Studies in Systems, Decision and Control

Volume 356

The series "Studies in Systems, Decision and Control" (SSDC) covers both new developments and advances, as well as the state of the art, in the various areas of broadly perceived systems, decision making and control–quickly, up to date and with a high quality. The intent is to cover the theory, applications, and perspectives on the state of the art and future developments relevant to systems, decision making, control, complex processes and related areas, as embedded in the fields of engineering, computer science, physics, economics, social and life sciences, as well as the paradigms and methodologies behind them. The series contains monographs, textbooks, lecture notes and edited volumes in systems, decision making and control spanning the areas of Cyber-Physical Systems, Autonomous Systems, Sensor Networks, Control Systems, Energy Systems, Automotive Systems, Biological Systems, Vehicular Networking and Connected Vehicles, Aerospace Systems, Automation, Manufacturing, Smart Grids, Nonlinear Systems, Power Systems, Robotics, Social Systems, Economic Systems and other. Of particular value to both the contributors and the readership are the short publication timeframe and the world-wide distribution and exposure which enable both a wide and rapid dissemination of research output.

Indexed by SCOPUS, DBLP, WTI Frankfurt eG, zbMATH, SCImago.

All books published in the series are submitted for consideration in Web of Science.

More information about this series at http://www.springer.com/series/13304

Dawei Shi · Yuan Huang · Junzheng Wang ·
Ling Shi

Event-Triggered Active Disturbance Rejection Control

Theory and Applications

Science Press
Beijing

Springer

Dawei Shi
School of Automation
Beijing Institute of Technology
Beijing, China

Yuan Huang
School of Automation
Beijing Institute of Technology
Beijing, China

Junzheng Wang
School of Automation
Beijing Institute of Technology
Beijing, China

Ling Shi
Department of Electronic and Computer Engineering
Hong Kong University of Science and Technology
Hong Kong, Hong Kong

ISSN 2198-4182 ISSN 2198-4190 (electronic)
Studies in Systems, Decision and Control
ISBN 978-981-16-0292-4 ISBN 978-981-16-0293-1 (eBook)
https://doi.org/10.1007/978-981-16-0293-1

Jointly published with Science Press
The print edition is not for sale in China (Mainland). Customers from China (Mainland) please order the print book from: Science Press.

This Springer imprint is published by the registered company Springer Nature Singapore Pte Ltd.
The registered company address is: 152 Beach Road, #21-01/04 Gateway East, Singapore 189721, Singapore

Preface

Control systems are composed of processes, sensors, actuators, and controllers. In the era of digital systems, we usually implement control algorithms through digital processors. This inevitably makes control systems to have a sampled-data nature, namely, the controlled processes operate in continuous time, while digital controllers provide control commands only at discrete sampling instants. The study of sampled-data control systems aims at dealing with the discrepancy between continuous evolution of physical system dynamics and the discrete update of computer control algorithms, and the primary idea is to analyze the effect of sampling. Earlier investigations focus on controller analysis and design based on periodic sampling, in which the process is sampled and controlled based on constant and equidistant sampling periods. Recently, to make efficient use of the limited communication and computation resources in networked control systems, event-based sampled-data control also received significant research attention. In an event-based control system, sampling and control are not performed until certain prespecified conditions are violated and sampling periods become event-triggered and nonuniform, which adds to the difficulty in analysis and design.

We started to work on active disturbance rejection control (ADRC) when we were trying to find a simple but theoretically sound approach to event-triggered control. At that time, event-triggered control had been investigated in the control community for a couple of years, with several interesting control frameworks considered (e.g., H_∞ control, model predictive control, to name a few), and different conditions were proposed to ensure the stability and performance of the event-triggered closed-loop control system. Our motivation, however, is that the theoretical efficiency and performance of event-triggered sampled-data control was not yet demonstrated for simple and easy to implement controllers—not even the PID controllers, and the effect of event-triggering mechanism on closed-loop control performance was not quantitatively characterized. In this regard, ADRC, a nontrivial generalization of the classic PID control method originally proposed and developed by Prof. Jingqing Han at the Chinese Academy of Sciences, seems to be an ideal candidate for our study due to its simple structure and its popularity and success in different industrial applications. On the other hand, the existing theoretic analysis of ADRC mainly focused on the continuous-time formulation of the controller and the effect of sampled-data implementation did not receive significant attention. In a certain sense, the study of event-triggered ADRC also provides a new possibility of analyzing the sampled-data performance of this controller.

Over the years, the theoretic developments and applications on event-triggered ADRC have been published in well-recognized journals; however, a comprehensive summary of the results is also necessary because the core ideas behind the design need to be synthesized to help the readers understand the development and motivate further study of the open problems, which is the purpose of this monograph.

As a prerequisite, an ideal reader would have knowledge of modern control theory and sampled-data systems, although we have included detailed proofs of the results in this book to make it self-contained.

The book starts with an introductory chapter, which introduces the basics of sampled-data systems, event-based sampled-data control, ADRC, and the related literature. The rest of the book is then divided into two parts. The first part covers theoretic developments, including the first attempt on discrete-time extended state observer (ESO), the developments of event-triggered ESO and ADRC, and the extensions to high-gain observer-based control. The second part presents a few application examples, including numerical application to attitude control of spacecraft system, and experimental results on DC motors and electrical cylinders. The last example is an artificial pancreas system that is used to achieve closed-loop glucose control for the patient with type 1 diabetes, the aim of which is not to purely illustrate event-based ADRC, but to motivate the future direction of event-based learning and control parameter adaptation.

This book would not have been possible without the help and support from many people and funding agencies. In particular, Dawei Shi wishes to thank his students, Jian Xue, Deheng Cai, Jiliang Song, Jing Chen, and Kaixin Cui, with whom he collaborated on event-triggered ADRC. Dawei Shi is also grateful to Prof. Tongwen Chen at the University of Alberta, Canada, who led him to the field of sampled-data systems and event-triggered control. Finally, he would also like to thank Prof. Francis J. Doyle III and Dr. Eyal Dassau at the Harvard University, USA, for offering the postdoc opportunity to work on artificial pancreas and closed-loop drug delivery, which made the discussions in Chap. 9 of the book possible. Financial support from National Natural Sciences Foundation of China (Grant Nos. 61973030 and 61503027) and Beijing National Natural Sciences Foundation (Grant No. 4192052) is gratefully acknowledged.

Yuan Huang would like to thank Prof. Shun-Ichi Azuma from Nagoya University, Japan, for hosting him as a visiting student, during which many interesting ideas on sampled-data systems were developed and investigated.

Ling Shi wishes to thank his Ph.D. students, Junfeng Wu, Chao Yang, Duo Han, and Yuzhe Li for a series of joint work on event-based state estimation and its applications. He would also like to thank his long-term collaborators on this topic: Prof. Karl H. Johansson (KTH), Dr. Yilin Mo (NTU), Prof. Bruno Sinopoli (CMU), Prof. Huanshui Zhang (Shandong U), and Prof. Qing-shan Jia (Tsinghua U). He is grateful for the financial support from the Hong Kong Research Grant Council and from the Hong Kong University of Science and Technology.

Junzheng Wang would like to thank his student Dongchen Liu for the joint work on event-based ADRC of electric cylinders. He gratefully acknowledges financial support from National Natural Sciences Foundation of China (Grant No. 51675041).

Finally, the authors would like to thank Lisa Fan from Springer and Ding Yu from China Science Publishing & Media Ltd. for their efficient handling of this joint book project from the early stage.

Beijing, China — Dawei Shi
Beijing, China — Yuan Huang
Beijing, China — Junzheng Wang
Hong Kong, Hong Kong — Ling Shi
December 2020

Contents

Part II Applications

Acronyms

ADC	Analog to digital converter
ADRC	Active disturbance rejection control
AP	Artificial Pancreas
BIBS	Bounded input bounded state
DAC	Digital to analog converter
DC	Direct current
ESO	Extended state observer
ET-ADRC	Event-triggered active disturbance rejection control
ET-ESO	Event-triggered extended state observer
HGO	High-gain observer
MPC	Model predictive control
PID	Propotional-integral-derivative
T1DM	Type 1 diabetes mellitus
TD	Tracking differentiator
ZOH	Zero-order hold

Chapter 1
Introduction

The scope of this book generally falls in event-based sampled-data systems and active disturbance rejection control (ADRC). In this introduction chapter, the background materials related to these topics will be discussed. We will start with the background introduction of event-triggered sampled-data control and more generally, sampled-data systems. We will then provide an introduction of ADRC, which is the main subject of the book, by presenting the essential elements in the controller. After that, we will give a brief overview of the literature related to the topics of the book, primarily focusing on event-based control, ADRC, and event-based ADRC. Finally, we present the organization of the book.

1.1 Event-Based Sampled-Data Control

Event-based sampled-data control has received significant attention in recent years, particularly after the pioneering work of Åström and Bernhardsson [8]. As a special form of sampled-data control, event-based approaches feature that the information transfer and controller updates are performed asynchronously according to the so-called event-triggering conditions. The main focus of event-triggered control design is to ensure the stability and performance given the nonuniform sampled-data control protocols. In the following, we first present an introduction of basic sampled-data systems and the general design approaches and then introduce the basics of event-based sampled-data control.

D. Shi et al., *Event-Triggered Active Disturbance Rejection Control*,
Studies in Systems, Decision and Control 356,
https://doi.org/10.1007/978-981-16-0293-1_1

1.1.1 Sampled-Data Systems

With the development of digital signal processing and large scale integrated circuits, the major functionalities of modern control systems are implemented through digital controllers using computers or microprocessors. This kind of control system is composed of two main modules: a generalized plant and a digital controller (see Fig. 1.1). We term the plant as a "generalized plant" as it contains all the physical components of a control system, including not only the process to be controlled, but also the sensors and actuators. The dynamics of the generalized plant usually evolve in continuous time, due to their physical nature. The digital controller incorporates the cyber components of a control system; as its functionalities operate based on periodic clock signals determined by the frequencies of the oscillators of the digital processors, the digital controller operates in discrete time. Generally, a digital controller consists of three basic components:

- Sampler: A sampler samples a continuous-time signal into a discrete-time signal based on certain sampling principles. This component is usually implemented through an analog-to-digital converter (ADC).
- Controller: A controller executes the control algorithm and computes the control commands. This component is implemented using computers or microprocessors (e.g., MCUs, DSPs, and FPGAs).
- Hold: A hold component converts the discrete-time control signal into a continuous-time signal according to certain principles, and is usually implemented using a digital-to-analog converter (DAC).

A more concise description of the structure of the above system is provided in Fig. 1.2. We use G, S, K, H to represent the generalized plant, the sampler, the controller, and the holder, respectively. A control system with the above structure is called a sampled-data system [10, 14]. Overall, a sampled-data system (including the physical and cyber modules) evolves in continuous time as it runs in our real world. Still, due to the effect of sampling, some of its signals are discrete-time signals. In this sense, a sampled-data system can be viewed as a hybrid system with

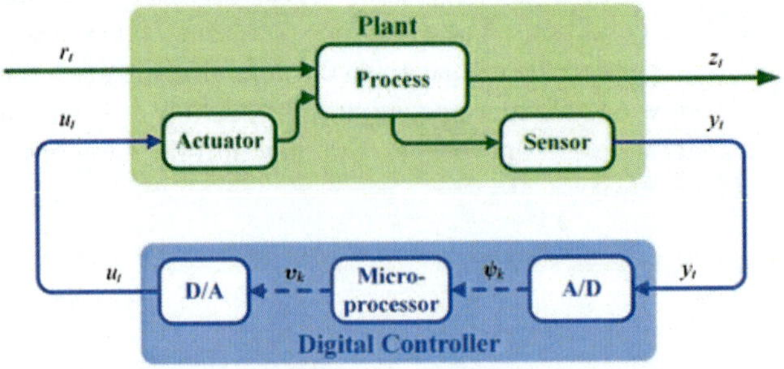

Fig. 1.1 Block diagram of a digital control system

Fig. 1.2 Sampled-data control system

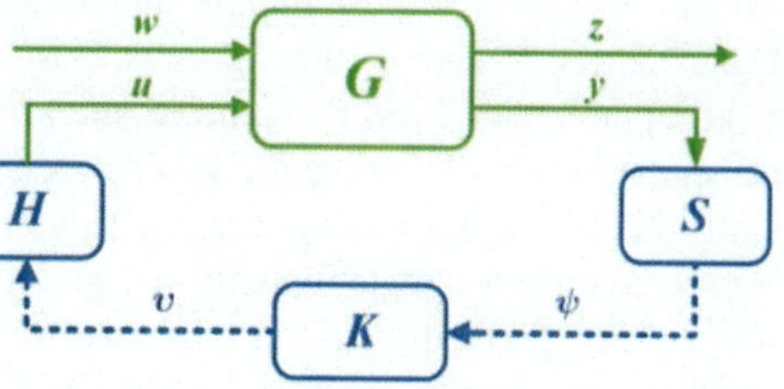

special structures. From this characteristic, three different approaches are normally employed in sampled-data control system design [14]:

- **Continuous-time design, discrete-time implementation:** For this approach, the effect of sampler and hold is first ignored and a continuous-time controller is designed based on the plant model. The obtained controller is then discretized based on a proper discretization method to obtain an approximate digital controller.
- **Discretize the plant and perform a discrete-time design:** First the plant is discretized to obtain a discrete-time model, and then a digital controller is designed based on discrete-time performance specifications.
- **Direct sampled-data design:** A digital controller is designed based on the continuous-time plant model, so that the performance specifications in continuous time can be satisfied.

The first two approaches are relatively straightforward in design, but it is difficult to ensure the performance of the sampled-data system in a continuous-time domain. In particular, the theoretic performance consistency in continuous time between the original continuous-time controller and the approximate discrete-time controller cannot be guaranteed in general for the first approach; although the second approach can ensure the performance in discrete time, namely, at the sampling instants, the inter-sample performance is not taken into account. In addition, for the second approach, standard methods exist to discretize linear time-invariant systems, but discretization method design remains challenging for general nonlinear plants.

In classic sampled-data control theory, the continuous-time signals are usually sampled periodically according to one or multiple fixed sampling periods, and correspondingly controllers update the control commands according to certain fixed sampling periods. As the controller runs in discrete time, a zero-order hold is used to generate the control commands during the inter-sampling time between two consecutive sampling instants, which means that the control command is kept constant before a new command is available. Since the operation of this type of sampled-data control systems is driven by periodic sampling instants, we call them time-triggered sampled-data systems or periodic sampled-data systems.

On the other hand, sampling instants can be determined in an adaptive fashion based on the value or characteristics of the target signal. For instance, we can consider a problem of remote estimation in a sampled-data system and we need to monitor a target signal based on its measurements. If the signal changes smoothly and slowly, it is not necessary to periodically update its value at a high frequency, and in fact, if the real-time value of the signal is close to the previously updated value within an acceptable range on the remote side, there is no need to update the value of the signal while achieving the goal of remote monitoring [84]. Similarly, for the controller, it is not always necessary to periodically update the control commands, and the controller update instants can be determined dynamically based on the real-time control performance (e.g., the size of tracking error). Basically, this type of sampled-data system operates on the basis of the occurrence of certain events, for instance, large changes in sensor measurements or obvious degradation in real-time control performance; such systems are called event-triggered sampled-data systems.

In addition, in many practical applications, the continuous signals are usually sampled using comparators built with digital circuits, and can only be measured during the sampling instants. Therefore, the events can only be triggered at the equidistant sampling instants. In digital controllers, control commands can only be updated at the discrete-time instants determined jointly by the maximum frequency of sensor measurements and the frequency of the oscillators of the microprocessors. As such, this type of event-triggered sampled system operates on the basis of the time-triggered sampling instants. For this reason, these systems are also called periodic event-triggered systems.

1.1.2 Event-Based Sampling and Control

In this subsection, we provide a further introduction of the event-triggered sampled-data systems. To make the discussions concrete and easier to understand, we focus on periodic event-triggered sampled-data systems, the schematic plot of which is provided in Fig. 1.3. Specifically, consider the following continuous-time plant:

$$\begin{aligned}\dot{x}(t) &= f(t, x(t), u(t)) \\ y(t) &= h(t, x(t), u(t)),\end{aligned}$$

where $x(t)$ denotes the system state, $u(t)$ denotes the control input and $y(t)$ represents the sensor measurements. Due to the sampled-data nature of the system, we use y_k to

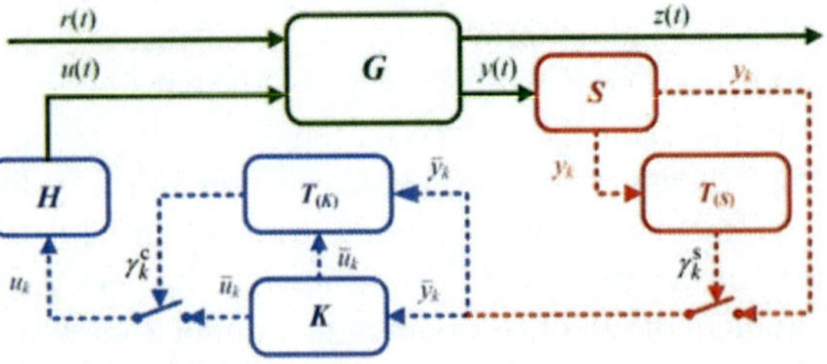

Fig. 1.3 Periodic event-triggered control system

represent the sampled version of $y(t)$ obtained by a sampler, and use u_k to represent the discrete version of $u(t)$ calculated by the digital controller; here $u(t)$ is obtained based on u_k using a zero-order hold. To make the relationships between $y(t), u(t)$ and y_k, u_k clear, we introduce the event triggers. The first event trigger, $T_{(s)}$, determines when the information of $y(t)$ will be sent to the controller K

$$T_{(S)} : \gamma_k^{\mathrm{S}} = \begin{cases} 0, \text{ if } \psi_k \in \Psi_k \\ 1, \text{ otherwise} \end{cases}$$

where ψ_k represents certain system information available to the digital controller and Ψ_k denotes the set of feasible ψ_k. Only when $\gamma_k^{\mathrm{S}} = 1$ holds will the value of y_k be transmitted to the controller K. We represent the information of y_k available to K as $\bar{y}_k$. Based on $\bar{y}_k$, the controller determines the control command $\bar{u}_k$, the transmission of which to the hold is determined by the second event trigger

$$T_{(K)} : \gamma_k^{\mathrm{C}} = \begin{cases} 0, \text{ if } \varphi_k \in \Phi_k \\ 1, \text{ otherwise} \end{cases}$$

where φ_k represents certain system information available to the digital controller and Φ_k denotes the set of feasible φ_k. Although the controller updates $\bar{u}_k$ at every sampling instant, the value of $\bar{u}_k$ is sent to the hold (which is u_k) only when $\gamma_k^{\mathrm{C}} = 1$. This completes the loop of how a periodic event-triggered control system works. We note again that periodic event-triggered systems is only a special type of event-triggered sampled-data systems, and other types of event-triggered control systems do exist but are not introduced here as the principles generally follow in a similar (and maybe easier) way.

Event-triggered sampled-data systems were developed in the background of networked control systems and wireless sensor networks, in which the digital controllers are implemented in a distributed manner. Usually, the sampler (or ADC) connects to a sensor, and the hold (or DAC) is connected to an actuator, but the sampler, controller, and the holder communicate with each other through wired or wireless communication networks, which motivates the structure in Fig. 1.3. In this situation, the effect of limited/restricted resources of communication, computation, and power on system performance becomes ignitable [49]. For instance, in many wireless monitoring applications, wireless sensors and actuators are powered by batteries, some of which are not even replaceable [2]. Therefore, it is necessary to consider energy consumption as an additional metric in control system design while considering control performance. As another motivating observation, the number of communication channels in a control system is limited, so that only one or a portion of the sensors or actuators can communicate with the controller to transmit the measurements or update the control signals.

Through performing the control actions and transmitting sensor measurement only when needed, event-triggered sampled-data control provides a new way of reducing the communication/computation burden between different components in a networked control system and has received significant attention in the control community in the recent years. Event-based sampled-data control is related with several other topics in control system design, including switched systems [63, 88], robust control [113], nonuniform sampled-data control [53, 77], set-membership estimation and control [3, 16, 60], and quantized control [24, 64]. However, compared with traditional time-triggered sampled-data control, the theoretic performance guarantee of event-triggered sampled-data control is more difficult to achieve, which provides new and unique challenges to analysis and design of sampled-data systems. The developments of event-triggered sampled-data control will be discussed in Sect. 1.3.1.

1.2 Active Disturbance Rejection Control

ADRC was an application-motivated model-free controller design method proposed by Professor Jingqing Han at the Chinese Academy of Sciences in the 1990s [45–48]. The original motivation was to develop an enhanced replacement of the classic PID controller that received dominance in different industrial applications but appears to have apparent drawbacks and may not be able to meet the performance requirements in advanced industrial applications. Specifically, as was mentioned in the seminar paper [47] that systematically introduced ADRC to the English audience, ADRC was designed to overcome the following four fundamental technical limitations of the PID framework:

- Step set-point changes lead to abrupt changes in the control signal and may not be appropriate for most dynamic systems;
- The derivative term of PID control is sensitive to measurement noises;
- The linear weighted sum of the proportional, integral, and derivative terms may not be optimal;
- The integral term introduces saturation problems and may reduce the stability margin.

An ADRC is normally built of three components: tracking differentiator, extended state observer, and nonlinear feedback (see Fig. 1.4). Before introducing these components, we first write the classic PID control law

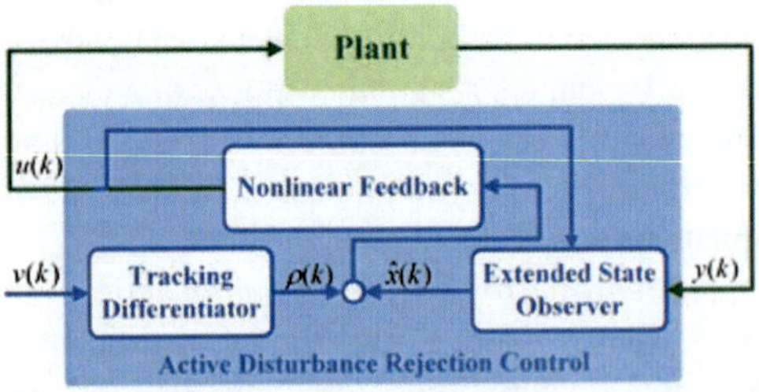

Fig. 1.4 Schematic of active disturbance rejection control (in discrete time)

$$u(t) = k_p e(t) + k_i \int_0^t e d\tau + k_d \frac{de}{dt}, \tag{1.1}$$

which would help explain how the components are developed. As a hindsight more than two decades after the development of the controller, the foundation of ADRC sits on the two major developments in modern control theory: the consideration of canonical form and the design of state observer. In a nutshell, these enable the model-free design of the controller, and the latter also helps enable offset-free tracking without explicitly using an integrator.

1.2.1 Functions **f han** *and* **f al**

Before introducing the Tracking Differentiator (TD), Extended State Observer (ESO), and nonlinear feedback modules, we first introduce the function $fhan$, which played an important role in the development of ADRC. This function was named after the inventor of this function (Professor Jingqing Han) and is mathematically defined as

$$\begin{cases} d = r\varepsilon^2, \ a_0 = \varepsilon n_2, \ y = n_1 + a_0, \\ a_1 = \sqrt{d(d + 8|y|)}, \\ a_2 = a_0 + \text{sign}(y)(a_1 - d)/2, \\ s_y = (\text{sign}(y + d) - \text{sign}(y - d))/2, \\ a = (a_0 + y - a_2)s_y + a_2, \\ s_a = (\text{sign}(a + d) - \text{sign}(a - d))/2, \\ fhan = -r(\frac{a}{d} - \text{sign}(a))s_a - r\text{sign}(a). \end{cases} \tag{1.2}$$

This function is a time-optimal control synthesis function for stabilizing two discrete-time integrators in cascade with a bounded input [47]

$$n_1(i + 1) = n_1(i) + T n_2(i), \tag{1.3}$$

$$n_2(i + 1) = n_2(i) + T u, \ |u| \le r, \tag{1.4}$$

where T is the sampling period, and u is the control law.

Another important function used in ADRC is fal, which is designed as

$$fal(e, \alpha, \delta) = \begin{cases} e/\delta^{1-\alpha}, & |e| \le \delta \\ |e|^\alpha \text{sign}(e), & |e| > \delta. \end{cases} \tag{1.5}$$

Note that this function determines the control strategy at two different phases. When tracking error e is within $|e| \le \delta$, e will approach 0 in infinite time by applying a linear feedback control law; otherwise, people select the nonlinear feedback control law in (1.5) such that the tracking error e can reach 0 quickly with $\alpha < 1$.

1.2.2 Tracking Differentiator

Let $v(t)$ denote a signal to be differentiated. The aim of designing a TD is to achieve the fastest tracking of $v(t)$ and $\dot{v}(t)$ subject to the acceleration limit of r. Now consider a second-order intergal plant (namely, double integrator)

$$\dot{x}_1 = x_2, \ \dot{x}_2 = u \tag{1.6}$$

with $|u| \leq r$ and v being the reference value for x_1, the time-optimal solution for u is

$$u = -r\text{sign}\left(x_1 - v + \frac{x_2|x_2|}{2r}\right). \tag{1.7}$$

With this input, the optimal tracking response is characterized by

$$\dot{x}_1 = x_2 \tag{1.8}$$

$$\dot{x}_2 = -r\text{sign}\left(x_1 - v + \frac{x_2|x_2|}{2r}\right), \tag{1.9}$$

which is the TD of $v(t)$. For the purpose of practical implementation, the discrete-time version in (9.3) is usually used to avoid unnecessary oscillations.

The motivation of designing TD is not only to design a desired transient profile for step changes but also to deal with the problem that the derivative action of PID control is sensitive to measurement noises. The rationale is explained as follows. Basically, a PID implementation normally calculate the differentiation of a signal v as

$$y = \frac{1}{\tau}\left(1 - \frac{1}{\tau s + 1}\right) v, \tag{1.10}$$

the time domain implementation of which is

$$y(t) = \frac{1}{\tau}(v(t) - v(t - \tau)). \tag{1.11}$$

To reduce the sensitivity to noises, an alternative way is to consider the following approximation:

$$\dot{v}(t) \approx \frac{v(t - \tau_1) - v(t - \tau_2)}{\tau_2 - \tau_1}, \tag{1.12}$$

which is shown to be able to avoid noise amplification in simulations and implies that a second-order process may lead to better performance. A particular second-order approximation of a differentiator is $s/(\tau s + 1)^2$, which corresponds to

$$\ddot{y} = -r^2(y - v) - 2r\dot{y} \tag{1.13}$$

in which y and $\dot{y}$ track v and y, respectively, and r determines the speed of tracking. TD is the differentiator that leads to fastest tracking.

1.2.3 Extended State Observer

A key component of ADRC is ESO, which is built on two important notions: total disturbance and its estimation. Consider a single-input-single-output process of the following form:

$$\begin{aligned}
\dot{x}_1 &= x_2 \\
&\vdots \\
\dot{x}_{n-1} &= x_n \\
\dot{x}_n &= f(x_1, \ldots, x_n, w, t) + bu \\
y &= x_1
\end{aligned}$$

where $x := [x_1, \ldots, x_n]^\top$ are the states, y is the output, u is the input and $f(x_1, \ldots, x_n, w, t)$ is an unknown multivariate function. The goal is to control the behavior of y using u. Here the unknown function $f(x_1, \ldots, x_n, w, t)$ describes the effect of unmodeled dynamics and external disturbances, and therefore, is named "total disturbance". Let $F(t) := f(x_1, \ldots, x_n, w, t)$ and $G(t) := \dot{F}(t)$ assuming that $F(t)$ is differentiable. If we treat $F(t)$ as an additional state variable $x_{n+1} = F(t)$, then the original dynamic system can be described as

$$\begin{aligned}
\dot{x}_1 &= x_2 \\
&\vdots \\
\dot{x}_{n-1} &= x_n \\
\dot{x}_n &= x_{n+1} + bu \\
\dot{x}_{n+1} &= G(t) \\
y &= x_1,
\end{aligned}$$

which is an observable system. For this system, an observer, which is named as the ESO, is proposed as

$$
\begin{aligned}
e &= z_1 - y \\
\dot{z}_1 &= z_2 - \beta_{01} e \\
&\vdots \\
\dot{z}_n &= z_{n+1} + bu - \beta_{0n} fe \\
\dot{z}_{n+1} &= -\beta_{0n+1} fe_1,
\end{aligned}
$$

where $fe = fal(e, 0.5, \delta)$ and $fe_1 = fal(e, 0.25, \delta)$. There are many ways to design the observer gains, and one crucial way is to design them using a pole assignment approach. For example, by replacing fe and fe_1 by e, one can conclude the system matrix N of the observation error dynamics as

$$
N = \begin{bmatrix} -\beta_{01} & 1 & 0 & \cdots & 0 \\ -\beta_{02} & 0 & 1 & \cdots & 0 \\ \ddots & \ddots & \ddots & \ddots & \vdots \\ -\beta_{0n} & 0 & 0 & 0 & 1 \\ -\beta_{0n+1} & 0 & \cdots & 0 & 0 \end{bmatrix}. \tag{1.14}
$$

Therefore, if we choose β_{0i} suitably such that N is Hurwitz, the observation performance of ESO can be guaranteed.

On the one hand, through introducing the viewpoint of ESO, an unknown and potentially nonlinear system is converted to a linear system with a canonical form and a disturbance term, which simplifies the control design problem and in particular, enables the design of a controller without the information of system dynamics—an important property inherited from PID control. Moreover, ESO has the form of a disturbance observer, which enables offset-free tracking without using an integrator. Thus, it helps avoid the effect of side effects (e.g., saturation) of using integral control. A similar idea has been adopted in offset-free design of model predictive control, see, e.g., [70, 76, 78] for the ideas and discussions. Discussions on the underlying internal model principle can be found in [22].

1.2.4 Nonlinear Feedback

The idea of using nonlinear feedback combination of different error terms for superior performance than PID control seems straightforward. In the seminar work of [47], three kinds of combinations were introduced

$$u = \frac{\beta_1 e_1 + \cdots + \beta_n e_n - z_{n+1}}{b_0}, \tag{1.15}$$

$$u = \frac{\beta_1 fal(e_1, \alpha_1, \delta) + \cdots + \beta_n fal(e_n, \alpha_n, \delta) - z_{n+1}}{b_0}, \tag{1.16}$$

$$u = -\frac{fhan(e_1, ce_2, r, h_1) + z_{n+1}}{b_0}, \tag{1.17}$$

with b_0 being an estimate of b. Note that the choices of the controller coefficients β_i are not determined by the specific plant model but the pre-designed poles of the closed-loop controlled system. By carefully selecting these coefficients, one can expect a suitable "timescale" for the controlled system. In many cases, the control law in (1.17) is used for $n = 2$.

1.3 Related Literature

In this section, we briefly discuss the literature that is related to the topic of this book. While the development of event-triggered ADRC started only a few years ago, ADRC and event-based control have been extensively studied in the control community during the past twenty years, which will be the main focus of this section.

1.3.1 Event-Based Sampled-Data Control

The investigation of event-triggering mechanisms (or event-triggers) in dynamic systems originated in the research on discrete-event systems in the 1980s [54]. The event-triggered estimation and control problems for general dynamic systems, however, did not start until recently [12, 31, 49, 61, 68]. In their pioneering work, Åström and Bernhardsson [9] compared the effect of event-triggered sampling and time-triggered sampling for first-order continuous-time stochastic systems and proved that event-based sampled-data control could effectively reduce output variance under the same average communication rate. Arzén [7] integrated the event-triggering mechanism with PID control, and showed that event-based sampling can effectively decrease the CPU occupation of the controller without affecting the closed-loop control performance. These two papers triggered the attention of the control community and to some extent promoted the developments on event-triggered sampled-data control. Although the theory of event-triggered sampled-data systems is still developing, a number of interesting results have been proposed in the literature. In this section, we offer a summary of the results developed along the three basic approaches to sampled-data control. Instead of providing a detailed literature review of all the published papers, we tend to focus on the general ideas and list a few typical developments. Comprehensive literature reviews on event-based control can be found in

[49, 68]. Also, note that we will focus on the developments on controller design in this section; for the developments on event-triggered state estimation, the interested readers can refer to the monograph of Shi et al. [84] and references therein.

1.3.1.1 Continuous-Time Design, Discrete-Time Implementation

The approach of continuous-time design and discrete-time implementation, which is also known as the "emulation approach", has been extensively adopted in designing event-triggered sampled-data controllers [1, 19, 21, 27, 66, 67, 69, 72, 80, 81, 92]. As the design of continuous-time controllers has been well studied, it is usually assumed that a continuous-time controller satisfying certain performance specifications has been designed, and the focus is to analyze the performance of the closed-loop sampled-data system after further incorporating event-triggering mechanisms. Specifically, the main research problems considered include (1) performance recovery of the original continuous-time control system through event-triggered sampled-data control, (2) analysis the Zeno phenomenon in the sampling patterns caused by introducing event-triggering mechanisms, the result of which could determine the effectiveness of the event-triggered control in reducing average/worst-case sampling rates. Currently, the majority of the studies along this direction focus on feedback control. Depending on the type of problems considered, we will further separately look into the cases of state and output feedback control.

State feedback: The majority of the existing investigations on event-triggered control focused on the state feedback case. For instance, Tabuada [92] considered the problem of event-based static state feedback control for nonlinear systems, and introduced an event-triggering condition of the form

$$\gamma(|e(t)|) \leq \sigma\alpha(|x(t)|), \tag{1.18}$$

where $e(t) = x(t_i) - x(t)$, t_i denotes the previous sampling instant, $\gamma(\cdot)$ and $\alpha(\cdot)$ are $\mathcal{K}_\infty$ functions and $\sigma > 0$. Under certain Lipschitz assumptions on the system dynamics and the feedback control law and the input-to-state stability (ISS) requirement of the continuous-time closed-loop system, it was proved that for any compact set S, there existed $\epsilon > 0$ and $\tau > 0$ such that the inter-triggering period is greater than τ (namely, $t_{i+1} - t_i \geq \tau$ hold for all $i \in \mathbb{N}$) provided the time required by controller update is less than ϵ. This result indicates that when the delay caused by controller update is sufficiently small, the simple event-triggering condition in (1.18) can avoid Zeno phenomenon. For the case of linear systems and $\Delta = 0$, τ can be determined through an implicit function. Wang and Lemmon [95] extended the work to consider event-triggering conditions with special logic structures. Compared with [92], the benefit of this result is that the energy function $V(t)$ does not need to be monotonically decreasing but rather only need to satisfy $V(t) \leq h(t, x_0)$. In addition, since the implementation of event-triggered control needs to continuously test whether the event-triggering conditions are satisfied, this leads to the challenges in algorithm implementation and consumption of communication resources. For this consideration, Anta and Tabuada [5] discussed self-triggered sampled-data control, in which the time interval between the next triggering instant t_{i+1} and the current triggering instant t_i can be predicted based on $x(t_i)$. Specifically, Anta and Tabuada [5] considered a self-triggered control protocol of the form

$$\tau(x(t_i)) := \min\left\{t > t_i \,\middle|\, |e(t)| = \sigma\frac{a}{b}|x(t)|\right\} \tag{1.19}$$

for state-dependent homogeneous systems and polynomial systems, and provided analytical results on the tradeoff between reachability and computation cost.

Mazo and Tabuada [73] discussed decentralized event-triggered control for the static state feedback case. In this case, each state x_i is separately measured and transmitted. To ensure the stability of the system through a triggering condition similar to (1.18), a set of parameters $\theta_1, \theta_2, \ldots, \theta_n \in \mathbb{R}$ satisfying $\sum_{i=1}^{n}\theta_i = 0$ are introduced. Obviously $|e(t)|^2 \leq \sigma|x(t)|^2$ is equivalent to

$$\sum_{i=1}^{n}(e_i^2(t) - \sigma x_i^2(t)) \leq 0 = \sum_{i=1}^{n}\theta_i. \tag{1.20}$$

So if $e_i^2(t) - \sigma x_i^2(t) \leq \theta_i$ holds for all i, we still have $|e(t)|^2 \leq \sigma|x(t)|^2$. Based on this observation, Mazo and Tabuada [73] proposed a series of heuristic decentralized event-triggered control methods. Mazo and Cao [72] further extended the work to the case of more general decentralized event-triggering conditions

$$t_{r_i}^i := \min\{t > t_{r_i-1}^i | e_i^2(t) = \eta_i\} \tag{1.21}$$

and discussed the uniformly globally practically asymptotic stability and uniformly globally asymptotic stability for the corresponding decentralized static feedback control design.

On the other hand, an event-triggered control system is interconnected by the original continuous-time control system and the event-triggering mechanism. Based on this structural property, Liu and Jiang [67] analyzed the event-triggered state feedback control problem for nonlinear systems using a small gain theorem approach, and proposed ISS conditions and the expression for the event-trigger that can ensure the closed-loop stability while avoiding Zeno behavior. Based on this result, the authors proposed self-triggering sampling strategies for disturbed nonlinear systems, and proved that with the knowledge of an upper bound on the disturbance, the closed-loop system was robust with respect to the external disturbance and that the system state would converge to the origin if the disturbance converged to zero.

The event-triggering conditions discussed in the above investigations are only related to the current state $x(t)$. In [27], Girard introduced an internal variable z in the event-triggering condition satisfying

$$\dot{z}(t) = -\beta(z(t)) + \sigma\alpha(\|z(t)\|) - \gamma(\|z(t)\|), z(0) = z_0, \tag{1.22}$$

where $\beta(\cdot)$, $\gamma(\cdot)$, $\alpha(\cdot)$ are $\mathcal{K}_\infty$ functions, $\sigma \in (0, 1)$ and z_0 are design parameters. Based on this variable, a dynamic event-triggering mechanism of the form

$$t_{i+1} = \inf\{t \in \mathbb{R} | t > t_i \cap z(t) + \theta \cdot (\sigma\alpha(\|x(t)\|) - \gamma(\|e(t^-)\|) \leq 0\} \tag{1.23}$$

was proposed, and the stability of the closed-loop system under static state feedback control was proved. A lower bound on the inter-triggering interval was proposed for linear systems.

With the development of network technology, the event-triggered implementation of nonlinear state feedback control was discussed in the background of networked control systems. Through connecting the communication bandwidth limit with audio signal processing theory, Premaratne et al. [81] proposed an event-triggering mechanism-based adaptive differential modulation, and proved the stability of the closed-loop systems under time delays, signal decoding errors, and disturbance inputs. The robustness against time delay and packet dropouts induced by the communication network was also analyzed. Abdelrahim et al. [1] considered a class of nonlinear networked systems with two timescales and discussed the problem of event-triggering mechanism design under communication constraints; in this scenario, the system was composed of a fast mode and a slow mode, so the event-triggering mechanism designed only considering the slow mode would not be able to guarantee the stability of the system and the boundedness of the time interval between two consecutive sampling instants. For this problem, the authors proposed a perturbed system approach to event-triggering condition design and the semi-global practical stability of the closed-loop system was guaranteed. A hybrid time- and event-driven control strategy was introduced to ensure that the inter-sampling interval is strictly greater than zero. Postoyan et al. [80] considered an event-triggered tracking control problem for unicycle mobile robots under communication limits and proved the practical convergence of the system for the prespecified reference trajectories under event-triggered control while guaranteeing the uniform boundedness of the triggering interval. Experimental results were obtained to verify the proposed results.

In addition, a number of investigations were performed for linear systems; compared with nonlinear systems, the systems considered have simpler and clearer structures and thus more delicate results were obtained. For instance, Lunze and Lehmann [69] discussed the case of linear systems with input disturbances, and proposed an event-triggering condition based on disturbance observation. Interestingly, an upper bound on the distance of the responses between the event-triggered implementation of the closed-loop system and the original time-triggered system was obtained analytically, together with a lower bound on the sampling intervals. Forni et al. [21] discussed the case without external disturbances and modeled the system into a hybrid system [28], and proposed an event-triggering condition design approach that ensured the global exponential convergence of the system.

Output feedback control: The problem of event-based output feedback control was also investigated, and a number of interesting attempts have been made in the recent years. For instance, for a linear time-invariant system under external disturbances, Heemels et al. [51] investigated the problem of event-triggering condition design and decentralized implementation of prespecified continuous-time linear time-invariant controllers. Through modeling the overall closed-loop system into an impulsive system [28] and utilizing a linear matrix inequality approach, sufficient conditions on closed-loop stability and $\mathcal{L}_\infty$ robustness were proposed, and it was shown that the inter-sampling interval was larger than that obtained by directly generalizing the results obtained in [92]. Through constructing a state estimator to assist event-triggering mechanism design, Forni et al. [21] extended the results to the case of output feedback, and obtained sufficient conditions to guarantee global asymptotic stability of the system. For general nonlinear systems, Liu and Jiang [66] generalized their results [67] to output feedback case through a Lyapunov approach, and proposed a class of event-triggering conditions with threshold parameters generated by a stable system; closed-loop stability and boundedness of sampling frequency were ensured through developing parameter design approaches for the event-triggering conditions.

1.3.1.2 Discretize the Plant and Perform a Discrete-Time Design

This approach starts with a discrete-time model, and the system performance is evaluated with discrete-time performance metrics. In the context of event-triggered sampled-data control, this approach can be divided into two sub-approaches. The first sub-approach still assumes that a discrete-time controller is designed for certain discrete-time performance specifications and mainly focus on event-triggered condition design, which is similar to the approach of continuous-time design and discrete-time implementation, but the key differences are that the utilization of discrete-time models and performance specifications leads to different problem description and challenges and that the discrete-time framework to some extent simplified the problems—for instance, discrete-time systems do not exhibit Zeno behaviors.[1] The second sub-approach mainly concerns the problem of controller design for prespecified event-triggering conditions or the problem of joint event-trigger and controller design. The problems considered are similar to the direct sampled-data design approach presented in the next section and are normally very difficult to solve systematically.

Compared with the study on continuous-time systems, the study on event-triggered control of discrete-time systems started relatively later. In the representative work of Heemels and Donkers [50], the authors considered discrete-time linear time-invariant systems and introduced the concept of periodic event-triggered control, for which sampling is still taken in a periodic time-triggered fashion but the events can only be triggered at the periodic sampling instants. The problem of output feedback control under the existence of separate event-triggers between the sensor and controller and between the controller and the actuator was considered. To solve this

[1]Note that due to the existence of communication constraints, the problem of analyzing average sampling or communication rate is also necessary.

problem, the authors proposed a perturbed linear system framework and a piecewise linear framework, and the global exponential stability and the ℓ_2 performance of the closed-loop system were analyzed by introducing linear matrix inequality conditions. Similar to the conclusions obtained in [69], it was shown in [50] that the performance obtained by a time-triggered discrete-time output feedback controller could be recovered to an arbitrary precision by periodic event-triggered control. Quevedo et al. [82] discussed the problem of anytime control for discrete-time nonlinear systems and a prespecified controller, and the convergence of the algorithm was ensured through introducing sufficient conditions on the event-triggering parameters.

Due to the adoption of the event-triggering mechanism, a number of properties no longer hold for linear systems. One typical and important example is the separation principle [49, 61]. Similar problems were noticed and studied in the optimal control of quantized systems [23, 103]. For discrete-time linear systems, Molin and Hirche [75] discussed the existence of certainty equivalence principle under event-triggered control, which was a general extension of the classic linear quadratic Gaussian regulation problem. Three classes of problems were considered. The first class incorporated the communication resource limitation into the cost function; the second considered resource limitation as constraints; while the third incorporated constraints on the effect of average resource limitation. Through analyzing the structures of the optimal solutions to the three types of optimization problems, the existence of certainty equivalence principle was proved. However, considering the similarity between event-triggered systems and quantized systems, the analysis of certainty equivalence and the design of optimal control strategy for general event-triggered systems remain very challenging.

The joint control and event-trigger design is another important problem investigated in event-triggered control of discrete-time systems. Meng and Chen [74] investigated the joint design problem for discrete-time linear systems under static and dynamic output feedback control. Similar to [50], this work considered the case that the sensor and the controller were controlled by separate event-triggers. For static output feedback control, the stability conditions were provided in terms of bilinear inequalities, which were converted into a linear matrix inequality related nonlinear optimization problem. For dynamic output feedback control, a sufficient stability condition was provided directly using linear matrix inequality. Al-Areqi et al. [4] discussed the joint design problem for linear time-invariant systems under transmission delay and communication constraints. By modeling the delay process as uncertainty and developing deterministic or stochastic models for the uncertainty, the joint design problem was converted to an optimization problem. Linear matrix inequality conditions were developed to design the controller and event-trigger that ensured the closed-loop stability.

1.3.1.3 Direct Sampled-Data Design

The main feature of this approach is that continuous-time performance metrics are employed in analysis and design. For event-triggered sampled-data control, the problems covered using this approach include both the joint controller and event-trigger design for continuous-time performance specifications, and the controller design for prespecified event-triggering conditions and performance requirements. In addition, with this approach, the event-triggering mechanism can be implemented following the periodic event-triggered control approach discussed above, but the essential difference is the performance of the system is analyzed in the continuous-time framework. Similar to the case of classic sampled-data control [14], the direct sampled-data design approach is much more challenging than the first two approaches.

Interestingly, earlier investigations on event-triggered sampled-data control were mostly performed along the idea of direct sampled-data design, including the work of Åström and Bernhardsson [9] and extensions that studied similar problems [52, 74, 96]. These investigations were normally developed using the theory of stochastic differential equations and discussed the optimal sampling protocols and control methods for certain performance metrics. The systems considered were normally first- or second-order systems, and the generalization of the results to general high-order systems is normally very challenging.

For continuous-time linear time-invariant systems under event-triggered sampling, Peng and Yang [79] discussed the joint controller and event-trigger design problem with periodic event-triggering strategies, and proposed linear matrix inequality conditions that ensured $\mathcal{H}_\infty$ performance under time delays and packet drops. Developments were also made for nonlinear systems. For instance, Marchand et al. [71] proposed a systematic event-triggered state feedback control approach for nonlinear systems by extending the nonlinear controller design theory [87]. Liu and Jiang [66] extended the results for state feedback using a small gain approach and proposed a nonlinear observer-based output control strategy, so that the closed-loop control system could be converted into a form of interconnected input-to-state-stable subsystems. Li and Shi [62] investigated the continuous-time event-triggered control problem in the framework of model predictive control and proposed sufficient conditions to ensure the recursive feasibility and closed-loop stability.

In addition, analyzing the impact of event-triggering mechanisms on system control performance is also an important problem. For instance, if the sampled-data performance cannot be guaranteed under external disturbance, the applicability of the proposed event-triggered control approach would become questionable. A few results for this problem were also developed. For instance, utilizing an impulsive system approach, Borgers and Heemels [11] analyzed the properties of the minimum inter-sampling intervals for different continuous-time dynamic systems (linear or nonlinear), different feedback control strategies (state or output feedback), and typical event-triggering mechanisms. Antunes and Heemels [6] adopted a performance metric of the form

$$\int_0^\infty \mathbf{E}[e^{-\alpha t} g_c(x(t), u(t))]dt, \tag{1.24}$$

and analyzed the performance of an event-triggered controller with a special structure, and mathematically proved that the performance of event-triggered control was superior to that of traditional time-triggered control.

1.3.2 Active Disturbance Rejection Control

ADRC was developed by Prof. Jingqing Han in his pioneering works [45–48]. Han discussed the construction of ADRC in detail [43, 47], suggesting that ADRC should be composed of an ESO, a TD, and a nonlinear feedback controller. The effectiveness of this unconventional control strategy has been verified in many practical applications [83, 93, 110]. This section will be devoted to the theoretical developments on the active disturbance rejection control scheme (especially the crucial component, ESO), as well as the practical applications of ADRC.

1.3.2.1 Theoretical Developments on ADRC, an ESO-Based Feedback Scheme

The basic idea of the ADRC strategy is to use an ESO to estimate the plant dynamics and unknown disturbances in real-time and dynamically compensate for them. Therefore, the ESO is an essential component of the active disturbance rejection control strategy [44]. The capability of ESO in estimating uncertainties was analyzed in Yang and Huang [38, 39, 101]. Following the enormous practical applications of ESO [55, 94, 101], the study of the convergence of ESO has progressed smoothly.

In modern control theory, ESO can be viewed as an extension of a standard state observer. A state observer is normally an auxiliary system that estimates a given real system's internal state from its input and output. ESO is a ground breakthrough of state observers where not only the states but also the "total disturbance" are jointly estimated. The "total disturbance" captures the composite effect of unmodeled system dynamics, uncertainties in system parameters, and external disturbances [42]. Since the "total disturbance" and state of the system are estimated simultaneously in ESO, we can design an output feedback control law which is not critically reliant on the mathematical model. Following this idea, Han first proposed the ESO in [44] for a single-input single-output (SISO) system, in which the unknown system function $f(\cdot, t)$ and the unknown external disturbance w were treated as the "total disturbance". Note that by suitably choosing the observer function and observer parameter, one can expect the observation performance that the observer state $\hat{x}$ will approach x.

To further reduce the difficulty of adjusting observer parameters, Gao [26] introduced the bandwidth parameter ω and ε into the linear extended state observer (LESO) design. Generally, if the total disturbance changes fast, the tuning parameter should be tuned to be large accordingly in order to track the states and the "total disturbance". Essentially, this parameter design method is similar to parameter design in a high-gain observer, although in the classical high-gain observer, only the system state is estimated and there is no "total disturbance" estimation. Also, the convergence of the proposed ESO was investigated and verified numerically in [26]. Note that a similar formation of LESO was presented in [59], and the stabilization of a high-gain observer[2] was rigorously proved. In [107], Zheng and Gao showed that for a given model plant, the LESO was asymptotically convergent. If the model was not given, the estimation error of the LESO was bounded; through these results, the convergence of ESO for a linear SISO system has been proved. Based on the obtained theoretical results in LESO, Guo, and Zhao proposed the nonlinear extended state observer (NLESO) of a multi-input multi-output (MIMO) system with uncertainty in [39]. The authors showed the design of the ESO could be flexible—for a given plant, the observer can be constructed either as a linear one or in a nonlinear form.

As a consequence of the theoretical results of ESO, significant developments of ADRC were made [34, 36]. As stated in Zheng et al. [107], the asymptotic stability of a linear ADRC could be established if the plant model was completely known. Furthermore, the tracking errors could be bounded even though the plant possessed considerable dynamic uncertainty. The convergence of a varying high-gain nonlinear extended state observer, a crucial part of the nonlinear ADRC, was investigated in [38] by Guo and Gao. They proved that if the high-gain parameter was set in a particular range, the convergence could be obtained. They also provided proof of the convergence of the tracking differentiator, another critical component of the ADRC in [37]. It was shown that the differentiator tracked the reference signal effectively. The convergence of the ADRC for minimum-phase plants with uncertainty was considered in the work of Zhao and Huang [105]. For SISO nonlinear system with uncertainty, the convergence of an ADRC was discussed by Guo and Zhao [33]. It was shown that the system state could track the given signal, and the observation error was bounded. Furthermore, Guo and Zhao also considered the ADRC for an MIMO system in [40]. For further theoretical properties of ADRC, we refer the readers to [32, 35, 41, 98–100]. An in-depth review of the methodology development of the ADRC was provided by Huang and Xue in [56].

1.3.2.2 Practical Applications of ADRC

Due to the advantages of ADRC on improving transient performance and enhancing system robustness in engineering systems, this control approach has been extensively applied in various engineering applications [110], e.g., two-mass drive system [20], delta robot system [13], and so on [30, 83, 86, 90]. In this section, we briefly discuss the effectiveness of ADRC in several common industrial control applications, including motion control, DC-DC power converters, continuous stirred tank reactor (CSTR), and micro-electro-mechanical systems (MEMS) gyroscope [34, 36, 111].

---

[2]The formulations of the high-gain observer and the LESO have the same theoretical meaning.

In [25, 109], Gao et al. proposed an active disturbance rejection controller for motion control systems. Motion control applications can be found in almost every industry sector, from factory automation and robotics to high-tech computer hard-disk drives. They are used to regulate mechanical motions in terms of position, velocity, acceleration, and to coordinate the motions of multiple axes or machine parts. By estimating the combined effects of internal dynamics and external disturbances, the ADRC controller was proved to be highly tolerant of dynamic variations and disturbances commonly found in manufacturing processes [29]. Through simulations, frequency response analysis, and hardware tests, it was shown that the proposed approach was superior to the current PID based technology, especially in handling set-point changes, large inertia and friction variations, and external torque disturbance, all of which could be seen as "disturbance" by ADRC and were actively compensated. The improvement in transient response and steady-state error was also quite evident.

In [89], Sun and Gao presented the design and implementation of an ADRC digital controller for an H-Bridge DC-DC converter, which was used for NASA to provide reliable, efficient, well-regulated DC-DC power conversion for on-board space electronics. The experimental comparison of ADRC with a well-tuned PI controller was carried out. ADRC clearly showed a marked improvement in terms of output voltage deviation from the set point and in terms of recovery time. In [102], the output voltage regulation problem of a PWM-based DC-DC buck converter under various sources of uncertainties and disturbances was investigated by Yang et al. via an optimized ADRC approach. It was shown that the proposed ADRC could guarantee the rigorous stability of the closed-loop system, for any bounded uncertainties of the circuit, by appropriately choosing the observer gains and the bandwidth factor. Experimental results illustrated that the proposed control solution was characterized by improved robustness performance against various disturbances and uncertainties compared with traditional ADRC and integral model predictive control approaches.

The CSTR is widely used in chemical and process industries [106]. Due to its highly nonlinear nature, it is a significant benchmark problem in process control [97]. In [15], ADRC was applied to the CSTR control problems. Numerical verification was performed on two CSTRs, and both demonstrated outstanding performance in the absence of accurate mathematical models of the processes. Design and tuning were shown to be simpler compared to the MPC strategy. More importantly, ADRC does not require detailed knowledge of the process dynamics, which makes it a promising approach in solving specific process control problems in which an accurate model is challenging to obtain.

As reported in [17, 108], ADRC was successfully applied to control the drive axis for MEMS gyroscope. Assuming that the resonant frequencies of the drive and sense axes are matched, the ADRC was employed to the drive axis to resonance and regulate the drive axis's output amplitude to the desired value. The controller was validated by software and analog hardware implementations on a vibration piezoelectric beam gyroscope [18, 65]. Since the control system design does not require exact information of system parameters, it was robust against the gyroscope's structural uncertainties. The robustness of the ADRC was analyzed in the frequency domain in [17]. A new control method based on ADRC to actuate the drive axis of a micro-electro-mechanical systems (MEMS) gyroscope to a fixed level was presented [112], where ADRC was shown to achieve high performance in tracking.

1.3.3 Event-Based Active Disturbance Rejection Control

Motivated by the demands in maintaining control performance under limited communication and computation resources, Huang et al. introduced the event-triggered sampling scheme in ADRC design [57, 58]. Generally, the proposed approach belongs to the first approach in sampled-data system design, as an underlying continuous-time ADRC that could stabilize the system is normally assumed. The work started in event-triggered ESO design. In [57], Huang et al. considered event-triggered ESO (ET-ESO) design for a continuous-time nonlinear system with uncertainty and disturbance. Since the convergence of the ET-ESO depends on the design of the triggering condition directly, the goal was to design an implementable event-triggering condition that did not rely on the state of the plant. For the extended state observer considered, an event-triggered transmission strategy was proposed such that the observation error was uniformly bounded. Furthermore, under mild conditions on the initial state of the system, the observation error could be guaranteed to be bounded at any time. It was shown that there was no Zeno behavior for the event-based transmission strategy. Based on the results of ET-ESO design, event-triggered active disturbance rejection (ET-ADRC) was introduced in [58]. A nonlinear continuous-time system with uncertainty and disturbance was considered, for which an event-triggered control problem based on an ADRC approach was formulated. The authors first presented the framework of the closed-loop system where the control signal was transmitted at each event instant, while the output measurements are transmitted in real-time. Based on this framework, an easy-to-implement event-triggering strategy that relied on the state observation generated by an extended state observer was proposed to guarantee the asymptotic performance of the closed-loop system. Moreover, the extension to a more general case in which separate event-triggering strategies determined the transmission of the measurements and control inputs, were also investigated. By applying the proposed ET-ADRC in DC motor systems, the effectiveness of ET-ADRC was verified in [85].

For discrete-time systems, a number of interesting attempts were made in ET-ADRC, which corresponds to the second approach to sampled-data system design. In [104], Zhang et al. investigated the problem of disturbance rejection control for discrete-time systems within an event-driven control framework. Through a predefined event-driven scheduler, both full- and reduced-order ESO-based output feedback controllers were designed. With the proposed ESOs, both the disturbance and the system states were estimated, based on which the controllers were constructed to ensure stability and disturbance attenuation. It was found that with the established event-driven control approaches, the updating frequency of the controller could be prominently reduced, and the disturbance could also be compensated in the output channels of the systems. In [91], Sun et al. developed an ET-ADRC for disturbed systems in a networked environment. By using disturbance/uncertainty estimation and attenuation technique, an event-based sampled-data composite controller was proposed together with a discrete-time ESO. The proposed ET-ADRC scheme was able to reduce the communication frequency while maintaining the closed-loop stability of the control system.

So far, a direct sampled-data design approach for ET-ADRC appears to be missing, although the sampled-data control framework in [58] seemed to provide some motivating ideas.

1.4 Organization of the Book

As indicated by the title, this book focuses on event-triggered ADRC, attempting to cover both the theoretic developments and applications. Specifically, we focus our analysis on deterministic systems, the benefit of which is that the analysis is easier to follow and thus the results are easier to be applied. Extensions to stochastic cases are definitely possible, although the derivations will be more involved. We have an intent to encourage the application of sampled-data ADRC design, and thus simulation examples and experimental results are presented and discussed, with an emphasis on how to design an event-triggered ADRC for a particular system or application.

The rest of the book is composed of two parts. Part I is composed of Chaps. 2–5, which will cover the theoretic developments. Before looking into event-triggered ADRC design, we first introduce the results on discrete-time ESO in Chap. 2, which is relatively independent of the other chapters and self-contained. The motivation of including this chapter is to show the difficulties in the analysis and design of sampled-data or discrete-time ADRC, even with an ESO that has a very special structure. The good news is that closed-form expressions of the observation performance can be obtained through using an ellipsoidal set method and guidelines on choosing the gain parameters of ESO are provided, but the problem would become very challenging to solve if the structure of the system or the ESO is changed.

The main theoretic results on event-triggered ADRC are summarized in Chaps. 3 and 4. In Chap. 3, we first introduce the design of event-triggered ESO, which is an important component of ADRC and was also the first step in our attempt of event-triggered ADRC design. We will discuss how the event-triggering conditions are designed to ensure the stability of the observation error dynamics. Based on the results in Chap. 3, we will close the ADRC control loop in Chap. 4. This will be done in two steps. In the first step, we will consider the case of event-triggered control with continuous measurements, and then move on to consider separate event-triggered observation and control (governed by different event-triggering mechanisms) in the second step. Motivated by the connections between ADRC and high-gain based control, we generalized our ideas in Chaps. 3 and 4 to deal with event-triggered high-gain control problems in Chap. 5.

Part II is composed of Chaps. 6–9. Although the example of DC motor is used in the theoretic chapters, we still devoted Chap. 6 to the application of DC motor control, and the aim is to show how the event-triggered ADRC can be applied to an engineering system with a theoretic performance guarantee. An application example of electrical valve control is provided in Chap. 7; the feature of this example is the introduction of a proportional-derivative event-triggering condition, which is introduced to improve the tracking performance. A simulation example of attitude control of a rigid spacecraft is presented in Chap. 8, in which we show that the tracking performance can be maintained at a reduced control updating rate. The example in Chap. 9 is the artificial pancreas, an engineering system developed for closed-loop glucose regulation. The event-triggered ADRC used in this example, however, is different from the examples in Chaps. 6–8. Instead of having the sensor measurements and control command updates in an event-triggered fashion, the event-triggering mechanisms used here are implicit and governs the updates of key controller parameters to ensure the safety and performance of the system. The aim here is to point out the idea of event-based learning, which is an interesting direction of event-triggered systems. Event-based learning will be further discussed in Chap. 10, which summarizes the results introduced in this book and points out the open problems on this topic.

References

1. Abdelrahim, M., Postoyan, R., Daafouz, J.: Event-triggered control of nonlinear singularly perturbed systems based only on the slow dynamics. Automatica **52**, 15–22 (2015)
2. Akyildiz, I., Su, W., Sankarasubramaniam, Y., Cayirci, E.: Wireless sensor networks: a survey. Comput. Netw. **38**, 393–422 (2002)
3. Alamo, T., Bravo, J., Camacho, E.: Guaranteed state estimation by zonotopes. Automatica **41**(6), 1035–1043 (2005)
4. Al-Areqi, S., Gorges, D., Liu, S.: Event-based control and scheduling codesign: Stochastic and robust approaches. IEEE Trans. Autom. Control **60**(99), 1291–1303 (2014)
5. Anta, A., Tabuada, P.: To sample or not to sample: self-triggered control for nonlinear systems. IEEE Trans. Autom. Control **55**(9), 2030–2042 (2010)
6. Antunes, D., Heemels, W.: Rollout event-triggered control: beyond periodic control performance. IEEE Trans. Autom. Control **59**(12), 3296–3311 (2014)
7. Arzén, K.E.: A simple event-based pid controller. In: Proceedings of the IFAC World Congress (1999)
8. Åström, K., Bernhardsson, B.: Comparison of periodic and event based sampling for first-order stochastic systems. In: Preprints 14th World Congress of IFAC (1999)
9. Åström, K., Bernhardsson, B.: Comparison of Riemann and Lebesgue sampling for first order stochastic systems. In: Proceedings of the 41st IEEE Conference on Decision and Control, vol. 2, pp. 2011–2016 (2002)
10. Åström, K.J., Wittenmark, B.: Computer-controlled systems: theory and design. Prentice-Hall, Upper Saddle River (1984)
11. Borgers, D., Heemels, W.: Event-separation properties of event-triggered control systems. IEEE Trans. Autom. Control **59**(10), 2644–2656 (2014)

12. Cassandras, C.G.: The event-driven paradigm for control, communication and optimization. J. Control Decis. **1**(1), 3–17 (2014)
13. Castaneda, L., Luviano-Juarez, A., Chairez, I.: Robust trajectory tracking of a delta robot through adaptive active disturbance rejection control. IEEE Trans. Control Syst. Technol. **23**(4), 1387–1398 (2015)
14. Chen, T., Francis, B.: Optimal Sampled-Data Control Systems. Springer, Berlin (1995)
15. Chen, Z., Zheng, Q., Gao, Z.: Active disturbance rejection control of chemical processes. In: 2007 IEEE International Conference on Control Applications, pp. 855–861 (2007)
16. Chisci, L., Garulli, A., Zappa, G.: Recursive state bounding by parallelotopes. Automatica **32**(7), 1049–1055 (1996)
17. Dong, L., Zheng, Q., Gao, Z.: A novel oscillation controller for vibrational mems gyroscopes. In: 2007 American Control Conference, pp. 3204–3209 (2007)
18. Dong, L., Avanesian, D.: Drive-mode control for vibrational mems gyroscopes. IEEE Trans. Ind. Electron. **56**(4), 956–963 (2009)
19. Donkers, M., Heemels, W.: Output-based event-triggered control with guaranteed $\mathcal{L}_\infty$-gain and improved and decentralized event-triggering. IEEE Trans. Autom. Control **57**(6), 1362–1376 (2012)
20. Erenturk, K.: Fractional-order $PI^\lambda D^\mu$ and active disturbance rejection control of nonlinear two-mass drive system. IEEE Trans. Ind. Electron. **60**(9), 3806–3813 (2013)
21. Forni, F., Galeani, S., Nesic, D., Zaccarian, L.: Event-triggered transmission for linear control over communication channels. Automatica **50**(2), 490–498 (2014)
22. Francis, B., Wonham, W.: The internal model principle of control theory. Automatica **12**(5), 457–465 (1976)
23. Fu, M.: Lack of separation principle for quantized linear quadratic gaussian control. IEEE Trans. Autom. Control **57**(9), 2385–2390 (2012)
24. Fu, M., Xie, L.: The sector bound approach to quantized feedback control. IEEE Trans. Autom. Control **50**(11), 1698–1711 (2005)
25. Gao, Z., Hu, S., Jiang, F.: A novel motion control design approach based on active disturbance rejection. In: Proceedings of the 40th IEEE Conference on Decision and Control (Cat. No.01CH37228), vol. 5, pp. 4877–4882 (2001)
26. Gao, Z.: Scaling and bandwidth-parameterization based controller tuning. Am. Control Conf. **6**, 4989–4996 (2003)
27. Girard, A.: Dynamic triggering mechanisms for event-triggered control. IEEE Trans. Autom. Control **60**(99), 1992–1997 (2014)
28. Goebel, R., Sanfelice, R., Teel, A.: Hybrid dynamical systems. IEEE Control Syst. Mag. **29**(2), 28–93 (2009)
29. Goforth, F.J.: On motion control design and tuning techniques. In: Proceedings of the 2004 American Control Conference, vol. 1, pp. 716–721 (2004)
30. Goforth, F., Gao, Z.: An active disturbance rejection control solution for hysteresis compensation. Am. Control Conf. **2008**, 2202–2208 (2008)
31. Grüne, L., Hirche, S., Junge, O., Koltai, P., Lehmann, D., Lunze, J., Molin, A., Sailer, R., Sigurani, M., Stöcker, C., Wirth, F.: Event-based control. In: J. Lunze (ed.) Networked Control Systems, Control Theory of Digitally Networked Dynamic Systems, pp. 169–261. Springer, Berlin (2014)
32. Guo, B.Z., Jin, F.F.: Output feedback stabilization for one-dimensional wave equation subject to boundary disturbance. IEEE Trans. Autom. Control **60**(99), 824–830 (2014)
33. Guo, B., Zhao, Z.: On convergence of nonlinear active disturbance rejection for SISO systems. In: 24th Chinese Control and Decision Conference (CCDC), 2012, pp. 3507–3512 (2012)
34. Guo, B., Jin, F.: Sliding mode and active disturbance rejection control to stabilization of one-dimensional anti-stable wave equations subject to disturbance in boundary input. IEEE Trans. Autom. Control **58**(5), 1269–1274 (2013)
35. Guo, B.Z., Jin, F.F.: The active disturbance rejection and sliding mode control approach to the stabilization of the Euler–Bernoulli beam equation with boundary input disturbance. Automatica **49**(9), 2911–2918 (2013)

36. Guo, B., Jin, F.: Output feedback stabilization for one-dimensional wave equation subject to boundary disturbance. IEEE Trans. Autom. Control **60**(3), 824–830 (2015)
37. Guo, B., Zhao, Z.: On convergence of tracking differentiator. Int. J. Control **84**(4), 693–701 (2011)
38. Guo, B.Z., Zhao, Z.L.: On the convergence of an extended state observer for nonlinear systems with uncertainty. Syst. Control Lett. **60**(6), 420–430 (2011)
39. Guo, B.Z., Zhao, Z.L.: On convergence of non-linear extended state observer for multi-input multi-output systems with uncertainty. IET Control Theory Appl. **6**(15), 2375–2386 (2012)
40. Guo, B., Zhao, Z.: On convergence of nonlinear active disturbance rejection for MIMO system. SIAM J. Control Optim. **51**(2), 1727–1757 (2013)
41. Guo, B.Z., Zhao, Z.L.: Weak convergence of nonlinear high-gain tracking differentiator. IEEE Trans. Autom. Control **58**(4), 1074–1080 (2013)
42. Guo, B.Z., Zhao, Z.: Active disturbance rejection control: theoretical perspectives. Commun. Inf. Syst. **15**, 361–421 (2015). https://doi.org/10.4310/CIS.2015.v15.n3.a3
43. Han, J.: Active Disturbance Rejection Control Technique-the Technique for Estimating and Compensating the Uncertainties. National Defense Industry Press, Arlington (2008)
44. Han, J.: A class of extended state observers for uncertain systems. Control Decis. **10**, 85–88 (1995)
45. Han, J.: Extended state observer for a class of uncertain plants. Control Decis. **10**(1), 85–88 (1995)
46. Han, J.: Auto disturbances rejection controller and its applications. Control Decis. **13**(1), 19–23 (1998)
47. Han, J.: From PID to active disturbance rejection control. IEEE Trans. Ind. Electron. **56**(3), 900–906 (2009)
48. Han, J., Wang, W.: Nonlinear tracking-differentiator. Syst. Sci. Math. **14**(2), 177–183 (1994)
49. Heemels, W., Johansson, K., Tabuada, P.: An introduction to event-triggered and self-triggered control. In: 2012 IEEE 51st Annual Conference on Decision and Control (CDC), pp. 3270–3285 (2012)
50. Heemels, W., Donkers, M.: Model-based periodic event-triggered control for linear systems. Automatica **49**(3), 698–711 (2013)
51. Heemels, W., Donkers, M., Teel, A.: Periodic event-triggered control for linear systems. IEEE Trans. Autom. Control **58**(4), 847–861 (2013)
52. Henningsson, T., Johannesson, E., Cervin, A.: Sporadic event-based control of first-order linear stochastic systems. Automatica **44**(11), 2890–2895 (2008)
53. Hetel, L., Fridman, E.: Robust sampled-data control of switched affine systems. IEEE Trans. Autom. Control **58**(11), 2922–2928 (2013)
54. Ho, Y.C., Cao, X.R.: Perturbation Analysis of Discrete-Event Dynamic Systems. Kluwer Academic Publisher, Amsterdam (1991)
55. Huang, Y., Xu, K., Han, J., Lam, J.: Flight control design using extended state observer and non-smooth feedback. In: Proceedings of the 40th IEEE Conference on Decision and Control, 2001, vol. 1, pp. 223–228 (2001)
56. Huang, Y., Xue, W.: Active disturbance rejection control: methodology and theoretical analysis. ISA Trans. **53**(4), 963–976 (2014)
57. Huang, Y., Wang, J., Shi, D., Shi, L.: Toward event-triggered extended state observer. IEEE Trans. Autom. Control **63**(6), 1842–1849 (2018)
58. Huang, Y., Wang, J., Shi, D., Wu, J., Shi, L.: Event-triggered sampled-data control: an active disturbance rejection approach. IEEE/ASME Trans. Mechatron. **24**(5), 2052–2063 (2019)
59. Khalil, H.K.: High-gain observers in nonlinear feedback control. In: Nijmeijer, H., Fossen, T. (eds.) New Directions in nonlinear observer design. Lecture Notes in Control and Information Sciences, vol. 244, pp. 249–268. Springer, London (1999)
60. Kosut, R., Lau, M., Boyd, S.: Set-membership identification of systems with parametric and nonparametric uncertainty. IEEE Trans. Autom. Control **37**(7), 929–941 (1992)
61. Lemmon, M.: Event-triggered feedback in control, estimation, and optimization. In: Bemporad, A., Heemels, M., Johansson, M. (eds.) Networked Control Systems. Lecture Notes in Control and Information Sciences, vol. 406, pp. 293–358. Springer, London (2010)

62. Li, H., Shi, Y.: Event-triggered robust model predictive control of continuous-time nonlinear systems. Automatica **50**(5), 1507–1513 (2014)
63. Liberzon, D.: Switching in Systems and Control. Birkhauser, Basel(2003)
64. Liberzon, D., Nesic, D.: Input-to-state stabilization of linear systems with quantized state measurements. IEEE Trans. Autom. Control **52**(5), 767–781 (2007)
65. Lili, D., Qing, Z., Avanesov, D.: The design and implementation of driving mode control for vibrational gyroscopes. In: 2008 American Control Conference, pp. 4419–4424 (2008)
66. Liu, T., Jiang, Z.P.: A small-gain approach to robust event-triggered control of nonlinear systems. IEEE Trans. Autom. Control **60**(99), 2072–2085 (2015)
67. Liu, T., Jiang, Z.P.: Event-based control of nonlinear systems with partial state and output feedback. Automatica **53**, 10–22 (2015)
68. Liu, Q., Wang, Z., He, X., Zhou, D.: A survey of event-based strategies on control and estimation. Syst. Sci. Control Eng. **2**, 90–97 (2014)
69. Lunze, J., Lehmann, D.: A state-feedback approach to event-based control. Automatica **46**(1), 211–215 (2010)
70. Maeder, U., Borrelli, F., Morari, M.: Linear offset-free model predictive control. Automatica **45**(10), 2214–2222 (2009)
71. Marchand, N., Durand, S., Castellanos, J.: A general formula for event-based stabilization of nonlinear systems. IEEE Trans. Autom. Control **58**(5), 1332–1337 (2013)
72. Mazo, M., Cao, M.: Asynchronous decentralized event-triggered control. Automatica **50**(12), 3197–3203 (2014)
73. Mazo, M., Tabuada, P.: Decentralized event-triggered control over wireless sensor/actuator networks. IEEE Trans. Autom. Control **56**(10), 2456–2461 (2011)
74. Meng, X., Chen, T.: Optimal sampling and performance comparison of periodic and event based impulse control. IEEE Trans. Autom. Control **57**(12), 3252–3259 (2012)
75. Molin, A., Hirche, S.: On the optimality of certainty equivalence for event-triggered control systems. IEEE Trans. Autom. Control **58**(2), 470–474 (2013)
76. Morari, M., Maeder, U.: Nonlinear offset-free model predictive control. Automatica **48**(9), 2059–2067 (2012)
77. Mustafa, G., Chen, T.: h_∞ filtering for nonuniformly sampled systems: a Markovian jump systems approach. Syst. Control Lett. **60**(10), 871–876 (2011)
78. Pannocchia, G., Rawlings, J.B.: Disturbance models for offset-free model-predictive control. AIChE J. **49**(2), 426–437 (2003)
79. Peng, C., Yang, T.C.: Event-triggered communication and control co-design for networked control systems. Automatica **49**(5), 1326–1332 (2013)
80. Postoyan, R., Bragagnolo, M.C., Galbrun, E., Daafouz, J., Nesic, D., Castelan, E.B.: Event-triggered tracking control of unicycle mobile robots. Automatica **52**, 302–308 (2015)
81. Premaratne, U., Halgamuge, S., Mareels, I.: Event triggered adaptive differential modulation: a new method for traffic reduction in networked control systems. IEEE Trans. Autom. Control **58**(7), 1696–1706 (2013)
82. Quevedo, D., Gupta, V., Ma, W.J., Yuksel, S.: Stochastic stability of event-triggered anytime control. IEEE Trans. Autom. Control **59**(12), 3373–3379 (2014)
83. Ramirez-Neria, M., Garcia-Antonio, J., Sira-Ramirez, H., Velasco-Villa, M., Castro-Linares, R.: On the linear active rejection control of thomson's jumping ring. Am. Control Conf. (ACC) **2013**, 6643–6648 (2013)
84. Shi, D., Shi, L., Chen, T.: Event-based State Estimation: A Stochastic Perspective. Springer, Berlin (2016). https://doi.org/10.1007/978-3-319-26606-0
85. Shi, D., Xue, J., Zhao, L., Wang, J., Huang, Y.: Event-triggered active disturbance rejection control of dc torque motors. IEEE/ASME Trans. Mechatron. **22**(5), 2277–2287 (2017)
86. Sira Ramirez, H., Linares Flores, J., Garcia Rodriguez, C., Contreras Ordaz, M.: On the control of the permanent magnet synchronous motor: an active disturbance rejection control approach. IEEE Trans. Control Syst. Technol. **22**(5), 2056–2063 (2014)
87. Sontag, E.D.: Mathematical Control Theory Deterministic Finite Dimensional Systems. Springer, Berlin (1998)

88. Sun, Z., Ge, S.: Stability Theory of Switched Dynamical Systems. Springer, Berlin (2011)
89. Sun, B., Gao, Z.: A dsp-based active disturbance rejection control design for a 1-kw h-bridge dc-dc power converter. IEEE Trans. Ind. Electron. **52**(5), 1271–1277 (2005)
90. Sun, B., Gao, Z.: A DSP-based active disturbance rejection control design for a 1-kw H-bridge DC-DC power converter. IEEE Trans. Ind. Electron. **52**(5), 1271–1277 (2005)
91. Sun, J., Yang, J., Li, S., Zheng, W.X.: Sampled-data-based event-triggered active disturbance rejection control for disturbed systems in networked environment. IEEE Trans. Cybern. **49**(2), 556–566 (2019)
92. Tabuada, P.: Event-triggered real-time scheduling of stabilizing control tasks. IEEE Trans. Autom. Control **52**(9), 1680–1685 (2007)
93. Tao, J., Sun, Q., Sun, H., Chen, Z., Dehmer, M., Sun, M.: Dynamic modeling and trajectory tracking control of parafoil system in wind environments. IEEE/ASME Trans. Mechatron. **22**(6), 2736–2745 (2017). https://doi.org/10.1109/TMECH.2017.2766882
94. Wang, H., Huang, H.: Properties and applications of ESO. Control Decis. **28**(7), 1078 (2013)
95. Wang, X., Lemmon, M.: On event design in event-triggered feedback systems. Automatica **47**(10), 2319–2322 (2011)
96. Wang, B., Meng, X., Chen, T.: Event based pulse-modulated control of linear stochastic systems. IEEE Trans. Autom. Control **59**(8), 2144–2150 (2014)
97. Wang, J., Tan, C., Wu, H.: Online shape modification of molecular weight distribution based on the principle of active disturbance rejection controller. IEEE Access **7**, 53163–53171 (2019)
98. Xue, W., Huang, Y., Yang, X.: What kinds of system can be used as tracking-differentiator. In: 29th Chinese Control Conference (CCC), 2010, pp. 6113–6120 (2010)
99. Xue, W., Huang, Y.: Comparison of the DOB based control, a special kind of PID control and ADRC. In: American Control Conference (ACC), pp. 4373–4379 (2011)
100. Xue, W., Huang, Y.: On performance analysis of ADRC for a class of MIMO lower-triangular nonlinear uncertain systems. ISA Trans. **53**(4), 955–962 (2014)
101. Yang, X., Huang, Y.: Capabilities of extended state observer for estimating uncertainties. In: American Control Conference, pp. 3700–3705 (2009)
102. Yang, J., Cui, H., Li, S., Zolotas, A.: Optimized active disturbance rejection control for dc-dc buck converters with uncertainties using a reduced-order gpi observer. IEEE Trans. Circuits Syst. I: Regul. Pap. **65**(2), 832–841 (2018)
103. Yuksel, S.: Jointly optimal lqg quantization and control policies for multi-dimensional systems. IEEE Trans. Autom. Control **59**(6), 1612–1617 (2014)
104. Zhang, J., Zheng, W.X., Xu, H., Xia, Y.: Observer-based event-driven control for discrete-time systems with disturbance rejection. IEEE Transactions on Cybernetics, pp. 1–11 (2019)
105. Zhao, C., Huang, Y.: ADRC based input disturbance rejection for minimum-phase plants with unknown orders and/or uncertain relative degrees. J. Syst. Sci. Complex **25**, 625–640 (2012)
106. Zheng, Q., Chen, Z., Gao, Z.: A practical approach to disturbance decoupling control. Control Eng. Pract. **17**(9), 1016–1025 (2009). https://doi.org/10.1016/j.conengprac.2009.03.005, http://www.sciencedirect.com/science/article/pii/S0967066109000550
107. Zheng, Q., Gao, L., Gao, Z.: On stability analysis of active disturbance rejection control for nonlinear time-varying plants with unknown dynamics. In: 46th IEEE Conference on Decision and Control, pp. 3501–3506 (2007)
108. Zheng, Q., Gao, Z.: Disturbance rejection in mems gyroscope: problems and solutions. In: Proceedings of the 30th Chinese Control Conference, pp. 6334–6339 (2011)
109. Zheng, Q., Gao, Z.: Motion control design optimization: problem and solutions. Int. J. Intell. Control Syst. **10** (2005)
110. Zheng, Q., Gao, Z.: On practical applications of active disturbance rejection control. In: 29th Chinese Control Conference (CCC), vol. 17, pp. 6095–6100 (2010)
111. Zheng, Q., Gao, Z.: On practical applications of active disturbance rejection control. In: Proceedings of the 29th Chinese Control Conference, pp. 6095–6100 (2010)
112. Zheng, Q., Dong, L., Lee, D.H., Gao, Z.: Active disturbance rejection control for MEMS gyroscopes. IEEE Trans. Control Syst. Technol. **17**(6), 1432–1438 (2009)
113. Zhou, K., Doyle, J.C., Glover, K.: Robust and Optimal Control. Prentice Hall, Upper Saddle River (1996)

Part I
Theoretic Developments

Chapter 2
Performance Assessment of Discrete-Time Extended State Observers

Due to the existence of the event-triggered sampling protocols, controller design is usually performed based on the nonuniform discrete sampled data for networked control systems [14, 21, 22]. Before discussing the properties of the event-triggered ADRC strategy, this chapter first considers a less challenging situation that an ESO is applied based on the system information obtained through uniform discrete sampling. The aim is to understand the main problems that may arise in designing a nonuniform sampling protocol and identify the key ideas that may help our further analysis.

Specifically, this chapter evaluates the discrete-time ESO for a disturbed nonlinear discrete-time nonlinear system with internal uncertainties and external disturbances. Since the choice of the gain parameter dominates the ESO's performance, the goal here is to provide the readers with an analysis of the observation error's behavior when the gain parameter varies from zero to infinity. It is shown that there exists a certain threshold such that if the designed gain parameter is smaller than this threshold, the observation error at steady state will become unbounded. Furthermore, an upper bound on the asymptotic observation error is also proposed by utilizing set-theoretic techniques. Moreover, a tighter upper bound on the asymptotic observation error is proposed for the case that the closed-loop matrix of the ESO is nonnegative. We evaluate the proposed theoretical results by numerical simulations and experimental verification on a DC motor loading platform.

2.1 Problem Formulation

2.1.1 Discrete-Time Nonlinear System

Consider a nonlinear discrete-time n dimensional control system with internal uncertainty and external disturbance (Fig. 2.1):

D. Shi et al., *Event-Triggered Active Disturbance Rejection Control*,
Studies in Systems, Decision and Control 356,
https://doi.org/10.1007/978-981-16-0293-1_2

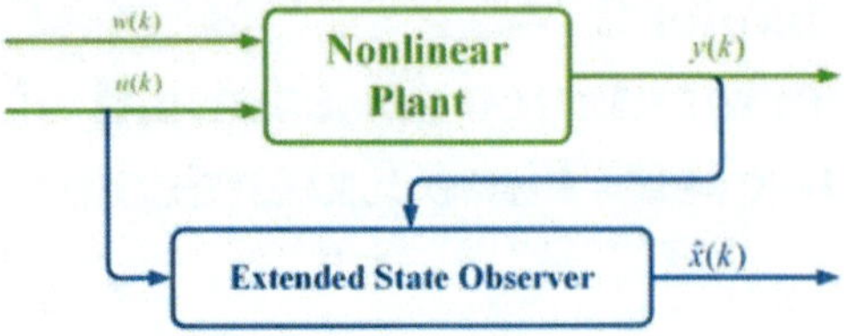

Fig. 2.1 Schematic of the extended state observer

$$
\begin{aligned}
x_i(k+1) &= x_{i+1}(k), \quad i \in \{1, 2, \ldots, n-1\}, \\
&\vdots \\
x_n(k+1) &= f(x(k), w(k)) + bu(k), \\
y(k) &= x_1(k),
\end{aligned} \tag{2.1}
$$

where $x := [x_1, \ldots, x_n]^{\mathrm{T}} \in \mathbb{R}^n$ is the system state, $w \in \mathbb{R}$ is the external disturbance, function $f : \mathbb{R}^n \times \mathbb{R} \to \mathbb{R}$ denotes the total disturbances including external disturbance and internal uncertainty, $y \in \mathbb{R}$ is the output signal, $u \in \mathbb{R}$ is the control signal and b is the unknown input channel parameter.

In the literature of ADRC (e.g., [2, 3, 20]), the core idea of ESO is to treat the internal uncertainty and external disturbance as an extended state, namely the $n+1$th state for an nth order system, and then observe this extended state by designing an observer. According to this strategy, the extended state is normally defined as

$$
x_{n+1}(k) := f(x(k), w(k)) + (b - b_0)u(k), \tag{2.2}
$$

where b_0 is the nominal parameter of b in (2.1). We note again that the main task of x_{n+1} is to represent these uncertainties and disturbances in the sense that neither their explicit expressions nor efficient estimates are available for controller design.

Before introducing ESO, we present the following assumption on the nonlinear uncertainty $f(\cdot, \cdot)$.

Assumption 2.1 With a bounded input u, there exists a constant $M_f > 0$ such that the solution x in (2.1) is bounded and the function $f(x, w)$ satisfies

$$
|f(x, w) + (b - b_0)u| \leq M_f. \tag{2.3}
$$

In the development of ADRC, ESO always serves as an important component, which was developed by Han and his co-workers in the 1990s [6, 7]. At that time, the motivation of developing this "new" control scheme was to extend the traditional PID controllers to a more delicate form, so that the performance of the control system can be improved; one fundamental principle in designing this scheme, however, is that the model-free properties of the PID controller should be inherited. With a large number of successful industrial applications (see, e.g., [1, 13]), the ADRC is easy to tune without requiring the information of the underlying process model. However, the explanation of this interesting model-free property is still unclear to control engineers, which motivates the recent investigations for theoretical support of the performance of ADRC [3, 5]. In such studies, one fundamental assumption we need to follow is that the structural properties of the process are assumed to be as unknown as possible, to be consistent with the "model-free" principle in ADRC design. One way of interpreting this black-box situation is to assume the process to be bounded, which explains why Assumption 2.1 is utilized in this chapter and is also a commonly used assumption to analyze the performance of ADRC [3, 5]. Similar assumptions will be adopted when we are designing and analyzing ET-ADRC in the next chapters.

2.1.2 Discrete-Time Extended State Observer

In this chapter, the discrete-time extended state observer is designed as

$$\begin{aligned} \hat{x}_i(k+1) &= \hat{x}_{i+1}(k) + a_i \cdot \varepsilon^{n-i} \cdot \frac{y(k) - \hat{x}_1(k)}{\varepsilon^n}, \\ &\vdots \\ \hat{x}_n(k+1) &= \hat{x}_{n+1}(k) + a_n \cdot \frac{y(k) - \hat{x}_1(k)}{\varepsilon^n} + b_0 u(k), \\ \hat{x}_{n+1}(k+1) &= a_{n+1} \cdot \frac{1}{\varepsilon} \cdot \frac{y(k) - \hat{x}_1(k)}{\varepsilon^n}, \end{aligned} \tag{2.4}$$

where $i \in \{1, \ldots, n-1\}$, $\hat{x} := [\hat{x}_1, \ldots, \hat{x}_{n+1}]^{\mathrm{T}} \in \mathbb{R}^{n+1}$ denotes the observer state, $\varepsilon > 0$ is the designed gain parameter, $\hat{x}_{n+1}$ denotes the observation of the extended state x_{n+1}. Note that $\{a_i\}$ are the observer parameters to be designed with $a_{n+1} \neq 0$. Since the nonlinear system function $f(\cdot)$ is unknown, the available information of $f(\cdot)$ is its bound as shown in Assumption 2.1. Actually, if an approximation $\hat{f}(\cdot)$ for the nonlinear part can be obtained such that

$$|f(x, w) - \hat{f}(\cdot) + (b - b_0)u| \leq M_f', \tag{2.5}$$

and $M_f' \leq M_f$, this approximation can be applied in the design of the extended state observer (e.g., this approximation can be added in the nth term of the ESO dynamics in (2.4)).

2.1.3 *Description of the Problem*

To further characterize the observation performance, we define the observation error variables $e_i(\cdot)$ and $\sigma_i(\cdot)$ by

$$e_i(k) = x_i(k) - \hat{x}_i(k), \tag{2.6}$$

$$\begin{aligned}\sigma_i(k) &= (x_i(k) - \hat{x}_i(k))/\varepsilon^{n+1-i}, \\ &= e_i(k)/\varepsilon^{n+1-i}, \quad i \in \{1, \ldots, n+1\},\end{aligned} \tag{2.7}$$

respectively, and write $e := [e_1, \ldots, e_{n+1}]^{\mathrm{T}}$, $\sigma := [\sigma_1, \ldots, \sigma_{n+1}]^{\mathrm{T}}$. In the literature of ESO [3], it is known that the observation performance can be adjusted by tuning the gain parameter ε; however, the quantitative relationship between the observation error at steady state and the high gain parameter ε is not clear for ESO in its discrete-time application. Specifically, the following problems will be considered in this chapter.

1. Would the observation error $e(k)$ remain bounded for all $\varepsilon \in (0, \infty)$ when $k \to \infty$? If not, what would be the design requirements for the gain parameter ε which guarantees the boundedness of $e(k)$?
2. If a well-designed ε guarantees the boundedness of $\lim_{k\to\infty} e(k)$, would it be possible to further provide upper bounds for $\lim_{k\to\infty} e(k)$ in terms of ε?

The discussions on the above problems in this chapter will provide interesting insights into discrete-time ESO and help better understand the relationship between the observation performance and the gain parameter design.

2.2 Observation Performance Analysis of Discrete-Time Extended State Observer

In this section, we evaluate the performance of the ESO and analyze the boundedness of the observation error.

2.2.1 *Gain Parameters Design Based on Spectral Radius Method*

The following result establishes the condition on the high gain parameter ε design such that the boundedness of $e(k)$ can be guaranteed.

Theorem 2.1 *Consider the plant model in* (2.1) *and the discrete-time extended state observer in* (2.4). *If Assumption* 2.1 *holds, then for any initial values of* x *and* $\hat{x}$, *there exist* $\varepsilon_0 > 0$ *and* $M_\sigma(\varepsilon) > 0$ *such that the following properties hold:*

1. $\forall \varepsilon \in (0, \varepsilon_0)$, $\limsup_{k\to\infty} |x_i(k) - \hat{x}_i(k)| \to \infty$,
2. $\forall \varepsilon \in [\varepsilon_0, \infty)$, $\limsup_{k\to\infty} |x_i(k) - \hat{x}_i(k)| \leq \varepsilon^{n+1-i} M_\sigma(\varepsilon)$,

where $1 \leq i \leq n+1$.

Proof According to the definition of $\sigma(k)$ in (2.7) and the dynamics in (2.1) and (2.4), we obtain that

$$\begin{aligned} \sigma_i(k+1) &= -\frac{a_i}{\varepsilon}\sigma_1(k) + \frac{1}{\varepsilon}\sigma_{i+1}(k),\ i \in \{1, 2, \ldots, n\}, \\ \sigma_{n+1}(k+1) &= -\frac{a_{n+1}}{\varepsilon}\sigma_1(k) + x_{n+1}(k+1). \end{aligned} \tag{2.8}$$

For notational brevity, we express (2.8) by

$$\sigma(k+1) = A\sigma(k) + Bx_{n+1}(k+1), \tag{2.9}$$

$$A = \frac{1}{\varepsilon}\begin{bmatrix} -a_1 & 1 & 0 & \cdots & 0 \\ \vdots & & & \ddots & \\ -a_n & 0 & \cdots & 0 & 1 \\ -a_{n+1} & 0 & \cdots & 0 & 0 \end{bmatrix}, B = \begin{bmatrix} 0 \\ \vdots \\ 0 \\ 1 \end{bmatrix}. \tag{2.10}$$

Define the matrix $\check{A}$ as

$$\check{A} := \varepsilon A,$$

and express the characteristic polynomial $D_{\check{A}}(\lambda)$ of $\check{A}$ as

$$D_{\check{A}}(\lambda) = \lambda^{n+1} + a_1\lambda^n + \cdots + a_n\lambda + a_{n+1}, \tag{2.11}$$

where the eigenvalues of $\check{A}$ are $\{\lambda_1, \lambda_2, \ldots, \lambda_{n+1}\}$. Therefore, the set of eigenvalues of A can be expressed as $\{\frac{\lambda_1}{\varepsilon}, \frac{\lambda_2}{\varepsilon}, \ldots, \frac{\lambda_{n+1}}{\varepsilon}\}$.

Note that the spectral radius $r(A)$ of matrices A and $\check{A}$ satisfy

$$r(A) = \max\left\{\frac{|\lambda_1|}{\varepsilon}, \frac{|\lambda_2|}{\varepsilon}, \ldots, \frac{|\lambda_{n+1}|}{\varepsilon}\right\},$$

$$r(\check{A}) = \max\{|\lambda_1|, |\lambda_2|, \ldots, |\lambda_{n+1}|\}.$$

Then we define ε_0 by $\varepsilon_0 := r(\check{A})$. If ε satisfies $\varepsilon < \varepsilon_0$ so that $r(A) > 1$, $\sigma(k)$ will be divergent. Consequently, to guarantee the performance of the ESO, the gain parameter ε should be designed to satisfy $\varepsilon > \varepsilon_0$.

Since $|\frac{\lambda_i}{\varepsilon}| < 1$ holds when $\varepsilon > \varepsilon_0$, system (2.9) is asymptotically stable and Bounded-Input-Bounded-State (BIBS) stable. Considering the boundedness of $x_{n+1}(k)$, namely, $|x_{n+1}(k)| < M_f$, the bounded property of the observation error $\sigma(k)$ can be obtained according to the BIBS stability of the dynamics in (2.9). Hence, there exists $M'_\sigma(\varepsilon) > 0$ such that $\|\sigma(k)\|_\infty < M'_\sigma(\varepsilon)$. Furthermore, from the definition of $\sigma(k)$ from (2.7), we have

$$\limsup_{k\to\infty}|x_i(k) - \hat{x}_i(k)| < \varepsilon^{n+1-i} M'_\sigma(\varepsilon). \tag{2.12}$$

Finally, we consider the case that $\varepsilon = \varepsilon_0$. In this case, the dynamics of observation error system is marginal stable. Considering the boundedness of x_{n+1} provided in Assumption 2.1, even though the observation error $\sigma(k)$ cannot decrease with the increase of time, we also can find a $M''_\sigma(\varepsilon)$ such that when $\varepsilon = \varepsilon_0$, we have

$$\limsup_{k\to\infty}|x_i(k) - \hat{x}_i(k)| < \varepsilon^{n+1-i} M''_\sigma(\varepsilon). \tag{2.13}$$

Therefore, we define M_σ by $M_\sigma := \max\{M'_\sigma, M''_\sigma\}$ to complete the proof. □

In Theorem 2.1, we observe that the gain parameter ε should be designed in a suitable range such that the observation performance of the proposed discrete-time extended state observer can be guaranteed. It is shown that the spectral radius ε_0 of matrix $\check{A}$ is the fence of this range. Note that when $\varepsilon = \varepsilon_0$, the observation error system is marginal stable since $r(\check{A}) = 1$. In this case, the observation error cannot be guaranteed to decrease with the increase of time. Thus, to guarantee the performance of this discrete-time ESO in a practical application, we need to set the gain parameter ε in (ε_0, ∞) excluding ε_0 and ∞.

2.2.2 *Observation Performance Analysis Through Ellipsoidal Set Approach*

In Theorem 2.1, we apply the infinity norm of $\sigma(k)$ to investigate the observation performance of the proposed discrete-time ESO. Thus the bound obtained is relatively conservative: when $\varepsilon \to \infty$, it is unclear whether the observation error will be infinity or not. On the other hand, the specific expression of $M_\sigma(\varepsilon)$ has not been presented in Theorem 2.1. To further solve these problems, an alternative technique is used to characterize the upper bound of the observation error; the core of our method is to bound the observation error with ellipsoidal sets and to calculate the sizes of the ellipsoids at steady state. To make this clear, we first introduce the concept of an ellipsoidal set introduced in [10].

2.2.2.1 Basic Introduction for Ellipsoidal Sets

Given $Y > 0$, an ellipsoidal set $\mathcal{Y} = \mathcal{E}(c, Y)$ is defined as

$$\mathcal{Y} := \{y \in \mathbb{R}^m | (y - c)^{\mathrm{T}} Y^{-1} (y - c) \leq 1, Y > 0\}. \tag{2.14}$$

If Y is singular and $Y \geq 0$, $\mathcal{Y} = \mathcal{E}(c, Y)$ is defined as

$$\mathcal{Y} := \{y \in \mathbb{R}^m | \langle l, y \rangle \leq \langle l, c \rangle + \langle l, Yl \rangle^{\frac{1}{2}}, \forall l \in \mathbb{R}^m\}. \tag{2.15}$$

In this definition, c is the center of the ellipsoidal set, which only describes the relative position of $\mathcal{Y}$. The size of the ellipsoidal set is determined by the shaping matrix $Y > 0$. Note that we define the size of $\mathcal{Y}$ as $\mathrm{Tr}Y$ which is equal to

$$\mathrm{Tr}Y = \chi_1^2 + \chi_2^2 + \cdots + \chi_m^2,$$

where $\{\chi_i\}$ is the set of the semi-major axes of the ellipsoid. According to the above definition, if $\mathrm{Tr}Y$ is bounded, then we say $\mathcal{Y}$ is bounded. In the following, the sets of possible observation errors caused by uncertainty and disturbance are captured by ellipsoidal sets. The upper bound of the observation error is obtained based on the sizes of these sets. To aid the analysis, we present the following lemmas on the properties of ellipsoidal sets.

Lemma 2.1 ([10]) *Let $p > 0$. We have*

$$\mathcal{E}(c_1, X_1) \oplus \mathcal{E}(c_2, X_2) \subseteq \mathcal{E}(c_1 + c_2, (1 + p^{-1})X_1 + (1 + p)X_2),$$

where $p = (\mathrm{Tr}X_1)^{1/2}/(\mathrm{Tr}X_2)^{1/2}$.

Lemma 2.2 ([10]) *Let $\mathcal{E}(a, Q) \subseteq \mathbb{R}^n$. Then $x \in \mathcal{E}(a, Q)$ is equivalent to $Ax + b \in \mathcal{E}(Aa + b, AQA^{\mathrm{T}})$.*

Lemma 2.3 (Lemma 2 in [15]) *Let $\mathcal{X}, \mathcal{Y} \subseteq \mathbb{R}^n$, and $T : \mathbb{R}^n \to \mathbb{R}^n$ be a linear transformation. Then $T(X \oplus Y) = (TX) \oplus (TY)$.*

Lemma 2.4 *Let P and Q be real-valued matrices with appropriate dimensions such that $P > 0$ and $Q \geq 0$. Then there exist $\overline{p} > 0$ and $\underline{p} > 0$ such that*

$$\underline{p}\mathrm{Tr}Q \leq \mathrm{Tr}QP \leq \overline{p}\mathrm{Tr}Q.$$

Proof Since P is real and Hermitian, according to Theorem 4.1.5 in [9], there exists U satisfying $U^{\mathrm{T}} = U^{-1}$ and $U^{\mathrm{T}}PU = \mathrm{diag}\{p_1, \ldots, p_n\}$. Then, denote $\underline{p}$ and $\overline{p}$ by

$$\underline{p} = \min\{p_1, \ldots, p_n\},$$

and

$$\overline{p} = \max\{p_1, \ldots, p_n\},$$

respectively. Through a similar line of arguments in the proof of Lemma 6 from [15], we obtain that

$$\underline{p}\mathrm{Tr}Q \leq \mathrm{Tr}QP = \mathrm{Tr}U^{\mathrm{T}}QU\mathrm{diag}\{p_i\} \leq \overline{p}\mathrm{Tr}Q. \tag{2.16}$$

This completes the proof. □

Define matrices $\overline{A}$ and $\overline{B}$ as

$$\overline{A} = P^{-1}AP, \overline{B} = P^{-1}B,$$

where

$$P = \mathrm{diag}\left\{\frac{1}{\varepsilon^n}, \frac{1}{\varepsilon^{n-1}}, \ldots, 1\right\}.$$

We directly obtain that

$$\overline{A} = \begin{bmatrix} \frac{-a_1}{\varepsilon} & 1 & 0 & \cdots & 0 \\ \vdots & & & \ddots & \\ \frac{-a_n}{\varepsilon^n} & 0 & \cdots & 0 & 1 \\ \frac{-a_{n+1}}{\varepsilon^{n+1}} & 0 & \cdots & 0 & 0 \end{bmatrix}, \overline{B} = \begin{bmatrix} 0 \\ \vdots \\ 0 \\ 1 \end{bmatrix}. \tag{2.17}$$

Write $\overline{A} := [\delta_1(\varepsilon)/\varepsilon, \delta_2, \ldots, \delta_{n+1}]$ and then the following properties can be identified.

Lemma 2.5 *Let $\overline{A}$ and $\overline{B}$ be defined in (2.17). Then*

1. $\overline{A}^i\overline{B} = \delta_{n+2-i}$, $1 \leq i \leq n$,
2. $\lim_{\varepsilon\to\infty} \overline{A}^{n+1} = 0$,
3. $\bigoplus_{i=0}^{n} \overline{A}^i\overline{B}\mathcal{U}_k \subseteq \mathcal{E}(0, (n+1)M_f^2 I)$ *holds where* $\mathcal{U}_k = \mathcal{E}(0, M_f^2)$.

Proof First, write $\overline{A}$ using its row or column vectors as

$$\overline{A} = [\delta_1(\varepsilon)/\varepsilon, \delta_2, \ldots, \delta_{n+1}] = \left[\gamma_{0,1}(\varepsilon), \ldots, \gamma_{0,n+1}(\varepsilon)\right]^{\mathrm{T}},$$

and define function $\phi_{\overline{A}} : \mathbb{R}^{n\times n} \times \mathbb{R}^{n\times n} \to \mathbb{R}^{n\times n}$ as

$$\phi_{\overline{A}}(X) = X\overline{A},$$

where $X \in \mathbb{R}^{n\times n}$. Then for any

$$X = [\alpha_1, \alpha_2, \ldots, \alpha_{n+1}] = \left[\beta_1, \beta_2, \ldots, \beta_{n+1}\right]^{\mathrm{T}}, \tag{2.18}$$

we have

$$\begin{aligned} X\overline{A} &= [\alpha_1^*/\varepsilon, \alpha_1, \alpha_2, \ldots, \alpha_n], \\ \alpha_1^* &= \left[\beta_1\delta_1(\varepsilon), \beta_2\delta_1(\varepsilon), \ldots, \beta_{n+1}\delta_1(\varepsilon)\right]^{\mathrm{T}}. \end{aligned} \tag{2.19}$$

Therefore, for $\phi_{\overline{A}}(\overline{A})$, we have

$$\phi_{\overline{A}}(\overline{A}) = \overline{A} \cdot \overline{A} = [\delta_1^*(\varepsilon)/\varepsilon, \delta_1(\varepsilon)/\varepsilon, \delta_2, \ldots, \delta_n],$$
$$\delta_1^*(\varepsilon) = [\gamma_{0,1}(\varepsilon)\delta_1(\varepsilon), \gamma_{0,2}(\varepsilon)\delta_1(\varepsilon), \ldots, \gamma_{0,n+1}(\varepsilon)\delta_1(\varepsilon)]^{\mathrm{T}}.$$

Write $\phi_{\overline{A}}{}^i(\overline{A})$ as

$$\phi_{\overline{A}}{}^i(\overline{A}) = \left[\gamma_{i,1}(\varepsilon), \ldots, \gamma_{i,n+1}(\varepsilon)\right]^{\mathrm{T}},$$

and then we will show that for $i \in [1, n-1]$,

$$\phi_{\overline{A}}{}^i(\overline{A}) = [\delta_i^*(\varepsilon)/\varepsilon, \ldots, \delta_1^*(\varepsilon)/\varepsilon, \delta_1(\varepsilon)/\varepsilon, \delta_2, \ldots, \delta_{n-i+1}],$$
$$\delta_i^*(\varepsilon) = [\gamma_{i-1,1}(\varepsilon)\delta_1(\varepsilon), \ldots, \gamma_{i-1,n+1}(\varepsilon)\delta_1(\varepsilon)]^{\mathrm{T}} \tag{2.20}$$

hold through mathematical induction.

First, when $i = 1$, the claim holds.

Next, suppose that the claim holds when $i = m$. We observe

$$\begin{aligned}
&\phi_{\overline{A}}{}^{m+1}(\overline{A}) = \phi_{\overline{A}}{}^m(\overline{A}) \cdot \overline{A} \\
&= [\delta_m^*(\varepsilon)/\varepsilon, \ldots, \delta_1^*(\varepsilon)/\varepsilon, \delta_1(\varepsilon)/\varepsilon, \delta_2, \ldots, \delta_{n-m+1}] \\
&\cdot [\delta_1(\varepsilon)/\varepsilon, \delta_2, \ldots, \delta_{n+1}] \\
&= [\delta_{m+1}^*(\varepsilon)/\varepsilon, \delta_m^*(\varepsilon)/\varepsilon, \ldots, \delta_1^*(\varepsilon)/\varepsilon, \delta_1(\varepsilon)/\varepsilon, \delta_2, \ldots, \delta_{n-m}],
\end{aligned}$$

holds, where $\delta_{m+1}^*(\varepsilon) = [\gamma_{m,1}(\varepsilon)\delta_1(\varepsilon), \ldots, \gamma_{m,n+1}(\varepsilon)\delta_1(\varepsilon)]^{\mathrm{T}}$. Thus when $i = m+1$, the claim also holds.

Since both the basis and the inductive step have been performed, the statement in (2.20) holds by mathematical induction and

$$\phi_{\overline{A}}{}^{i-1}(\overline{A}) \cdot \overline{B} = \overline{A}^i \overline{B} = \delta_{n+2-i}, \ 1 \le i \le n. \tag{2.21}$$

Then, for $\phi_{\overline{A}}{}^{(n)}(\overline{A})$, we have

$$\begin{aligned}
\phi_{\overline{A}}{}^{(n)}(\overline{A}) &= \overline{A}^n \cdot \overline{A} = \overline{A}^{n+1} \\
&= [\delta_n^*(\varepsilon)/\varepsilon, \ldots, \delta_1^*(\varepsilon)/\varepsilon, \delta_1(\varepsilon)/\varepsilon],
\end{aligned}$$
$$\delta_n^*(\varepsilon) = [\gamma_{n-1,1}(\varepsilon)\delta_1(\varepsilon), \ldots, \gamma_{n-1,n+1}(\varepsilon)\delta_1(\varepsilon)]^{\mathrm{T}}.$$

Since the order of ε in $\gamma_{i,j}(\varepsilon)\delta_1(\varepsilon)$ is smaller than zero, it is shown that

$$\lim_{\varepsilon \to \infty} \overline{A}^{n+1} = 0. \tag{2.22}$$

According to Lemma 2.2, we can define $\mathcal{X}_i$ as the following form:

$$\begin{aligned}
\mathcal{X}_1 &= \overline{B}\mathcal{U}_k = \mathcal{E}(0, X_1), X_1 = \overline{B}M_f^2\overline{B}^{\mathrm{T}}, \\
\mathcal{X}_2 &= \overline{A}\cdot\overline{B}\mathcal{U}_k = \mathcal{E}(0, X_2), X_2 = \overline{A}\cdot\overline{B}M_f^2(\overline{A}\cdot\overline{B})^{\mathrm{T}}, \\
&\vdots \\
\mathcal{X}_{n+1} &= \overline{A}^n\overline{B}\mathcal{U}_k = \mathcal{E}(0, X_{n+1}), X_{n+1} = \overline{A}^n\overline{B}M_f^2(\overline{A}^n\overline{B})^{\mathrm{T}}.
\end{aligned}$$

From (2.21), we have

$$\overline{A}^i\overline{B} = \delta_{n+2-i}.$$

Considering the structural properties of $\mathcal{X}_i$, it is easy to obtain that

$$\mathcal{X}_i = \delta_{n+3-i}\mathcal{U}_k = \mathcal{E}(0, X_i), X_i = \delta_{n+3-i}M_f^2\delta_{n+3-i}^{\mathrm{T}}.$$

Then, based on the definition of δ_i in (2.17), we obtain

$$\mathrm{Tr}X_i = M_f^2.$$

Now, we define $\mathcal{Y}_i, i \in [1, n]$ by

$$\begin{aligned}
\mathcal{Y}_1 &:= \mathcal{E}(0, Y_1) = \mathcal{E}(0, X_1), \\
Y_1 &:= M_f^2 \cdot \mathrm{diag}\{0, \ldots, 0, 1\}, \\
\mathcal{Y}_{i+1} &:= \mathcal{E}(0, Y_{i+1}) \supseteq \mathcal{Y}_i \oplus \mathcal{X}_{i+1}, \\
Y_{i+1} &:= (1 + 1/p_i)Y_i + (1 + p_i)X_{i+1}, \ 1 \le i \le n,
\end{aligned} \tag{2.23}$$

where $p_i = (\mathrm{Tr}Y_i)^{1/2}/(\mathrm{Tr}X_{i+1})^{1/2}$, and we will prove that for $1 \le i \le n$,

$$Y_{i+1} = M_f^2 \ldots \mathrm{diag}\{0, \cdots, 0, \underbrace{i+1, \ldots, i+1}_{i+1 \text{ times}}\}, \tag{2.24}$$

holds through mathematical induction.

First, when $i = 1$, we have

$$\begin{aligned}
\mathcal{Y}_2 &= \mathcal{E}(0, Y_2), \\
Y_2 &= (1 + 1/p_1)Y_1 + (1 + p_1)X_2 \\
&= M_f^2 \cdot \mathrm{diag}\{0, \ldots, 0, 2, 2\},
\end{aligned} \tag{2.25}$$

where $p_1 := (\mathrm{Tr}Y_1)^{1/2}/(\mathrm{Tr}X_2)^{1/2} = 1$. Thus, the claim holds.
Next, assume that when $i = m$ the claim holds such that we have for $\mathcal{Y}_{m+1}$,

$$\begin{aligned} p_m &= (\mathrm{Tr}Y_m)^{1/2}/(\mathrm{Tr}X_{m+1})^{1/2} \\ &= \sqrt{m^2 M_f^2/M_f^2} = m, \\ Y_{m+1} &= (1+1/p_m)Y_m + (1+p_m)X_{m+1} \\ &= (1+1/m)M_f^2 \cdot \mathrm{diag}\{0,\ldots,0,\underbrace{m,\ldots,m}_{m \text{ times}}\} + (1+m)M_f^2 \delta_{n+2-m}\delta_{n+2-m}^{\mathrm{T}} \\ &= M_f^2 \cdot \mathrm{diag}\{0,\ldots,0,\underbrace{m+1,\ldots,m+1}_{m+1 \text{ times}}\}. \end{aligned} \tag{2.26}$$

Thus when $i = m + 1$, the claim holds.
Since both the basis and the inductive step have been performed, by mathematical induction, the statement (2.24) holds. Therefore, when $i = n + 1$, we have

$$\bigoplus_{j=1}^{n+1} \mathcal{X}_j \subseteq \mathcal{Y}_{n+1} = \mathcal{E}(0, Y_{n+1}),$$

and

$$\bigoplus_{j=0}^{n} \overline{A}^j \overline{B} \mathcal{U}_k \subseteq \mathcal{E}(0, (n+1)M_f^2 I). \tag{2.27}$$

Combining the results in (2.21), (2.22), (2.27), we complete the proof. □

In Lemmas 2.1, 2.2, 2.3, 2.4, the properties of ellipsoidal sets and matrices are provided to support the essential deduction in this chapter. Specifically, according to Lemmas 2.1, 2.2, 2.3, the outer ellipsoidal approximation of the Minkowski sum between two ellipsoids can be obtained. Moreover, we present the upper/lower bound for a matrix with certain multiplication factors in Lemma 2.4. Considering the specific properties of the proposed discrete-time ESO, the first two items of Lemma 2.5 are introduced to discuss the properties of $\overline{A}$ and $\overline{B}$ defined in (2.17) which represent the state matrix and input matrix of the observation error dynamics in (2.28), respectively. In particular, when $\varepsilon \to \infty$, we show that $\overline{A}^{n+1} = 0$. This result will be used in the analysis of the performance of the observer when $\varepsilon \to \infty$. Based on the simplified expression of $\overline{A}^j \overline{B}$, a critical preliminary result for the ellipsoidal set of the observation error (2.30) is introduced in the third item of Lemma 2.5. Specifically, according to Lemma 2.5, we can obtain an upper bound of the Minkowski sum of error ellipsoidal sets, which enables the analysis of the observation error.

2.2.2.2 Analysis of Observation Performance for Discrete-Time ESO

Based on the results in Lemmas 2.1, 2.2, 2.3, 2.4, 2.5, we are ready to introduce the results on the analysis of the observation error for the proposed discrete-time ESO using ellipsoidal sets. In Theorem 2.1, it is shown that when $\varepsilon > \varepsilon_0$, the boundedness of $\lim_{k\to\infty} e(k)$ can be guaranteed. Next, an upper bound of the observation error will be provided based on ε in the following theorem:

Theorem 2.2 *Consider the plant model in (2.1) and the extended state observer in (2.4). If Assumption 2.1 holds, then for any initial values of x and $\hat{x}$, there exist $\varepsilon_0 > 0$, $0 < \alpha < 1$ and $\beta > 0$ so that when $\varepsilon > \varepsilon_0$, the following properties hold:*

1. $\lim_{k\to\infty} |x_i(k) - \hat{x}_i(k)| \leq \frac{M_f \beta \sqrt{n+1}}{1-\sqrt{\alpha}}$,
2. $\lim_{\varepsilon\to\infty, k\to\infty} |x_i(k) - \hat{x}_i(k)| \leq (n+1)M_f$,

where $1 \leq i \leq n+1$.

Proof The proof will be splitted into three steps.
Step 1: According to Theorem 2.1, we should design $\varepsilon > \varepsilon_0$ to guarantee the observation performance of the proposed discrete-time ESO. From the definition of $\sigma(k)$ in Eq. (2.7), we obtain that

$$e_i = \sigma_i \cdot \varepsilon^{n+1-i}.$$

Therefore, we design a matrix

$$P = \text{diag}\left\{\frac{1}{\varepsilon^n}, \frac{1}{\varepsilon^{n-1}}, \ldots, 1\right\}$$

such that

$$\sigma = Pe.$$

For notational brevity, we denote $\mu(k) := x_{n+1}(k+1)$, and then from (2.1)–(2.4), it is shown that

$$e(k+1) = \overline{A}e(k) + \overline{B}\mu(k), \tag{2.28}$$

where $\overline{A} = P^{-1}AP$, and $\overline{B} = P^{-1}B$. Since $\overline{A} = P^{-1}AP$, the eigenvalues of $\overline{A}$ are $\{\frac{\lambda_1}{\varepsilon}, \frac{\lambda_2}{\varepsilon}, \ldots, \frac{\lambda_{n+1}}{\varepsilon}\}$. First, we analyze the boundedness property of observation error at time instants

$$\{0, (n+1), 2(n+1), \ldots\}.$$

Define

$$\zeta(k) := e(nk+k), \tilde{A} := \overline{A}^{n+1}.$$

From (2.28), we have

$$\zeta(k+1) = \tilde{A}\zeta(k) + \textstyle\sum_{j=0}^{n} \overline{A}^j \overline{B}\mu((n+1)k - j + n). \tag{2.29}$$

Considering $|\mu(k)| \leq M_f$ and the definition of the ellipsoidal set, we can obtain that the observation error ellipsoidal set $\zeta(k)$ has the following form:

$$\mathcal{Z}_{k+1} = \tilde{A}\mathcal{Z}_k \oplus \left(\textstyle\bigoplus_{j=0}^{n} \overline{A}^j \overline{B}\mathcal{U}_k\right), \tag{2.30}$$

where $\mathcal{Z}_k = \mathcal{E}(\zeta(k), Z_k)$ and $\mathcal{U}_k = \mathcal{E}(0, M_f^2)$.

Step 2: We propose the upper bound of the observation error in the case that $\|\tilde{A}\|_2 > 1$. Since $\varepsilon > \varepsilon_0$, $\tilde{A}$ is stable. Furthermore, there exists a nonsingular matrix T defined by $T = P_s^{\frac{1}{2}}$ where $P_s > I > 0$ such that

$$\tilde{A}^{\mathrm{T}} P_s \tilde{A} - P_s + I = 0.$$

Define $\hat{A} := T\tilde{A}T^{-1}$, and we have

$$\begin{aligned}\hat{A}^{\mathrm{T}}\hat{A} &= (T^{-1})^{\mathrm{T}}\tilde{A}T^{\mathrm{T}}T\tilde{A}T^{-1} = T^{-1}(P_s - I)T^{-1} \\ &= I - P_s^{-1} < I.\end{aligned} \tag{2.31}$$

Therefore, $\|\hat{A}\|_2 \leq 1$ holds. Denote

$$\check{\zeta}(k) := T\zeta(k).$$

It is not difficult to show that

$$\check{\zeta}(k+1) = \hat{A}\check{\zeta}(k) + T\textstyle\sum_{j=0}^{n} \overline{A}^j \overline{B}\mu((n+1)k - j + n).$$

In this case, $\check{\zeta}(k)$ satisfies

$$\check{\mathcal{Z}}_{k+1} = \hat{A}\check{\mathcal{Z}}_k \oplus \left(\bigoplus_{j=0}^{n} T\overline{A}^j\overline{B}\mathcal{U}_k\right), \tag{2.32}$$

where $\check{\mathcal{Z}}_k = \mathcal{E}(\check{\zeta}(k), \check{Z}_k)$ and $\mathcal{U}_k = \mathcal{E}(0, M_f^2)$. According to Lemmas 2.3 and 2.5, let $\hat{\mathcal{Z}}_k := \mathcal{E}(\hat{\check{\zeta}}(k), \hat{Z}_k)$ and then the approximate observation error will be given by

$$\begin{aligned}\hat{\mathcal{Z}}_{k+1} &= \mathcal{E}(\hat{A}\hat{\check{\zeta}}(k), \hat{Z}_{k+1}) \\ \hat{Z}_{k+1} &= \left(1 + \frac{\sqrt{\mathrm{Tr}(n+1)M_f^2TT^{\mathrm{T}}}}{\sqrt{\mathrm{Tr}\hat{A}\hat{Z}_k\hat{A}^{\mathrm{T}}}}\right)\hat{A}\hat{Z}_k\hat{A}^{\mathrm{T}} \\ &\quad + \left(1 + \frac{\sqrt{\mathrm{Tr}\hat{A}\hat{Z}_k\hat{A}^{\mathrm{T}}}}{\sqrt{\mathrm{Tr}(n+1)M_f^2TT^{\mathrm{T}}}}\right)(n+1)M_f^2TT^{\mathrm{T}}.\end{aligned} \tag{2.33}$$

Since $\hat{Z}_k \geq 0$, we have $\mathrm{Tr}\hat{Z}_k \geq 0$ and

$$\begin{aligned}
\sqrt{\mathrm{Tr}\hat{Z}_{k+1}} &= \sqrt{\mathrm{Tr}\hat{A}\hat{Z}_k\hat{A}^{\mathrm{T}}} + \sqrt{\mathrm{Tr}(n+1)M_f^2TT^{\mathrm{T}}} \\
&= \sqrt{\mathrm{Tr}\hat{Z}_k\hat{A}^{\mathrm{T}}\hat{A}} + \sqrt{\mathrm{Tr}(n+1)M_f^2TT^{\mathrm{T}}} \\
&\leq \sqrt{\alpha}\sqrt{\mathrm{Tr}\hat{Z}_k} + \sqrt{\mathrm{Tr}(n+1)M_f^2P_s},
\end{aligned} \tag{2.34}$$

where $\alpha = \|\hat{A}\|_2^2$. Hence, when $k \to \infty$, we have

$$\lim_{k\to\infty} \sqrt{\mathrm{Tr}\check{Z}_k} \leq \lim_{k\to\infty} \sqrt{\mathrm{Tr}\hat{Z}_k} \leq \frac{\sqrt{\mathrm{Tr}(n+1)M_f^2P_s}}{1-\sqrt{\alpha}}, \tag{2.35}$$

and the boundedness of $\check{\mathcal{Z}}$ is obtained. According to the definition of Minkowski sum, $\check{\mathcal{Z}} = T\mathcal{Z}$ holds. Thus, the boundedness of $\hat{\mathcal{Z}}$ is equivalent to the boundedness of $\mathcal{Z}$. Since $T > 0$, from Lemmas 2.2 and 2.4, we have

$$\mathrm{Tr}\check{Z}_k = \mathrm{Tr}TZ_kT^{\mathrm{T}} = \mathrm{Tr}Z_kT^{\mathrm{T}}T = \mathrm{Tr}Z_kP_s \geq \beta\mathrm{Tr}Z_k,$$

where $\beta' = \min\{p_1, \ldots, p_{n+1}\}$ and p_i is the eigenvalue of P_s. Therefore

$$\lim_{k\to\infty} \sqrt{\mathrm{Tr}Z_k} \leq \frac{\sqrt{\mathrm{Tr}(n+1)M_f^2P_s}}{\sqrt{\beta'}(1-\sqrt{\alpha})}. \tag{2.36}$$

As $\varepsilon > \varepsilon_0$, $\lim\limits_{k\to\infty} \hat{A}^k = 0$. Thus, when $k \to \infty$, the center of $\mathcal{Z}_k$ is zero. From the definition of ellipsoidal set size and (2.36), we have

$$\lim_{k\to\infty} |x_i(k) - \hat{x}_i(k)| \leq \frac{M_f\beta\sqrt{n+1}}{1-\sqrt{\alpha}}, \tag{2.37}$$

where $\beta = \sqrt{\mathrm{Tr}P_s}/\sqrt{\min\{p_1, \ldots, p_{n+1}\}}$ and $\alpha = \|P_s^{1/2}\tilde{A}P_s^{-1/2}\|_2^2$.

Step 3: If $\|\tilde{A}\|_2 < 1$, let $P_s = I$ and we will obtain the same result as stated in (2.37). Further, we have $\lim_{\varepsilon\to\infty} \|\tilde{A}\|_2 = 0$ and

$$\lim_{k\to\infty} |x_i(k) - \hat{x}_i(k)| \leq (n+1)M_f. \tag{2.38}$$

Therefore, the upper bound of the observation error can be obtained at time instants $\{0, (n+1), 2(n+1), \ldots\}$. The analysis for the bound at other instants, for example,

$$\{1, (n+1)+1, 2(n+1)+1, \ldots\},$$

will follow the similar lines of arguments to those in (2.29)–(2.38). This completes the proof. □

In Theorem 2.2, we propose the upper bound in terms of ε, which converges to $(n+1)M_f$ when $\varepsilon \to \infty$. Due to the difficulty in calculating the Minkowski sum of ellipsoids, outer ellipsoidal approximations are utilized to obtain the upper bound of the observation error with the aid of some technical lemmas presented in [12]. Based on these lemmas, we further present two lemmas to show that an outer approximation of the error ellipsoidal set can be obtained, which helps investigate the upper bound of the error. Then an upper bound in terms of the size of the observation error is proposed. Next, we look into a special case and present the following results.

Theorem 2.3 *Consider the plant model in* (2.1) *and the extended state observer in* (2.4). *If Assumption 2.1 holds and* $a_i \leq 0$, *for any initial values of* x *and* $\hat{x}$, *there exists* $\varepsilon_0 > 0$ *such that*

$$\lim_{k\to\infty} |x_i(k) - \hat{x}_i(k)| \leq M_f \left| \frac{\varepsilon^{n+1} + \sum_{j=1}^{i-1} a_j \varepsilon^{n-j+1}}{\varepsilon^{n+1} + a_1\varepsilon^n \cdots + a_n\varepsilon + a_{n+1}} \right|,$$

$$\lim_{\varepsilon\to\infty, k\to\infty} |x_i(k) - \hat{x}_i(k)| \leq M_f,$$

for $1 \leq i \leq n+1$ *and* $\varepsilon > \varepsilon_0$.

Proof First, we rewrite (2.9) as

$$\sigma(k+1) = A^{k+1}\sigma(0) + \textstyle\sum_{j=0}^{k} A^{k-j} B x_{n+1}(j+1).$$

When $a_i \leq 0$, each element of $A^k B$ is not negative, and thus $A^k B$ is a nonnegative matrix. In this way, we have

$$\begin{aligned} &|\textstyle\sum_{j=0}^{k} A^{k-j} B x_{n+1}(j+1)| \\ \leq &B|x_{n+1}(k+1)| + AB|x_{n+1}(k)| + \cdots + A^k B|x_{n+1}(1)| \\ \leq &M_f(B + AB \cdots + A^k B). \end{aligned} \tag{2.39}$$

If $\varepsilon > \varepsilon_0$, then $\lim_{k\to\infty} A^k = 0$ and we can obtain the upper bound of $\sigma(k)$ by

$$\begin{aligned} \lim_{k\to\infty} |\sigma(k+1)| &= \lim_{k\to\infty} \left|\textstyle\sum_{j=0}^{k} A^{k-j} B x_{n+1}(j+1)\right| \\ &\leq \lim_{k\to\infty} M_f B \textstyle\sum_{i=0}^{k} A^i. \end{aligned} \tag{2.40}$$

Since $r(A) < 1$, $\lim\limits_{k\to\infty} \sum_{i=0}^{k} A^i$ is absolutely convergent and

$$\lim_{k\to\infty} \sum_{i=0}^{k} A^i = (I - A)^{-1}$$

based on Theorem 6.17 in [19]. In this way, we have

$$\lim_{k\to\infty} |\sigma(k+1)| \leq M_f(I-A)^{-1}B.$$

From the definition of $\sigma_i(k)$, $2 \leq i \leq n+1$, we have

$$\lim_{k\to\infty} |x(k)-\hat{x}(k)| \leq M_f \mathrm{diag}\{\varepsilon^n, \varepsilon^{n-1}, \ldots, 1\}(I-A)^{-1}B \tag{2.41}$$

$$= M_f \begin{bmatrix} \frac{\varepsilon^{n+1}}{\varepsilon^{n+1}+a_1\varepsilon^n\cdots+a_n\varepsilon+a_{n+1}} \\ \vdots \\ \frac{\varepsilon^{n+1}+\sum_{j=1}^{i-1}a_j\varepsilon^{n-j+1}}{\varepsilon^{n+1}+a_1\varepsilon^n\cdots+a_n\varepsilon+a_{n+1}} \\ \vdots \\ \frac{\varepsilon^{n+1}+a_1\varepsilon^n\cdots+a_n\varepsilon}{\varepsilon^{n+1}+a_1\varepsilon^n\cdots+a_n\varepsilon+a_{n+1}} \end{bmatrix}. \tag{2.42}$$

With the above analysis, the upper bound of the observation error can be obtained as

$$\lim_{k\to\infty,\varepsilon\to\infty} |x_i(k)-\hat{x}_i(k)| \leq M_f.$$

This completes the proof. □

2.2.3 *Extensions and Special Cases*

In this section, we discuss a few interesting extensions, which can be obtained based on the results introduced above.

2.2.3.1 Observation Performance with Measurement Noises

In many application scenarios, the effect of measurement noises is not negligible. In this section, we show that the obtained results can be extended to the case where the measurement noise exists. To do this, we define the output measurement by $y(k) := x_1(k)+d(k)$, where $d(k)$ denotes the measurement noise with the assumption that there exists $M_d > 0$ such that $|d(k)| \leq M_d$. Due to the existence of $d(k)$, the dynamics of observer states satisfy for $i = 1, \ldots, n-1$,

$$\hat{x}_i(k+1) = \hat{x}_{i+1}(k) + a_i \cdot \varepsilon^{n-i} \cdot \frac{x_1(k)-\hat{x}_1(k)+d(k)}{\varepsilon^n},$$

$$\hat{x}_n(k+1) = \hat{x}_{n+1}(k) + a_n \cdot \frac{x_1(k)-\hat{x}_1(k)+d(k)}{\varepsilon^n} + b_0u(k),$$

$$\hat{x}_{n+1}(k+1) = a_{n+1} \cdot \frac{1}{\varepsilon} \cdot \frac{x_1(k)-\hat{x}_1(k)+d(k)}{\varepsilon^n}.$$

Defining $\tilde{d}(k) := d(k)/\varepsilon^{n+1}$, we express (2.9) as

$$\sigma(k+1) = A\sigma(k) + Bx_{n+1}(k+1) + C\tilde{d}(k), \tag{2.43}$$

where C is defined as $C := [-a_1, \ldots, -a_{n+1}]^{\mathrm{T}}$. It is shown that the choice of ε_0 is independent of $d(k)$, suggesting the existence of measurement noise does not effect the basic conclusion that we should set $\varepsilon \in (\varepsilon_0, \infty)$. Next, we express (2.30) as

$$\mathcal{Z}_{k+1} = \tilde{A}\mathcal{Z}_k \oplus \left(\bigoplus_{j=0}^{n} \overline{A}^j \overline{B}\mathcal{U}_k\right) \oplus \left(\bigoplus_{j=0}^{n} \overline{A}^j \overline{C}\mathcal{D}_k\right),$$

where $\overline{C} := [-\frac{a_1}{\varepsilon}, \ldots, -\frac{-a_{n+1}}{\varepsilon^{n+1}}]$ and $\mathcal{D}_k = \mathcal{E}(0, M_d^2)$. For $\overline{A}^j\overline{C}$, through simple computation, we obtain

$$\overline{A}^j\overline{C} = \begin{bmatrix} \frac{f_{j,1}}{\varepsilon^{j+1}} \\ \vdots \\ \frac{f_{j,n}}{\varepsilon^{j+n}} \\ \frac{f_{j,n+1}}{\varepsilon^{j+n+1}} \end{bmatrix}, \tag{2.44}$$

for some constants $f_{j,k}$ which are determined by $[a_1, a_2, \ldots, a_{n+1}]$ but independent of ε. Thus, from Lemma 2.5, we obtain that

$$\bigoplus_{j=0}^{n} \overline{A}^j\overline{C}\mathcal{D}_k = \bigoplus_{j=0}^{n} \mathcal{E}(0, D_j), \tag{2.45}$$

where D_j is of the form

$$D_j = \frac{M_d^2}{\varepsilon^{2j+2}} \begin{bmatrix} f_{j,1}^2 & \frac{f_{j,1}f_{j,2}}{\varepsilon} & \cdots & \frac{f_{j,1}f_{j,n+1}}{\varepsilon^n} \\ \frac{f_{j,1}f_{j,2}}{\varepsilon} & \frac{f_{j,2}^2}{\varepsilon^2} & \cdots & \frac{f_{j,2}f_{j,n+1}}{\varepsilon^{n+1}} \\ \vdots & \vdots & \ddots & \vdots \\ \frac{f_{j,1}f_{j,n+1}}{\varepsilon^n} & \frac{f_{j,2}f_{j,n+1}}{\varepsilon^{n+1}} & \cdots & \frac{f_{j,n+1}^2}{\varepsilon^{2n}} \end{bmatrix}.$$

Then through a similar line of arguments in the proof of Theorem 2.2, we can analyze the observation performance combined with the measurement noise, although the bound obtained might not be as simple as that for the measurement noise-free case in Theorem 2.2.

2.2.3.2 Multi-input-Multi-output Systems

In this chapter, although a SISO system with disturbances is considered, the obtained results can be extended to certain types of MIMO systems. For instance, consider a MIMO system of the form

$$
\begin{aligned}
x_i(k+n_i) &= f_i(x_1(k), \ldots, x_m(k)) + b_i u_i(k), \\
y_i(k) &= x_{n_i,1}(k), i = 1, 2, \ldots, m,
\end{aligned}
$$

which can be transformed into a set of first-order difference equations as

$$
\begin{aligned}
x_{i,1}(k+1) &= x_{i,2}(k), \\
x_{i,2}(k+1) &= x_{i,3}(k), \\
&\vdots \\
x_{i,n_i}(k+1) &= f_i(x_1(k), \ldots, x_m(k)) + b_i u_i(k), \\
y_i(k) &= x_{i,1}(k), i = 1, 2, \ldots, m.
\end{aligned}
$$

By this way, an ESO for the considered MIMO system can be designed as

$$
\begin{aligned}
\hat{x}_{i,1}(k+1) &= \hat{x}_{i,2}(k) + a_{i,1} \cdot \varepsilon^{n_i-1} \cdot \frac{y_i(k) - \hat{x}_{i,1}(k)}{\varepsilon^{n_i}}, \\
&\vdots \\
\hat{x}_{i,n_i}(k+1) &= \hat{x}_{i,n+1}(k) + a_{i,n_i} \cdot \frac{y_i(k) - \hat{x}_{i,1}(k)}{\varepsilon^{n_i}} + b_{i,0} u_i(k), \\
\hat{x}_{i,n_i+1}(k+1) &= a_{i,n_i+1} \cdot \frac{1}{\varepsilon} \cdot \frac{y_i(k) - \hat{x}_{i,1}(k)}{\varepsilon^{n_i}}.
\end{aligned}
\tag{2.46}
$$

It is shown that the ESO in (2.46) bears a similar structure as that of the ESO in Eq. (2.4). Based on this observation, the theoretic results developed for the SISO system can be extended to this MIMO case under assumptions similar to Assumption A1 in [4], following a similar line of arguments as those in the proofs of Theorems 2.1, 2.2.

2.2.3.3 Systems with Unmatched Uncertainties

On the other hand, it is also possible to extend the results obtained in this chapter so that certain systems with unmatched uncertainties can be considered. Consider a system of the form

$$
\begin{aligned}
q(k+1) &= \Delta_q(x(k), q(k)), \\
x_i(k+1) &= x_{i+1}(k), \quad i \in \{1, 2, \ldots, n-1\}, \\
z(k+1) &= \Delta_z(z(k), u(k), q(k)), \\
x_n(k+1) &= g(x(k), w_1(k)) + z(k), \\
y(k) &= x_1(k),
\end{aligned}
\tag{2.47}
$$

where z is the auxiliary state, q is the unmeasurable part in the system, g is the nonlinear function, Δ_z and Δ_q describe the dynamics of z and q, respectively. In this case, the system state is composed of x and z. If we denote $w := [w_1^\top \; z^\top]^\top$ and define

$$f(x, w) := g(x, w_1) + z, \tag{2.48}$$

we are able to express the system in (2.47) into the form of Eq. (2.1) so that the developed results can be applied. The sacrifice of doing this is that the state observation for z cannot be precisely obtained, since from (2.48), the fine structural properties of Δ_z in (2.47) are not exploited.

2.2.3.4 Nonlinear Control Input Signal

In this chapter, the performance assessment of discrete-time extended state for a class of nonlinear systems with linear control input is considered. The obtained results, however, can be extended to case with the nonlinear control input terms by modifying the definition of extended state and reconstructing the observer accordingly. To show this, consider a system $x_i(k+1) = x_{i+1}(k)$, $x_n(k+1) = f(x(k), w(k)) + \eta(u(k))$, where $\eta(\cdot)$ is a continuous nonlinear function. In the case that $\eta(\cdot)$ is known, define the extended state by $x_{n+1}(k) := f(x(k), w(k))$ instead of that in (2.2), and replace inequality (2.3) by $|f(x, w)| \le M_f$. Then the proposed results can be extended to this case by considering an extended state observer of the same form as (2.4) but with its nth state of the form

$$\hat{x}_n(k+1) = \hat{x}_{n+1}(k) + a_n \cdot \frac{y(k) - \hat{x}_1(k)}{\varepsilon^n} + \eta(u(k)). \tag{2.49}$$

If $\eta(\cdot)$ is unknown, we can define the extended state in this case by $x_{n+1}(k) := f(x(k), w(k)) + \eta(u(k))$ and replace (2.3) by $|f(x, w) + \eta(u)| \le M_f$. Accordingly, we construct the nth term in (2.4) by

$$\hat{x}_n(k+1) = \hat{x}_{n+1}(k) + a_n \cdot \frac{y(k) - \hat{x}_1(k)}{\varepsilon^n}. \tag{2.50}$$

With these, the proposed results can be generalized to cover this case through similar lines of arguments.

2.3 Numerical Example

In this section, we illustrate the obtained results through the following numerical example. Consider a plant model of the form:

$$\begin{aligned} x_1(k+1) &= x_2(k), \\ x_2(k+1) &= x_3(k), \\ x_3(k+1) &= \text{sat}[\sin(x_1(k)) + e^{w(k)}] + u(k), \\ y(k) &= x_1(k), \end{aligned} \tag{2.51}$$

where $u(k) = 60\sin(1.3k) + 30\sin(2.2k) + 18\sin(3.1k)$, $w(k) = 20\cos(k)$, the extended state $x_4(k)$ is defined by

$$x_4(k) = \text{sat}(\sin(x_1(k)) + e^{w(k)}) + (1 - 0.95)u(k)$$

and $\text{sat}(x)$ is defined as

$$\text{sat}(x) = \begin{cases} \text{sgn}(x), & |x| > 1, \\ x, & |x| \leq 1. \end{cases} \tag{2.52}$$

Thus $|f(x, w) + (1 - 0.95)u| \leq 6.4$ holds such that Assumption 2.1 can be satisfied. For the considered system, the discrete-time ESO is designed of the form:

$$\begin{aligned} \hat{x}_1(k+1) &= \hat{x}_2(k) + a_1 \cdot (y(k) - \hat{x}_1(k))/\varepsilon, \\ \hat{x}_2(k+1) &= \hat{x}_3(k) + a_2 \cdot (y(k) - \hat{x}_1(k))/\varepsilon^2, \\ \hat{x}_3(k+1) &= \hat{x}_4(k) + a_3 \cdot (y(k) - \hat{x}_1(k))/\varepsilon^3 + b_0 u(k), \\ \hat{x}_4(k+1) &= a_4 \cdot (y(k) - \hat{x}_1(k))/\varepsilon^4. \end{aligned} \tag{2.53}$$

To perform the simulations, we take $b_0 = 0.95$, $x_1(0) = x_2(0) = x_3(0) = 20$ and $\hat{x}_1(0) = \hat{x}_2(0) = \hat{x}_3(0) = \hat{x}_4(0) = 0$. Next we select two groups of parameters to validate the introduced theoretic results.

Case 1: $[a_1, a_2, a_3, a_4] = [1, 1, 1, 1]$.

In this case, the characteristic polynomial of $\breve{A}$ is

$$D_{\breve{A}}(\lambda) = \lambda^4 + \lambda^3 + \lambda^2 + \lambda + 1.$$

The eigenvalues of A are

$$\left\{ \frac{0.31 + 0.95i}{\varepsilon}, \frac{0.31 - 0.95i}{\varepsilon}, \frac{-0.809 + 0.588i}{\varepsilon}, \frac{-0.809 - 0.588i}{\varepsilon} \right\}.$$

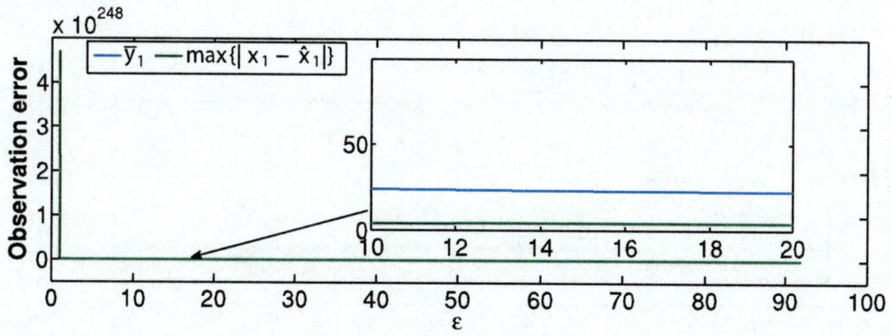

Fig. 2.2 The trend of the maximum observation error when ε varies from 0.98 to 96

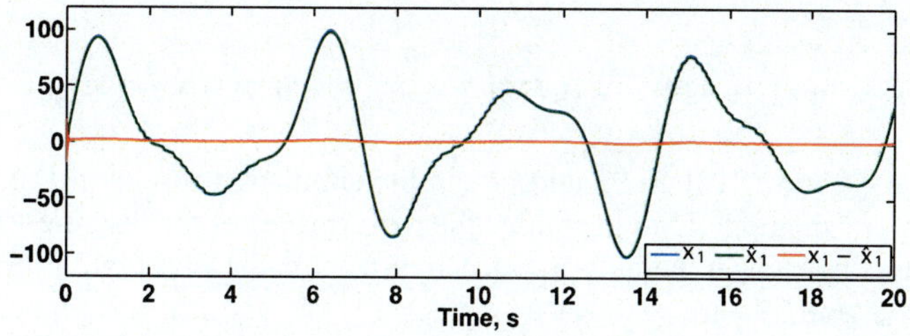

Fig. 2.3 The performance of the ESO (5.116) for the system (2.51) with $[a_1, a_2, a_3, a_4] = [1, 1, 1, 1]$

Therefore, let

$$\varepsilon_0 = \max\{|0.31 \pm 0.95i|, |-0.809 \pm 0.588i|\} = 1.$$

First, we vary ε from 0.98 to 96, and we illustrate the trend of maximum observation error in Fig. 2.2, where the upper bound $\overline{y}_1$ obtained from Theorem 2.2 is also provided. When $\varepsilon < \varepsilon_0$, it is shown that the upper bound of the observation error is divergent. Now, we set $\varepsilon = 1.05 > \varepsilon_0$, and the observation performance is shown in Fig. 2.3, where the ESO achieves a satisfactory performance in terms of a small and bounded observation error.

Case 2: $[a_1, a_2, a_3, a_4] = [-1, -1, -1, -1]$.

In this case, the characteristic polynomial of $\breve{A}$ is obtained as

$$D_{\breve{A}}(\lambda) = \lambda^4 - \lambda^3 - \lambda^2 - \lambda - 1.$$

The eigenvalues of A are

$$\left\{\frac{1.928}{\varepsilon}, \frac{-0.775}{\varepsilon}, \frac{-0.706 + 0.815i}{\varepsilon}, \frac{-0.706 - 0.815i}{\varepsilon}\right\}.$$

Thus we have

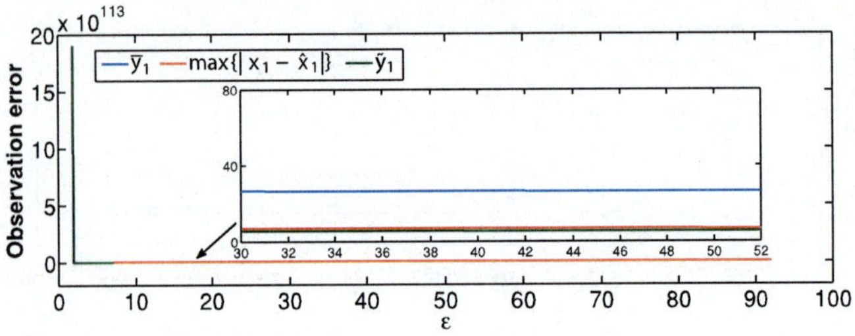

Fig. 2.4 The trend of the upper bound of observation error when ε changes from 1.91 to 92

$$\varepsilon_0 = \max\{|1.928|, |-0.775|, |-0.706 \pm 0.815i|\} = 1.928.$$

Then we vary ε from 1.91 to 92 and obtain the simulation results in Fig. 2.4. We find that $\varepsilon > \varepsilon_0$ should hold to guarantee the performance of the discrete-time ESO. For this case, two upper bounds $\overline{y}_1(\varepsilon)$ and $\tilde{y}_1(\varepsilon)$ for the observation error from Theorems 2.2 and 2.3 with

$$\overline{y}_1(\varepsilon) = \frac{M_f \beta \sqrt{n+1}}{1-\sqrt{\alpha}},$$

$$\tilde{y}_1(\varepsilon) = \frac{M_f \varepsilon^4}{\varepsilon^4 + a_1\varepsilon^3 + a_2\varepsilon^2 + a_3\varepsilon + a_4}$$

can be calculated, which are provided in Fig. 2.4. The numerical results show that (1) when $a_i < 0$, the upper bound $\tilde{y}_1$ is tighter than $\overline{y}_1$, since $\overline{y}_1$ is obtained based on the size of an ellipsoid that characterizes the sum of observation errors along different dimensions; (2) when $\varepsilon \to \infty$, the maximum observation error converges to M_f. Finally, we set $\varepsilon = 15 > \varepsilon_0$, and the performance of the observer is shown in Fig. 2.5.

On the other hand, due to the high gain properties of the ESO, we consider the high gain observer

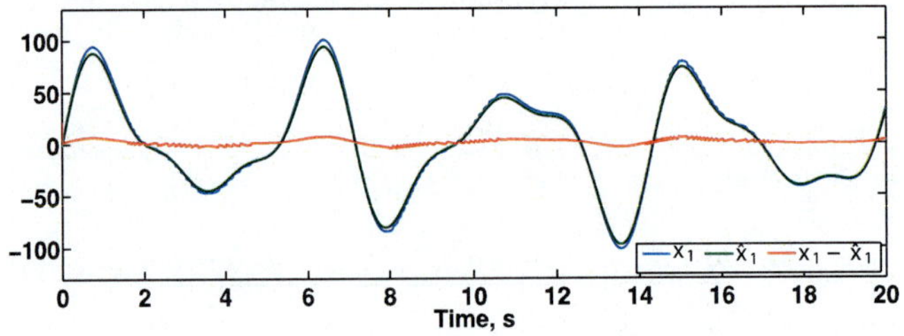

Fig. 2.5 The performance of the ESO (5.116) for the system (2.51) with $[a_1, a_2, a_3, a_4] = [-1, -1, -1, -1]$

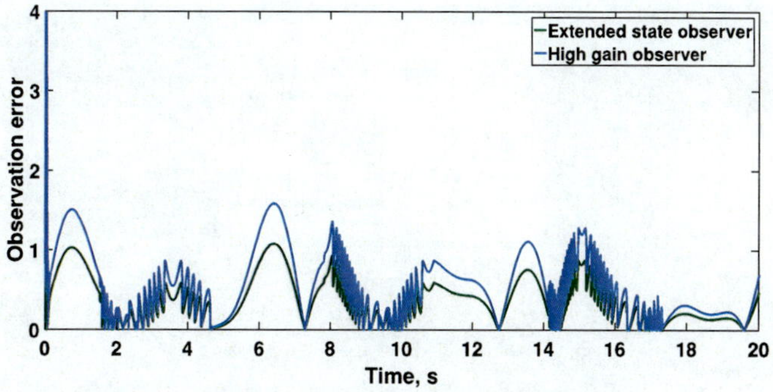

Fig. 2.6 The comparison of the performance between the extended state observer and the high gain observer

$$
\begin{aligned}
z_1(k+1) &= z_2(k) + h_1 \cdot (y(k) - z_1(k))/\varepsilon_h, \\
z_2(k+1) &= z_3(k) + h_2 \cdot (y(k) - z_1(k))/\varepsilon_h^2, \\
z_3(k+1) &= z_4(k) + h_3 \cdot (y(k) - z_1(k))/\varepsilon_h^3 + b_0 u(k),
\end{aligned} \tag{2.54}
$$

to compare the performance of the discrete-time ESO with that of the other existing observers. Set $z_1(0) = z_2(0) = z_3(0) = 0$, $[h_1, h_2, h_3] = [1, 1, 1]$ and $\varepsilon_h = 1.05$, and then illustrate the performance in Fig. 2.6. We find that for a large portion of simulation period, the observation error of discrete-time ESO is smaller than that of the high gain observer. The reason is that the extended state $\hat{x}_4$, which represents the observation of the nonlinear part and the disturbance, can be utilized in ESO to improve the performance.

2.4 Experimental Results

Throughout this book, a DC motor loading system (see Fig. 2.7) is used to validate the obtained theoretic results. The purpose of illustrating the results through experiments here is to tell whether the theoretic results obtained can indeed capture the properties of the ESO in practice; the reason for choosing DC motor is that successful applications of ADRC and ESO on motor systems have been extensively reported in the literature (e.g., [16, 17]), implying the potential feasibility of verifying the theoretical properties developed on this type of systems.

The experimental platform consists of two main parts: the motor loading platform and the operation platform (Fig. 2.8). The operation platform (the right part in Fig. 2.7) is mainly utilized to process the data and the signals obtained from the loading platform. The specific structure of the DC motor is shown in Fig. 4.3. It consists of a speed-torque sensor, an optical encoder, a magnetic brake, an inertial loading mechanism, and a DC motor. The DC motor is mounted on the bracket and connected

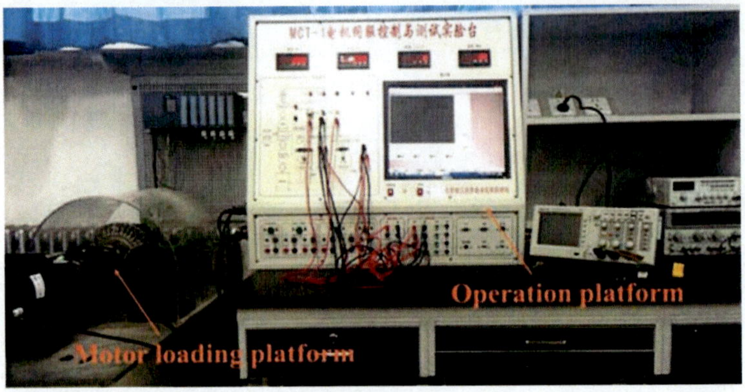

Fig. 2.7 The motor loading platform and the operation platform

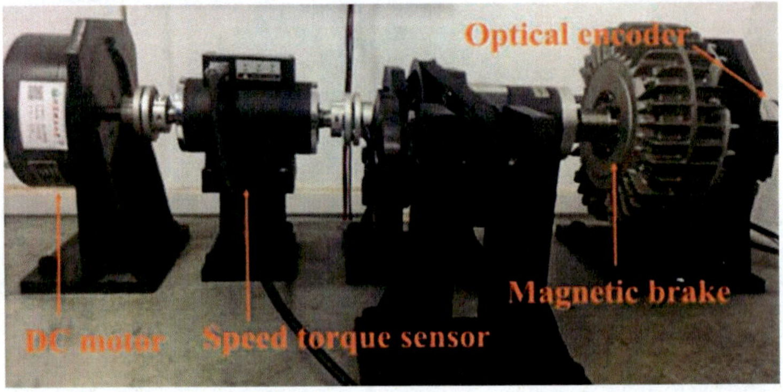

Fig. 2.8 The motor loading platform

with the speed-torque sensor. To apply inertia loads to the DC motor, the other end of the speed-torque sensor is connected with a timing belt pulley via a torque coupling. The output terminal of the belt pulley is connected with the magnetic brake, which is used to apply sliding friction load to the DC motor in this chapter.

The platform can load the motor using the magnetic brake with a maximum output torque of 20 N·m, a driving voltage of 24VDC, and a maximum driving current of 1210 mA. The control of the sliding friction torque is achieved by controlling the driving current. The platform offers sensors to measure the output torque and the speed of the DC motor which is with a maximum output speed of 325 r/min. The ranges of the torque and the speed are ±20 N·m and ±8000 r/min, respectively.

To verify the applicability of the proposed method to a DC motor system, we analyze the structure of this system as follows. First, the dynamics of the DC torque system can be described as

$$LJ\tau\dddot{y} + (LJ + RJ\tau)\ddot{y} + (RJ + 1.5p^2k_e^2\tau)\dot{y} + 1.5p^2k_e^2y$$
$$+ L\tau\ddot{F}_{fri} + (L + R\tau)\dot{F}_{fri} + RF_{fri} = 1.5pk_e\mu u_c.$$

where y denotes the rotor angle speed, u_c is the actual control signal, f_{fri} denotes the friction disturbance, and J, p, k_e, R, μ are the rotor inertia, the number of pole pairs, the phase-voltage coefficient, the resistance between any two phases and the equivalent driving gain, respectively. According to [18], F_{fri} is a nonlinear function of y such that in this place we can express it as $F_{fri}(y)$. Then, we discretize this system through finite-difference methods that

$$\frac{dz}{dt} \approx \frac{z[(n+1)T] - z(nT)}{T},$$
$$\frac{d^2z}{dt^2} \approx \frac{z[(n+2)T] - 2z[(n+1)T] + z(nT)}{T^2},$$
$$\frac{d^3z}{dt^3} \approx \frac{z[(n+3)T] - 3z[(n+2)T] + 3z[(n+1)T] - z(nT)}{T^3},$$

where T is the sample interval. It is not difficult to show that the discrete-time model of this system can be presented as

$$a_y y(k+3) + b_y y(k+2) + c_y y(k+1) + d_y y(k)$$
$$+ a_d \overline{F}_{fri}(y(k+2), y(k+1), y(k)) = a_u u_c(k),$$

for some parameters $a_y, b_y, c_y, d_y, a_d, a_u$, where $\overline{F}_{fri}(y(k+2), y(k+1), y(k))$ is deduced from F_{fri} and its derivatives. Therefore, if we choose

$$x_1(k) := y(k), \quad x_2(k) := y(k+1), \quad x_3(k) := y(k+2),$$

then we can describe the dynamics of the motor in the form of equations in (2.1). Considering the saturation phenomenon caused by hard constraints in engineering systems (e.g., operation restrictions and energy limitations [8, 11]), Assumption 2.1 is also satisfied. Note that $x_2(k), x_3(k)$ are the two delayed states of $x_1(k)$, and we can verify the observation performance of the proposed ESO by testing the output tracking performance for the DC motor.

For the DC motor platform, we design the ESO based on (2.4) with $n = 3$. First, we test the observer performance in the no-load case. Here, the initial values and observer parameters are set by

$$\hat{x}(0) = 0, \ [a_1, a_2, a_3, a_4] = [1, 1, 1, 1], \ \varepsilon = 1.05, \text{ and } b_0 = 0.223,$$

In Figs. 2.9 and 2.10, we show the performance of the ESO on observing the output of the DC motor. Besides, ESO is also tested in the loaded case with the same parameter setting. Figure 2.11 shows the experimental results of the observation performance when the DC motor is loaded with a 6 N·m torque through the magnetic brake. From

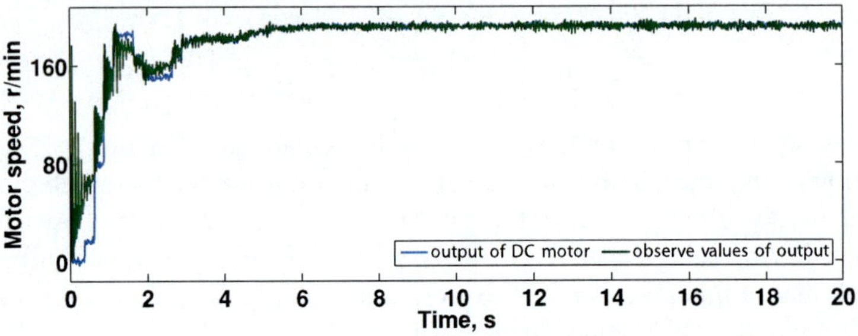

Fig. 2.9 The observation performance on the step output of DC motor with no-load

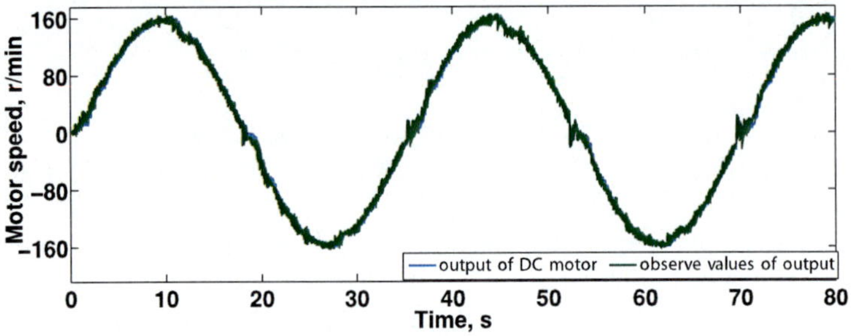

Fig. 2.10 The observation performance on the sine output of DC motor with no-load

Fig. 2.11, a steady-state error can be observed. The reason is that when the motor is loaded with torque, the system dynamics are changed, which leads to the change of the bound of the total disturbance. When the plant is changed, the parameters of the ESO should be modified to maintain good performance. To do this, we adjust the observer parameters to

$$[a_1, a_2, a_3, a_4] = [1, 1, 1, 1],\ \varepsilon = 1.203 \text{ and } b_0 = 0.125.$$

Then, as shown in Fig. 2.12, observation performance is improved.

In order to validate the results in Sect. 2.2.2, we set

$$[a_1, a_2, a_3, a_4] = [-1, -1, -1, -1] \text{ and } b_0 = 0.125,$$

and vary ε from 0.98 to 98. The experimental results also show that ε should satisfy $\varepsilon > \varepsilon_0$, otherwise the observation error will be unbounded. In Fig. 2.13, we observe that when $\varepsilon \to \infty$, the observation error converges to 11.104, which validates the convergence of the observation error. In Sect. 2.2.2, the upper bounds of the observation error obtained are based on the bound M_f of the extended state. Since not all the parameters of the DC motor are available, it is hard to find an accurate M_f. In this chapter, our choice is to choose M_f as the steady-state value of $|x_1(k) - \hat{x}_1(k)|$, which is known to be bounded by M_f according to Theorem 2.3, and is shown to

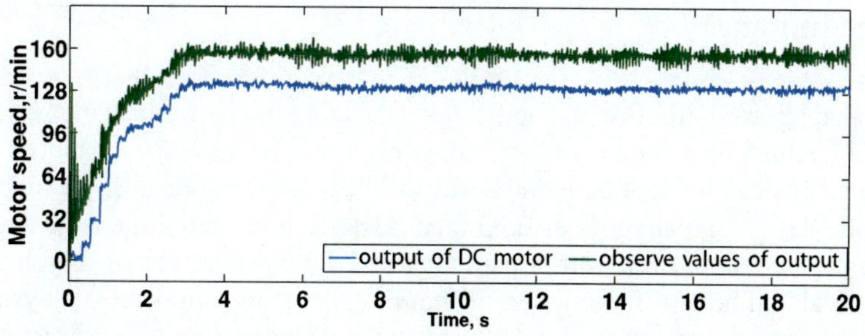

Fig. 2.11 The observation performance of the output of DC motor with 6 N·m torque when the ESO parameters are set as $[a_1, a_2, a_3, a_4] = [1, 1, 1, 1]$, $\varepsilon = 1.05$ and $b_0 = 0.223$

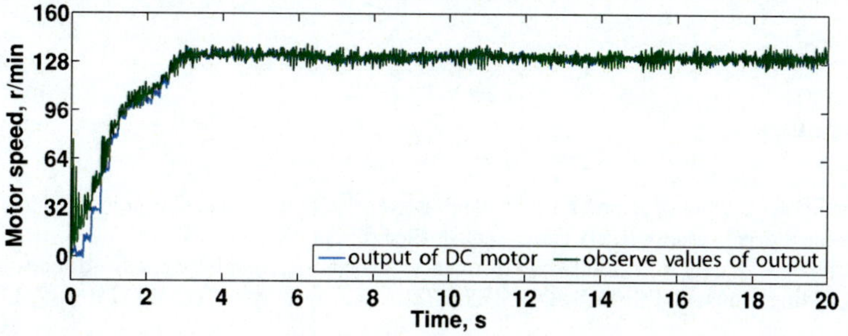

Fig. 2.12 The observation performance of the output of DC motor with 6 N·m torque when the ESO parameters are set as $[a_1, a_2, a_3, a_4] = [1, 1, 1, 1]$, $\varepsilon = 1.203$ and $b_0 = 0.125$

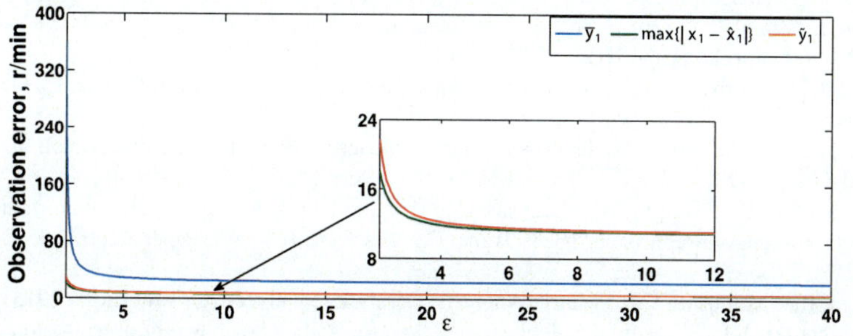

Fig. 2.13 The trend of the upper bound of observation error

be effective in the experimental results obtained. With this choice, the upper bounds $\overline{y}$ and $\tilde{y}_1$ obtained from Eqs. (2.37) and (2.42) are further illustrated in Fig. 2.13, respectively.

From the above experimental results, we have verified the main results developed in this chapter: (1) if $\varepsilon > \varepsilon_0$, the observation error of the ESO will be bounded; (2) the upper bounds effectively describe the behavior of steady-state observation error for different values of ε; (3) in particular, if $\varepsilon \to \infty$, the convergence of the upper bounds on the observation error can be observed.

2.5 Summary

In this chapter, we introduce an approach to analyzing the performance of discrete-time ESO. Although a number of interesting results can be developed in closed-form and the results can be verified by numerical and experimental results, the hindsight is that the design and analysis of ESO (and ADRC) in discrete-time is generally a challenging problem, even with a particular and simple structure of the observer. The general and helpful finding here is that the high gain parameter ϵ plays a key role in handling the effect of sampled data, and a different way of constructing the Lyapunov functions would be necessary due to the consideration of the ESO/ADRC structure. As will be shown in the next few chapters, these observations will be of significant importance in analyzing the stability of ET-ESO and ET-ADRC.

References

1. Erenturk, K.: Fractional-order $PI^{\lambda}D^{\mu}$ and active disturbance rejection control of nonlinear two-mass drive system. IEEE Trans. Indust. Electron. **60**(9), 3806–3813 (2013)
2. Godbole, A., Kolhe, J., Talole, S.: Performance analysis of generalized extended state observer in tackling sinusoidal disturbances. IEEE Trans. Control Syst. Technol. **21**(6), 2212–2223 (2013)
3. Guo, B., Zhao, Z.: On the convergence of an extended state observer for nonlinear systems with uncertainty. Syst. Control Lett. **60**(6), 420–430 (2011)
4. Guo, B., Zhao, Z.: On convergence of non-linear extended state observer for multi-input multi-output systems with uncertainty. IET Control Theory Appl. **6**(15), 2375–2386 (2012)
5. Guo, B., Zhao, Z.: On convergence of nonlinear active disturbance rejection for MIMO system. SIAM J. Control Optim. **51**(2), 1727–1757 (2013)
6. Han, J.: Active disturbance rejection controller and its applications. Control Decis. **13**(1), 19–23 (1998)
7. Han, J.: From PID to active disturbance rejection control. IEEE Trans. Indust. Electron. **56**(3), 900–906 (2009)
8. He, N., Shi, D.: Event-based robust sampled-data model predictive control: A non-monotonic lyapunov function approach. IEEE Trans. Circuits Syst. I: Regul. Paper **62**(10), 2555–2564 (2015)
9. Horn, R.A., Johnson, C.R.: Matrix Analysis. Cambridge University, Cambridge (2013)
10. Kurzhanski, V.I.: Ellipsoidal Calculus for Estimation and Control. Birkhäuser, Boston (1997)
11. Magni, L., Scattolini, R.: Model predictive control of continuous-time nonlinear systems with piecewise constant control. IEEE Trans. Autom. Control **49**(6), 900–906 (2004)
12. Noack, B., Klumpp, V., Hanebeck, U.: State estimation with sets of densities considering stochastic and systematic errors. In: 12th International Conference on Information Fusion, 2009, pp. 1751–1758 (2009)
13. Ramirez Neria, M., Garcia Antonio, J., Sira Ramirez, H., Velasco Villa, M., Castro Linares, R.: On the linear active rejection control of Thomson's jumping ring. In: American Control Conference (ACC), vol. 2013, pp. 6643–6648 (2013)
14. Shen, B., Wang, Z., Liu, X.: A stochastic sampled-data approach to distributed h_{∞} filtering in sensor networks. IEEE Trans. Circuits Syst. I: Regul. Papers **58**(9), 2237–2246 (2011)
15. Shi, D., Chen, T., Shi, L.: On set-valued Kalman filtering and its application to event-based state estimation. IEEE Trans. Autom. Control **60**(5), 1275–1290 (2015)

16. Sira Ramirez, H., Linares Flores, J., Garcia Rodriguez, C., Contreras Ordaz, M.: On the control of the permanent magnet synchronous motor: An active disturbance rejection control approach. IEEE Trans. Control Syst. Technol. **22**(5), 2056–2063 (2014)
17. Su, Y.X., Zheng, C.H., Duan, B.Y.: Automatic disturbances rejection controller for precise motion control of permanent-magnet synchronous motors. IEEE Trans. Indust. Electron. **52**(3), 814–823 (2005)
18. de Wit, C.C., Olsson, H., Astrom, K.J., Lischinsky, P.: A new model for control of systems with friction. IEEE Trans. Autom. Control **40**(3), 419–425 (1995)
19. Wu, Q.: Linear Algebra with Applications in Automatic Control. National Defense Industry Press (2011)
20. Zheng, Q., Gao, L., Gao, Z.: On stability analysis of active disturbance rejection control for nonlinear time-varying plants with unknown dynamics. In: 46th IEEE Conference on Decision and Control, pp. 3501–3506 (2007)
21. Zhou, B., Li, Z.Y., Lin, Z.: Stabilization of discrete-time systems with multiple actuator delays and saturations. IEEE Trans. Circuits Syst. I: Regul. Papers **60**(2), 389–400 (2013)
22. Zhou, B., Lin, Z.: Parametric Lyapunov equation approach to stabilization of discrete-time systems with input delay and saturation. IEEE Trans. Circuits Syst. I: Regul. Papers **58**(11), 2741–2754 (2011)

Chapter 3
Event-Triggered Extended State Observer

Based on the performance analysis on discrete-time ESO and the gain parameter design provided in Chap. 2, we consider the event-triggered extended state observer (ET-ESO) design for a continuous-time nonlinear system with uncertainty and disturbance in this chapter. The aim of the design is to guarantee the observation performance while saving communication source. Since the convergence of the ET-ESO directly depends on the design of the event-triggering condition, the primary task is to ensure an implementable event-triggering condition design that does not rely on the unknown plant state. For the ET-ESO introduced, we perform a rigorous analysis on its asymptotic performance. The obtained theoretical results are validated through numerical simulations.

3.1 Problem Formulation

3.1.1 Continuous-Time Nonlinear System

As depicted in Fig. 3.1, we consider an n-dimensional nonlinear continuous-time system with disturbance and uncertainty

$$\begin{aligned} \dot{x}_1(t) &= x_2(t), x_1(t_0) = x_{10}, \\ \dot{x}_2(t) &= x_3(t), x_2(t_0) = x_{20}, \\ &\vdots \\ \dot{x}_n(t) &= f(x(t), t) + u(t) + w(t), x_n(t_0) = x_{n0}, \\ y(t) &= x_1(t), \end{aligned} \tag{3.1}$$

D. Shi et al., *Event-Triggered Active Disturbance Rejection Control*,
Studies in Systems, Decision and Control 356,
https://doi.org/10.1007/978-981-16-0293-1_3

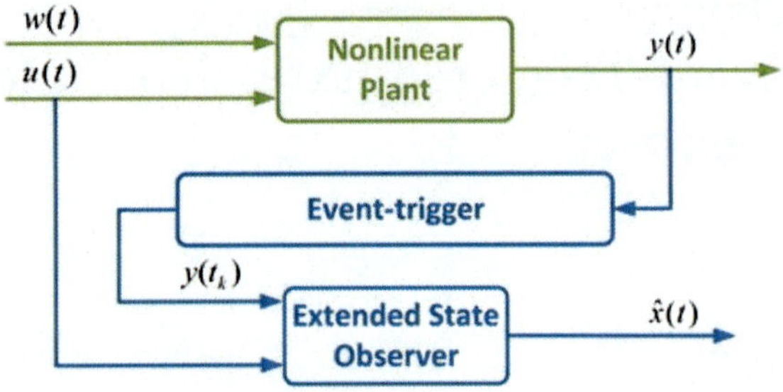

Fig. 3.1 Schematic of the extended state observer with event-trigger

where $x := [x_1, \ldots, x_n]^{\mathrm{T}} \in \mathbb{R}^n$ denotes the state of the plant, $w \in \mathbb{R}$ is the external disturbance, the function $f : \mathbb{R}^n \times \mathbb{R}_0^+ \rightarrow \mathbb{R}$ is the unknown nonlinear part, $y \in \mathbb{R}$ is the output signal, $u \in \mathbb{R}$ is the control signal and $x(t_0) := [x_{10}, \ldots, x_{n0}]^{\mathrm{T}} \in \mathbb{R}^n$ is the initial value of the state. Following the literature of ESO, an extended state is normally defined as

$$x_{n+1} := f(x, t) + w.$$

The system introduced above is composed of a chain of integrators with matched disturbance. A large number of systems (e.g., antenna pointing control systems [19], fast tool servo systems [33], and cavity dynamics of the superconducting RF cavities [31]) can be constructed into this form (3.1) if the relative degree conditions in [3] are satisfied. In this chapter, we still assume that the nonlinear function $f(x, t)$ is unknown in the sense that neither $f(x, t)$ nor its approximation $\hat{f}(x, t)$ is available in observer design.

Before introducing the ET-ESO, we first present the following assumptions for system (3.1), which are typically employed in the analysis of continuous-time ESO [7].

Assumption 3.1 The functions $f(x, t)$ and $w(t)$ are continuously differentiable with respect to their variables, and $f(x, t)$, $w(t)$, and $u(t)$ satisfy

$$|u(t)| + |f(x, t)| + |\dot{w}(t)| + \left|\frac{\partial f}{\partial t}\right| + \left|\frac{\partial f}{\partial x_i}\right| \leq c_0 + \sum_{j=1}^{n} c_j |x_j|^{\kappa},$$

for $i = 1, \ldots, n$, and some positive constants c_j and κ.

Assumption 3.2 There exist a constant $B > 0$ and a domain $\mathcal{X}$ such that with u and w satisfying Assumption 3.1, the solution $x_i(t)$ of system (3.1) and the external disturbance w are bounded by

$$|x_i(t)| + |w(t)| \leq B, \text{ for } i = 1, \ldots, n, \tag{3.2}$$

for initial value $x(t_0) \in \mathcal{X}$.

Note that in Assumption 3.1, we also allow u to be state-dependent such that $u(t) := u(x; t)$; this does not affect the implementation of the designed observer as only the numerical value of u is needed. A numerical example in simulation section is presented to show Assumptions 3.1, 3.2 can be easily satisfied.

3.1.2 *Event-Triggered Extended State Observer*

In this chapter, we consider an event-triggered extended state observer of the following form:

$$
\begin{aligned}
\dot{\hat{x}}_1(t) &= \hat{x}_2(t) + \varepsilon^{n-1} g_1\left(\frac{\xi(t) - \hat{x}_1(t)}{\varepsilon^n}\right), \hat{x}_1(t_0) = \hat{x}_{10}, \\
\dot{\hat{x}}_2(t) &= \hat{x}_3(t) + \varepsilon^{n-2} g_2\left(\frac{\xi(t) - \hat{x}_1(t)}{\varepsilon^n}\right), \hat{x}_2(t_0) = \hat{x}_{20}, \\
&\ \vdots \\
\dot{\hat{x}}_n(t) &= \hat{x}_{n+1}(t) + g_n\left(\frac{\xi(t) - \hat{x}_1(t)}{\varepsilon^n}\right) + u(t), \hat{x}_n(t_0) = \hat{x}_{n0}, \\
\dot{\hat{x}}_{n+1}(t) &= \frac{1}{\varepsilon} g_{n+1}\left(\frac{\xi(t) - \hat{x}_1(t)}{\varepsilon^n}\right), \hat{x}_{n+1}(t_0) = \hat{x}_{(n+1)0},
\end{aligned}
\tag{3.3}
$$

where $\hat{x} := [\hat{x}_1, \ldots, \hat{x}_{n+1}]^{\mathrm{T}} \in \mathbb{R}^{n+1}$ is the state of the observer, ε is a high gain parameter, $\hat{x}(t_0) := [\hat{x}_{10}, \ldots, \hat{x}_{(n+1)0}]^{\mathrm{T}} \in \mathbb{R}^{n+1}$ is the initial value of the observer state, $g_i(\cdot)$ are some chosen functions which are Lipschitz with $L_i > 0$, and $\xi(t)$ is given by

$$
\xi(t) = \begin{cases} y(t_k), & \text{if } r(t) = 0, \\ y(t), & \text{otherwise}, \end{cases}
\tag{3.4}
$$

where $r(t)$ is the event-trigger to be designed (see Eq. (3.9) in Sect. 3.2.1), t_k denotes the previous event-triggering time instant and k denotes the total number of triggered events during the time period $[t_0, t)$. Note that the differences of the ET-ESO from standard ESO include

- An event-triggering mechanism is applied to control the transmission of the output measurement $y(t)$ to the ET-ESO such that only when an event happens will the output measurement be transmitted;
- A zero-order-hold module is utilized at a no-event instant which leads to the different structural properties of the observer.

Throughout the chapter, we make the following assumption for ET-ESO.

Assumption 3.3 There exist continuously differentiable positive semi-definite functions $V(\cdot) : \mathbb{R}^{n+1} \to \mathbb{R}_0^+$ and $W(\cdot) : \mathbb{R}^{n+1} \to \mathbb{R}_0^+$ satisfying

1. $\lambda_1\|z\|^2 \leq V(z) \leq \lambda_2\|z\|^2, \lambda_3\|z\|^2 \leq W(z) \leq \lambda_4\|z\|^2$,
2. $\sum_{i=1}^{n} \frac{\partial V}{\partial z_i}(z_{i+1} - g_i(z_1)) - \frac{\partial V}{\partial z_{n+1}} g_{n+1}(z_1) \leq -W(z)$,
3. $\|\frac{\partial V}{\partial z}\| \leq \beta\|z\|$,

where $\beta, \lambda_1, \lambda_2, \lambda_3, \lambda_4$ are positive constants.

Note that we can adjust the ET-ESO scheme through the choice of functions g_i. As shown in Theorems 5.1 and 5.2 in [17], Assumption 3.3 is often used to ensure the stability of the nominal system considered. In this chapter, Assumption 3.3 helps guarantee the exponential stability of the dynamics of observation error e presented in next section when $h(t) = 0$ and $\sigma(t) = 0$, and can be satisfied if g_i are designed properly (see Sect. 3.3 for an example).

3.1.3 ET-ESO Design Problem

To describe the performance of the ET-ESO more precisely, we introduce the definition of the convergence of the ET-ESO.

Definition 3.1 Let x and $\hat{x}$ be the solution of system (3.1) and the ET-ESO in (3.3) with bounded initial values $x(t_0)$ and $\hat{x}(t_0)$, respectively. Let x_{n+1} be the extended state defined by $x_{n+1} := f(x, t) + w$. The ET-ESO is said to be *convergent* if for any given $E > 0$, there exists an $\varepsilon^* > 0$ such that for $\varepsilon \in (0, \varepsilon^*)$, x_i and $\hat{x}_i$ satisfy

$$\limsup_{t \to \infty} |x_i(t) - \hat{x}_i(t)| \leq E\varepsilon^{n+1-i}, \quad 1 \leq i \leq n+1.$$

The convergence property of the continuous-time ESO was investigated in [7]. However, due to the existence of the event-trigger scheme in ET-ESO, two problems need to be addressed in this chapter

1. Does there exist an event-triggering condition $r(t)$ that can ensure Zeno-free behavior of an ET-ESO?
2. Given the event-triggering condition, can we guarantee the convergence of the ET-ESO in the sense of Definition 3.1?

3.2 Main Theoretic Results

In this section, we will discuss the problems stated previously in detail. To analyze the observation performance, we first define the observation error variables $\tilde{x}_i$ and e_i as

$$\tilde{x}_i(t) := x_i(t) - \hat{x}_i(t), \tag{3.5}$$

$$e_i(t) := \tilde{x}_i(t)/\varepsilon^{n+1-i}, \quad i \in \{1, \ldots, n+1\}. \tag{3.6}$$

3.2.1 Event-Triggering Condition Design

To introduce the design of event-triggering condition, we define the sampling error $\sigma(t)$ here as

$$\sigma(t) := (y(t_k) - y(t))/\varepsilon^n, \tag{3.7}$$

where $t \in [t_k, t_{k+1})$, and t_k is the sample instant determined by the triggering condition. For notational brevity, we write α_i as

$$\alpha_i(e_1(t), \sigma(t)) := g_i(e_1(t) + \sigma(t)) - g_i(e_1(t)). \tag{3.8}$$

From Assumptions 3.1 and 3.2, there exists $M > 0$ such that

$$\left| \frac{d}{dt}(f(x(t), t) + w(t)) \right| < M.$$

Then, based on the definition of the sampling error, we introduce the following event-triggering mechanism:

$$r(t) = \begin{cases} 0, \text{ if } \frac{\beta\lambda_2}{\lambda_1\lambda_3} \sum_{i=1}^{n+1} L_i|\sigma(t)| \leq E - \frac{\beta\lambda_2 M\varepsilon}{\lambda_1\lambda_3}, \\ 1, \text{ otherwise,} \end{cases} \tag{3.9}$$

where $E > 0$ is a design parameter.

3.2.2 Non-existence of Zeno Phenomenon

In this section, we will show the nonexistence of Zeno phenomenon in the designed ET-ESO to guarantee the existence of the solution when $t \to \infty$. The key observations are summarized in the following result:

Proposition 3.1 *Consider the plant model in (3.1), the ESO in (3.3), the event-triggering condition in (3.9) and $\hat{x}(t_0) \in \hat{\mathcal{X}}$, where $\hat{\mathcal{X}}$ is any compact set of $\mathbb{R}^{n+1}$. Suppose that Assumptions 3.1, 3.2 hold, then there exist $\varepsilon^* > 0$ and $\tau > 0$ such that when $\varepsilon \in (0, \varepsilon^*)$,*

$$\min\{t_{k+1} - t_k\} \geq \tau. \tag{3.10}$$

for any $k > 0$.

Proof Define

$$\varepsilon^* := \min\{\frac{E\lambda_1\lambda_3}{\lambda_2\beta M}, 1\}$$

such that for any $\varepsilon \in (0, \varepsilon^*)$, we have

$$E - \frac{\beta\lambda_2 M\varepsilon}{\lambda_1\lambda_3} > 0.$$

Then, based on the event-triggering condition

$$r(t) = \begin{cases} 0, \text{ if } \frac{\beta\lambda_2}{\lambda_1\lambda_3}\sum_{i=1}^{n+1} L_i|\sigma(t)| \leq E - \frac{\beta\lambda_2 M\varepsilon}{\lambda_1\lambda_3}, \\ 1, \text{ otherwise,} \end{cases} \tag{3.11}$$

and from the definition of σ in (3.7), $r(t)$ is equivalent to

$$r(t) = \begin{cases} 0, \text{ if } |y(t) - y(t_k)| \leq (E - \frac{\beta\lambda_2 M\varepsilon}{\lambda_1\lambda_3})\varepsilon^n/(\frac{\beta\lambda_2}{\lambda_1\lambda_3}\sum_{i=1}^{n+1} L_i), \\ 1, \text{ otherwise.} \end{cases}$$

For the sampling error $y(t) - y(t_k)$, when $t \in [t_k, t_{k+1})$, we have

$$y(t) - y(t_k) = \int_{t_k}^{t} \dot{x}_1(\tau)d\tau = \int_{t_k}^{t} x_2(\tau)d\tau. \tag{3.12}$$

Considering the boundedness of x as shown in Assumption 3.2, we have

$$|y(t) - y(t_k)| \leq \int_{t_k}^{t} |x_2(\tau)|d\tau \leq (t - t_k)B \tag{3.13}$$

for some $B > 0$. Thus, there exists

$$\tau := \left(E - \frac{\beta\lambda_2 M\varepsilon}{\lambda_1\lambda_3}\right)\varepsilon^n / \left(\frac{\beta\lambda_2 B}{\lambda_1\lambda_3}\sum_{i=1}^{n+1} L_i\right) \tag{3.14}$$

such that

$$|y(t) - y(t_k)| \leq \left(E - \frac{\beta\lambda_2 M\varepsilon}{\lambda_1\lambda_3})\varepsilon^n/(\frac{\beta\lambda_2}{\lambda_1\lambda_3}\sum_{i=1}^{n+1} L_i\right)$$

holds for $t = t_k + \tau$. This completes the proof. □

In Proposition 3.1, it is shown that there is no Zeno phenomenon in this ET-ESO based on the event-triggering condition in (3.9), in the sense that there exists a nonzero dwell-time interval between two sampling instants. Moreover, in the literature, certain necessary conditions were provided to avoid Zeno phenomenon, e.g., [1, 2]. Actually, for the event-triggering condition proposed in this chapter, when $\sigma = 0$ we have

$$0 \neq E - \frac{\beta\lambda_2 M\varepsilon}{\lambda_1\lambda_3},$$

which is consistent with the requirements proposed in [1, 2].

3.2.3 Convergence Analysis of Event-Triggered Extended State Observer

In the previous section, we show that Zeno phenomenon does not arise for the event-triggering condition considered, based on which we investigate the performance of the ET-ESO in the following theorem:

Theorem 3.1 *Consider the plant model in (3.1) and the ESO in (3.3) under the event-triggering condition in (3.9). Suppose that Assumptions 3.1–3.3 hold. For $E > 0$, $x(t_0) \in \mathcal{X}$ and $\hat{x}(t_0) \in \hat{\mathcal{X}}$, $\hat{\mathcal{X}}$ being a compact set in $\mathbb{R}^{n+1}$, there exists $\varepsilon^* > 0$ such that for any $\varepsilon \in (0, \varepsilon^*)$, the trajectory of the ET-ESO is uniformly bounded and the ET-ESO is convergent, namely*

$$\limsup_{t\to\infty} |x_i(t) - \hat{x}_i(t)| \le E\varepsilon^{n+1-i}, 1 \le i \le n+1. \tag{3.15}$$

Proof From (3.1) and (3.3), we have for $\tilde{x}_i$ and $t \in [t_k, t_{k+1})$,

$$\begin{aligned} \dot{\tilde{x}}_i(t) &= x_{i+1}(t) - \hat{x}_{i+1}(t) - \varepsilon^{n-i} g_i\left(\frac{y(t_k) - \hat{x}_1(t)}{\varepsilon^n}\right) \\ &= \tilde{x}_{i+1} - \varepsilon^{n-i} g_i\left(\frac{y(t) - \hat{x}_1(t)}{\varepsilon^n} + \frac{y(t_k) - y(t)}{\varepsilon^n}\right), \ i \in \{1, \ldots, n\}, \end{aligned} \tag{3.16}$$

$$\begin{aligned} \dot{\tilde{x}}_{n+1}(t) &= \frac{d}{dt}(f(x(t), t) + w(t)) - \frac{1}{\varepsilon} g_{n+1}\left(\frac{y(t_k) - \hat{x}_1(t)}{\varepsilon^n}\right) \\ &= \frac{d}{dt}(f(x(t), t) + w(t)) - \frac{1}{\varepsilon} g_{n+1}\left(\frac{y(t) - \hat{x}_1(t)}{\varepsilon^n} + \frac{y(t_k) - y(t)}{\varepsilon^n}\right) \end{aligned} \tag{3.17}$$

holds. For notational brevity, we denote

$$h(t) := \frac{d}{dt}(f(x(t), t) + w(t)). \tag{3.18}$$

Then, from (3.6) and (3.8), the dynamics of e_i can be expressed as

$$\begin{aligned} \dot{e}_i(t) &= \frac{1}{\varepsilon} e_{i+1}(t) - \frac{1}{\varepsilon} g_i(e_1(t) + \sigma(t)) \quad i \in \{1, \ldots, n\}, \\ &= \frac{1}{\varepsilon} e_{i+1}(t) - \frac{1}{\varepsilon} g_i(e_1(t)) - \frac{1}{\varepsilon}\alpha_i(e_1(t), \sigma(t)), \\ \dot{e}_{n+1}(t) &= h(t) - \frac{1}{\varepsilon} g_{n+1}(e_1(t)) - \frac{1}{\varepsilon}\alpha_{n+1}(e_1(t), \sigma(t)). \end{aligned} \tag{3.19}$$

For a positive semi-definite function V, we observe through a simple computation that

$$\begin{aligned}\frac{d}{dt}V(t) &= \sum_{i=1}^{n+1}\frac{\partial V}{\partial e_i}\dot{e}_i(t)\\ &= \frac{1}{\varepsilon}\sum_{i=1}^{n}\frac{\partial V}{\partial e_i}(e_{i+1}(t) - g_i(e_1)) - \frac{1}{\varepsilon}\frac{\partial V}{\partial e_{n+1}}g_{n+1}(e_1(t))\\ &\quad - \frac{1}{\varepsilon}\sum_{i=1}^{n+1}\frac{\partial V}{\partial e_i}\alpha_i(e_1(t), \sigma(t)) + \frac{\partial V}{\partial e_{n+1}}h(t). \end{aligned} \tag{3.20}$$

According to Assumptions 3.1 and 3.2, there exists $M > 0$ such that

$$|h(t)| < M.$$

Then, from Assumption 3.3, we can construct an upper bound on $\dot{V}$ as

$$\begin{aligned}\dot{V} &\le -\frac{1}{\varepsilon}W + \beta M\|e\| - \frac{1}{\varepsilon}\sum_{i=1}^{n+1}\frac{\partial V}{\partial e_i}\alpha_i(e_1, \sigma)\\ &\le -\frac{\lambda_3}{\lambda_2\varepsilon}V + \frac{\sqrt{\lambda_1}}{\lambda_1}\beta M\sqrt{V} - \frac{1}{\varepsilon}\sum_{i=1}^{n+1}\frac{\partial V}{\partial e_i}\alpha_i(e_1, \sigma)\\ &\le -\frac{\lambda_3}{\lambda_2\varepsilon}V + \frac{\beta M\sqrt{\lambda_1}}{\lambda_1}\sqrt{V} + \frac{\beta\sqrt{\lambda_1}}{\lambda_1\varepsilon}\sum_{i=1}^{n+1}|\alpha_i(e_1, \sigma)|\sqrt{V}.\end{aligned}$$

To obtain a linear differential inequality, we take $Q(t) = \sqrt{V(t)}$ such that

$$\frac{dV}{dt} = \frac{dQ^2}{dt} = 2Q\frac{dQ}{dt}, \tag{3.21}$$

$$\sqrt{\lambda_1}\|e\| \le Q(e) \le \sqrt{\lambda_2}\|e\|. \tag{3.22}$$

Therefore, when $Q > 0$, we have

$$2Q\frac{dQ}{dt} \le -\frac{\lambda_3}{\lambda_2\varepsilon}Q^2 + \frac{\beta M\sqrt{\lambda_1}}{\lambda_1}Q + \frac{\beta\sqrt{\lambda_1}}{\lambda_1\varepsilon}\sum_{i=1}^{n+1}|\alpha_i(e_1, \sigma)|Q$$

which is equal to

$$\frac{dQ}{dt} \le -\frac{\lambda_3}{2\lambda_2\varepsilon}Q + \frac{\beta M\sqrt{\lambda_1}}{2\lambda_1} + \frac{\beta\sqrt{\lambda_1}}{2\lambda_1\varepsilon}\sum_{i=1}^{n+1}|\alpha_i(e_1, \sigma)|. \tag{3.23}$$

According to item 1 and item 3 in Assumption 3.3, we observe that

$$\lambda_1\|z\|^2 \le V(z) \le \frac{\beta}{2}\|z\|^2,$$

such that $\beta \geq 2\lambda_1$. Thus, for $Q = 0$

$$D^+Q(t) \leq \frac{\beta M\sqrt{\lambda_1}}{2\lambda_1} + \frac{\beta\sqrt{\lambda_1}}{2\lambda_1\varepsilon}\sum_{i=1}^{n+1}|\alpha_i(e_1,\sigma)|, \tag{3.24}$$

where D^+Q is the upper right-hand derivative of Q. Thus, $D^+Q(t)$ satisfies (3.23) for all $V \geq 0$. From the comparison lemma in [17], we obtain that

$$\begin{aligned} Q(t) \leq \phi(t,t_0)Q(t_0) &+ \frac{\beta M\sqrt{\lambda_1}}{2\lambda_1}\int_{t_0}^{t}\phi(t,\tau)d\tau \\ &+ \frac{\beta\sqrt{\lambda_1}}{2\lambda_1\varepsilon}\int_{t_0}^{t}\sum_{i=1}^{n+1}|\alpha_i(e_1(\tau),\sigma(\tau))|\phi(t,\tau)d\tau, \end{aligned} \tag{3.25}$$

where $\phi(t,t_0)$ is defined by

$$\phi(t,t_0) = \exp\left(-\frac{\lambda_3}{2\lambda_2\varepsilon}(t-t_0)\right).$$

From (3.22), for e we have

$$\begin{aligned} \|e(t)\| \leq& \sqrt{\frac{\lambda_2}{\lambda_1}}\|e(t_0)\|\phi(t,t_0) + \frac{\beta M}{2\lambda_1}\int_{t_0}^{t}\phi(t,\tau)d\tau \\ &+ \frac{\beta}{2\lambda_1\varepsilon}\int_{t_0}^{t}\sum_{i=1}^{n+1}|\alpha_i(e_1(\tau),\sigma(\tau))|\phi(t,\tau)d\tau \\ \leq& \sqrt{\frac{\lambda_2}{\lambda_1}}\|e(t_0)\|\phi(t,t_0) + \frac{\beta M}{2\lambda_1}\int_{t_0}^{t}\phi(t,\tau)d\tau \\ &+ \frac{\beta}{2\lambda_1\varepsilon}\sup_{t>t_0}\sum_{i=1}^{n+1}|\alpha_i(e_1(t),\sigma(t))|\int_{t_0}^{t}\phi(t,\tau)d\tau \\ \leq& \sqrt{\frac{\lambda_2}{\lambda_1}}\|e(t_0)\|\phi(t,t_0) + \int_{t_0}^{t}\phi(t,\tau)d\tau \\ &\cdot\left(\frac{\beta M}{2\lambda_1} + \frac{\beta}{2\lambda_1\varepsilon}\sup_{t>t_0}\sum_{i=1}^{n+1}|\alpha_i(e_1(t),\sigma(t))|\right) \\ \leq& \sqrt{\frac{\lambda_2}{\lambda_1}}\|e(t_0)\|\phi(t,t_0) + (1-\phi(t,t_0)) \\ &\cdot\left(\frac{\beta\lambda_2 M\varepsilon}{\lambda_1\lambda_3} + \frac{\beta\lambda_2}{\lambda_1\lambda_3}\sup_{t>t_0}\sum_{i=1}^{n+1}|\alpha_i(e_1(t),\sigma(t))|\right). \end{aligned} \tag{3.26}$$

Then, for a given $E > 0$, we define

$$\varepsilon^* = \min\left\{\frac{E\lambda_1\lambda_3}{\lambda_2\beta M}, 1\right\}.$$

From Proposition 3.1, as no Zeno behavior exists, $\hat{x}_i(t)$ will be well-defined when $t \to \infty$. Moreover, considering the fact that the functions g_i are supposed to be Lipschitz with $L_i > 0$, we have

$$|\alpha_i(e_1, \sigma)| = |g_i(e_1 + \sigma) - g_i(e_1)| \leq L_i|e_1 + \sigma - e_1| = L_i|\sigma|.$$

Thus, when $t \to \infty$ such that $\phi(t, t_0) \to 0$,

$$\|e(t)\| \leq \frac{\beta\lambda_2 M\varepsilon}{\lambda_1\lambda_3} + \frac{\beta\lambda_2}{\lambda_1\lambda_3}\sup_{t>t_0}\sum_{i=1}^{n+1} L_i|\sigma(t)|,$$

holds. Considering the event-triggering condition $r(t)$ in (3.9), for any $\varepsilon \in (0, \varepsilon^*)$, if

$$\frac{\beta\lambda_2}{\lambda_1\lambda_3}\sum_{i=1}^{n+1} L_i|\sigma| > E - \frac{\beta\lambda_2 M\varepsilon}{\lambda_1\lambda_3}, \tag{3.27}$$

the transmission task will be executed such that $\sigma = 0$. Then, it is guaranteed that

$$\limsup_{t\to\infty} \|e(t)\| \leq E. \tag{3.28}$$

Finally, according to the definition of $\tilde{x}$ in (3.5) and using the fact that

$$|x_i - \hat{x}_i| = |\tilde{x}_i| \leq \|e(t)\|\varepsilon^{n+1-i}, \tag{3.29}$$

we have

$$\limsup_{t\to\infty} |x_i(t) - \hat{x}_i(t)| \leq E\varepsilon^{n+1-i}, 1 \leq i \leq n+1. \tag{3.30}$$

Thus, the ET-ESO is convergent in the sense of Definition 3.1 and the trajectory of ET-ESO is uniformly bounded. This completes the proof. □

In Theorem 3.1, we show that for a continuous-time system and the ESO considered, namely, (3.1) and (3.3), we can design a suitable event-triggering condition $r(t)$ according to (3.9) to guarantee the convergence of the ET-ESO. Moreover, the practical stability of the proposed ET-ESO also can be obtained directly according to the proof of Theorem 3.1. Note that the precision of this observer can be adjusted by selecting E. Specifically, from (3.15), for a smaller E, the event-triggering condition is tighter which leads to a tighter bound of observation error, at the cost of a larger number of measurement transmissions. Therefore, even though the measurements are only available in an intermittent fashion, the observation performance of the ET-ESO can be guaranteed by properly choosing E. Moreover, according to the proof of Theorem 3.1, the selection of E will affect the choice of ε^* in the sense that for a smaller E, the value of ε^* becomes smaller; consequently, since $\varepsilon \in (0, \varepsilon^*)$, we can only select ε within a smaller range with the decrease of E.

3.2.4 Other Extensions

3.2.4.1 Performance Analysis with Weaker Assumption

In the previous section, we investigate the performance of the ET-ESO based on Assumption 3.3. Alternatively, based on a weaker assumption, for instance, Assumption H4 in [7], the problem of ET-ESO design and performance analysis can also be investigated if we replace item 3 in Assumption H4 with

$$\left\| \frac{\partial V}{\partial y} \right\| \le \alpha W(y)$$

for $\|y\| > R$ due to the existence of sampling error in all $n+1$ channels. We note that Assumption H4 in [7] does not require $\lambda_1\|z\|^2 \le V(z) \le \lambda_2\|z\|^2$, $\lambda_3\|z\|^2 \le W(z) \le \lambda_4\|z\|^2$ but only requires that $\{y|V(y) \le d\}$ is bounded for any $d > 0$. Accordingly, if we refine the event-triggering condition as

$$r(t) = \begin{cases} 0, \text{ if } \sum_{i=1}^{n+1} L_i|\sigma| \le \overline{B}\varepsilon, \\ 1, \text{ otherwise}, \end{cases}$$

for some $\overline{B} > 0$, the convergence in the sense of that in Theorem 2.2 in [7] can be pursued by combining the ideas in the proof of Theorem 3.1 with the ideas of proving Theorem 2.2. We note that following the derivations in the proof, the observation error can be ensured to converge to zero without requiring $y(t)$ to converge to zero.

3.2.4.2 Other Event-Triggering Conditions

In this chapter, the correction term depends on $\hat{x}_1(t) - y(t_k)$ and a similar technique was applied in [23] for the design of the time-triggered state estimation in networked systems. A different way is to use the term $\hat{x}_1(t_k) - y(t_k)$ to correct the error between the observer and the system state under consideration [24]. Motivated by this observation, an alternative event-triggering condition that depends on $\hat{x}_1(t_k) - y(t_k)$ can be introduced as

$$r(t) = \begin{cases} 0, \text{ if } \frac{\beta\lambda_2}{\lambda_1\lambda_3}\sum_{i=1}^{n+1} L_i|\sigma_y - \sigma_{\hat{x}}| \le E - \frac{\beta\lambda_2 M\varepsilon}{\lambda_1\lambda_3}, \\ 1, \text{ otherwise}, \end{cases}$$

where $\sigma_y(t) := [y(t_k) - y(t)]/\varepsilon^n$ and $\sigma_{\hat{x}}(t) := [\hat{x}_1(t_k) - \hat{x}_1(t)]/\varepsilon^n$, which can be obtained by setting a local copy of the observer on the sensor side [11, 26, 32, 34]; the performance of the ET-ESO also can be analyzed following a similar line of arguments as those in the proof of Theorem 3.1.

Another possible event-triggering condition can be provided as

$$r(t) = \begin{cases} 0, \text{ if } \frac{\beta\lambda_2}{\lambda_1\lambda_3}\sum_{i=1}^{n+1} |\alpha_i(e_1(t), \sigma(t))| \le E - \frac{\beta\lambda_2 M\varepsilon}{\lambda_1\lambda_3}, \\ 1, \text{ otherwise}. \end{cases}$$

The convergence results of ET-ESO can be obtained following the same line of ideas in this chapter. This condition can also be implemented by setting a local observer on the sensor side.

3.2.5 *Event-Triggered Extended State Observer with Designed Initial Value*

In Theorem 3.1, we obtain the convergence of the ET-ESO without restrictions on the initial values of the ESO. In many practical applications, the initial values can be set in advance according to the specific operating conditions. Next, we will present a corollary of Theorem 3.1 with the knowledge of initial conditions to obtain stronger results.

Corollary 3.1 *Consider the plant model in* (3.1) *and the ESO in* (3.3)*. Suppose that Assumptions 3.1–3.3 hold. For any given $E > 0$, there exist $\varepsilon^* > 0$ and a designed event-triggering condition $r(t)$ in the form of* (3.9)*, such that for any $\varepsilon \in (0, \varepsilon^*)$, the ET-ESO satisfies*

$$\max_{t>t_0} |x_i(t) - \hat{x}_i(t)| \le E\varepsilon^{n+1-i}, \ 1 \le i \le n+1, \tag{3.31}$$

if $e(t_0)$ defined in (3.6) *satisfies*

$$\sqrt{\frac{\lambda_2}{\lambda_1}}\|e(t_0)\| \le E. \tag{3.32}$$

Proof Since $\phi(t, t_0) \in (0, 1)$, we have for $A, B > 0$

$$A\phi(t, t_0) + B(1 - \phi(t, t_0)) \le \max\{A, B\}. \tag{3.33}$$

From (3.26), the upper bound on $\|e\|$ can be obtained as

$$\|e(t)\| \leq \max\left\{\sqrt{\frac{\lambda_2}{\lambda_1}}\|e(t_0)\|, \frac{\beta\lambda_2 M\varepsilon}{\lambda_1\lambda_3} + \frac{\beta\lambda_2}{\lambda_1\lambda_3}\sup_{t>t_0}\sum_{i=1}^{n+1} L_i|\sigma(t)|\right\}.$$

Since

$$\sqrt{\frac{\lambda_2}{\lambda_1}}\|e(t_0)\| \leq E,$$

$e(t)$ satisfies

$$\|e(t)\| \leq \max\left\{E, \frac{\beta\lambda_2 M\varepsilon}{\lambda_1\lambda_3} + \frac{\beta\lambda_2}{\lambda_1\lambda_3}\sup_{t>t_0}\sum_{i=1}^{n+1} L_i|\sigma(t)|\right\}.$$

Therefore, through a similar lines of arguments as in the proof of Theorem 3.1, we have that for any $\varepsilon \in (0, \varepsilon^*)$,

$$\max_{t>t_0}|x_i(t) - \hat{x}_i(t)| \leq E\varepsilon^{n+1-i}, 1 \leq i \leq n+1, \tag{3.34}$$

which completes the proof. □

3.3 Numerical Example

In this section, we illustrate the obtained theoretic results through a numerical example. Consider the following system plant:

$$\begin{aligned}
\dot{x}_1(t) &= x_2(t), x_1(0) = 1, \\
\dot{x}_2(t) &= f(x_1(t), x_2(t)) + u(t) + w(t), x_2(0) = 1, \\
y(t) &= x_1(t),
\end{aligned} \tag{3.35}$$

where

$$f(x_1, x_2) := -3x_1 - 2x_2 - \sin^3(x_1 + x_2),$$

and

$$w(t) = \begin{cases} 0, & t < 4\pi \text{ or } t > 6\pi, \\ -0.4\cos(2t) + 0.4, & 4\pi \leq t \leq 6\pi. \end{cases} \tag{3.36}$$

The control signal u is chosen as

$$u = 2x_1 + x_2,$$

and the extended state x_3 is defined as

$$x_3 = -3x_1 - 2x_2 - \sin^3(x_1 + x_2) + w.$$

For this system, the ET-ESO is designed as

$$\begin{aligned}
\dot{\hat{x}}_1(t) &= \hat{x}_2(t) + \frac{3}{\varepsilon} \cdot (\xi(t) - \hat{x}_1(t)) + \varepsilon \cdot \varphi\left(\frac{\xi(t) - \hat{x}_1(t)}{\varepsilon^2}\right), \\
\dot{\hat{x}}_2(t) &= \hat{x}_3(t) + \frac{3}{\varepsilon^2} \cdot (\xi(t) - \hat{x}_1(t)) + u(t), \\
\dot{\hat{x}}_3(t) &= \frac{1}{\varepsilon^3} \cdot (\xi(t) - \hat{x}_1(t)), \\
\xi(t) &= y(t_k), \ t \in [t_k, t_{k+1}),
\end{aligned} \tag{3.37}$$

where $\hat{x}(0) = [0, -1, 0]^{\mathrm{T}}$ and $\varphi(\theta)$ is selected as

$$\varphi(\theta) = \begin{cases} -\frac{1}{4}, & \theta < -\frac{\pi}{2}, \\ \frac{1}{4}\sin\theta, & -\frac{\pi}{2} \leq \theta \leq \frac{\pi}{2}, \\ \frac{1}{4}, & \theta > \frac{\pi}{2}, \end{cases} \tag{3.38}$$

which means that g_i's are given as

$$g_1(z_1) = 3z_1 + \varphi(z_1), \ g_2(z_1) = 3z_1, \ g_3(z_1) = z_1$$

in this ET-ESO. Through simple computations, it is easy to verify that

$$|u| + |f(x,t)| + |\dot{w}| + \left|\frac{\partial f}{\partial t}\right| + \left|\frac{\partial f}{\partial x_i}\right| \leq 5|x_1| + 3|x_2| + 7.8. \tag{3.39}$$

Moreover, from the expression of w and according to Exercise 4.6 in [17], we can obtain the boundedness of x. In this way, Assumptions 3.1, 3.2 hold for this example. Next, define the matrix N and $P > 0$ as

$$N = \begin{bmatrix} -3 & 1 & 0 \\ -3 & 0 & 1 \\ -1 & 0 & 0 \end{bmatrix}, \tag{3.40}$$

and

$$PN + N^{\mathrm{T}}P = -I, \tag{3.41}$$

respectively, in which $I \in \mathbb{R}^{3\times 3}$ is an identity matrix. It is clear that

$$\lambda_{\min}(P)\|z\|^2 \le \langle Pz, z\rangle \le \lambda_{\max}(P)\|z\|^2. \tag{3.42}$$

Moreover, for any $\theta \in \mathbb{R}$, we have

$$\int_0^\theta \varphi(s)ds \le \frac{1}{4}\int_0^\theta s ds = \frac{1}{8}\theta^2. \tag{3.43}$$

Hence, the function V can be designed as

$$V(z) = \langle Pz, z\rangle + \int_0^{z_1} \varphi(s)ds,$$

where P is written as

$$P = \begin{bmatrix} 1 & -0.5 & -1 \\ -0.5 & 1 & -0.5 \\ -1 & -0.5 & 4 \end{bmatrix} \tag{3.44}$$

according to (3.41). For function V, we can verify that

$$\lambda_{\min}(P)\|z\|^2 \le V(z) \le \lambda_{\max}(P)\|z\|^2 + \frac{1}{8}\|z\|^2 \tag{3.45}$$

and

$$\begin{aligned} &\sum_{i=1}^{2} \frac{\partial V}{\partial z_i}(z_{i+1} - g_i(z_1)) - \frac{\partial V}{\partial z_3} g_3(z_1) \\ &\quad \le -\left(\frac{z_1^2}{8} + \frac{7z_2^2}{8} + \frac{3z_3^2}{4}\right) \le -\frac{1}{8}\|z\|^2. \end{aligned} \tag{3.46}$$

For $\frac{\partial V}{\partial z}$, we have

$$\left\|\frac{\partial V}{\partial z}\right\| \le \|2z^{\mathrm{T}}P\| + \frac{1}{4}\|z\| \le \left(2\lambda_{\max}(P) + \frac{1}{4}\right)\|z\|. \tag{3.47}$$

Hence, Assumption 3.3 can also be satisfied if we choose the parameters as

$$\lambda_1 = 0.1966, \lambda_2 = 4.4622, \lambda_3 = 0.125, \beta = 9.1745.$$

In this way, Assumptions 3.1–3.3 are verified for the system considered.

3.3.1 Observation Performance Verification for ET-ESO

To implement the event-triggering condition

$$r(t) = \begin{cases} 0, \text{ if } \frac{\beta\lambda_2}{\lambda_1\lambda_3}\sum_{i=1}^{3} L_i|\sigma(t)| \le E - \frac{\beta\lambda_2 M\varepsilon}{\lambda_1\lambda_3}, \\ 1, \text{ otherwise,} \end{cases}$$

we choose

$$L_1 = 3.25,\ L_2 = 3,\ L_3 = 1,\ E = 6484,\ M = 6,\ \varepsilon = 0.3$$

and the observation performance of ET-ESO is shown in Fig. 3.2. Moreover, we also plot event-trigger $r(t)$ in Fig. 3.2 to show the event instants at which the measurements $y(t)$ are transmitted to the observer. According to Fig. 3.2, we observe that the condition $r(t)$ is triggered only at certain discrete-time instants, while satisfactory performance is guaranteed by the proposed ET-ESO. The steady-state observation errors of x_1 and x_2 are less than 0.0068 and 0.0586, respectively, which are also much smaller than $\varepsilon^2 E$ and εE.

On the other hand, if we set the initial values of x and $\hat{x}$ as $x_i(t_0) = \hat{x}_i(t_0)$, the maximum observation errors will become 0.0977 and 0.6724, which are much smaller than $\varepsilon^2 E$ and εE.

From the figures, we also observe that when the system output y is oscillatory (namely, $t \in [12, 25]$), more events are observed than the case when the system output approaches a steady state. These events are inevitable for the proposed event-triggering scheme, since the nonlinear dynamics of the system (namely, $f(x, t)$) is supposed to be unknown in event-triggering condition design.

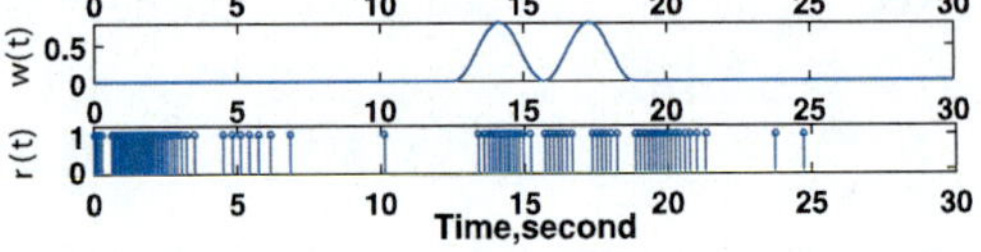

Fig. 3.2 Performance of the proposed ET-ESO

3.3.1.1 Comparison with Standard ESO

For this example, we also compare the observation error between ET-ESO and standard ESO [7] in Fig. 3.3. Compared with the standard ESO, the performance of the ET-ESO is only degraded within a tolerable level when the event-triggered output measurements are utilized. Furthermore, we quantitatively investigate the relationship between the observation error and the number of events, which is defined as the total number of triggered events within 30 s. To do this, we vary parameter E from 4000 to 9070, and the simulation results are provided in Fig. 3.4a, b.

As shown in Fig. 3.4, the general trend is that the observation error decreases with the increase of the number of events (caused by the decrease of parameter E (see Fig. 3.4a); with the increase of the number of events (or transmitted measurements), the performance of ET-ESO approaches that of the standard ESO.

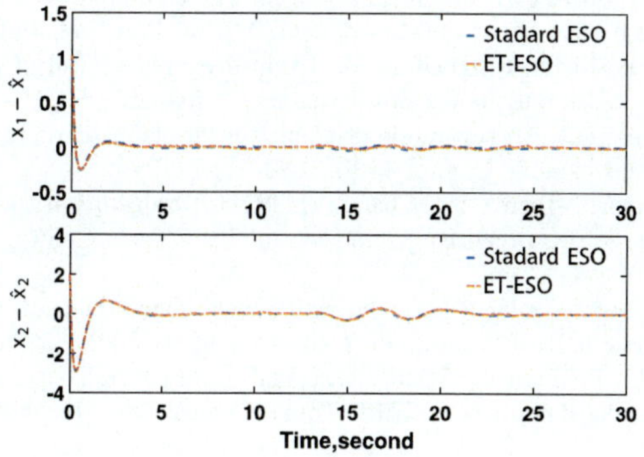

Fig. 3.3 Comparison on the observation error of standard ESO and ET-ESO

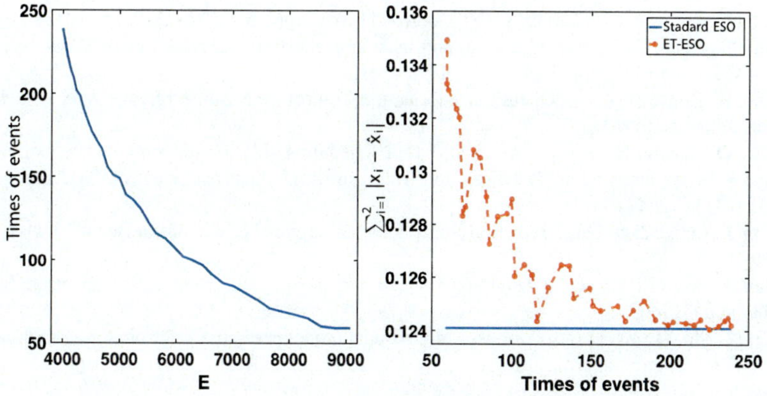

Fig. 3.4 The trend of the average observation error with number of events in 30 s

3.4 Summary

In this chapter, we consider the design of ET-ESO for a continuous-time nonlinear system with disturbance and uncertainty. Based on the main results, several extensions are also discussed, and a numerical example is adopted to illustrate the performance of the ET-ESO. These results were originally proposed in [16] and were later generalized to consider other problems (e.g., [36]). Based on the obtained theoretic results of ET-ESO, we will move on to introduce the design for ET-ADRC scheme in the next chapter.

References

1. Abdelrahim, M., Postoyan, R., Daafouz, J.: Event-triggered control of nonlinear singularly perturbed systems based only on the slow dynamics. IFAC Proc. Vol. **46**(23), 347–352 (2013)
2. Abdelrahim, M., Postoyan, R., Daafouz, J.: Event-triggered control of nonlinear singularly perturbed systems based only on the slow dynamics. Automatica **52**, 15–22 (2015)
3. Atassi, A.N., Khalil, H.K.: A separation principle for the stabilization of a class of nonlinear systems. IEEE Trans. Autom. Control **44**(9), 1672–1687 (1999)
4. Castaneda, L., Luviano-Juarez, A., Chairez, I.: Robust trajectory tracking of a delta robot through adaptive active disturbance rejection control. IEEE Trans. Control Syst. Technol. **23**(4), 1387–1398 (2015)
5. Erenturk, K.: Fractional-order $\mathrm{PI}^{\lambda}\mathrm{D}^{\mu}$ and active disturbance rejection control of nonlinear two-mass drive system. IEEE Trans. Indust. Electron. **60**(9), 3806–3813 (2013)
6. Godbole, A., Kolhe, J., Talole, S.: Performance analysis of generalized extended state observer in tackling sinusoidal disturbances. IEEE Trans. Control Syst. Technol. **21**(6), 2212–2223 (2013)
7. Guo, B., Zhao, Z.: On the convergence of an extended state observer for nonlinear systems with uncertainty. Syst. Control Lett. **60**(6), 420–430 (2011)
8. Guo, B., Zhao, Z.: On convergence of non-linear extended state observer for multi-input multi-output systems with uncertainty. IET Control Theory Appl. **6**(15), 2375–2386 (2012)
9. Guo, B., Zhao, Z.: On convergence of nonlinear active disturbance rejection for MIMO system. SIAM J. Control Optim. **51**(2), 1727–1757 (2013)
10. Guo, B., Zhao, Z.: Active Disturbance Rejeciton Control for Nonlinear Systems: An Introduction. Wiley Inc. (2016)
11. Han, D., Zhang, H., Shi, L.: An event-based scheduling solution for remote state estimation of two lti systems under bandwidth constraint. In: 2013 American Control Conference, pp. 3314–3319 (2013)
12. Han, J.: Active disturbance rejection controller and its applications. Control Decis. **13**(1), 19–23 (1998)
13. Han, J.: From PID to active disturbance rejection control. IEEE Trans. Indust. Electron. **56**(3), 900–906 (2009)
14. He, N., Shi, D.: Event-based robust sampled-data model predictive control: A non-monotonic lyapunov function approach. IEEE Trans. Circuits Syst. I: Regul. Papers **62**(10), 2555–2564 (2015)
15. Horn, R.A., Johnson, C.R.: Matrix Analysis. Cambridge University, Cambridge (2013)
16. Huang, Y., Wang, J., Shi, D., Shi, L.: Toward event-triggered extended state observer. IEEE Trans. Autom. Control **63**, 1842–1849 (2018)
17. Khalil, H.K.: Nonlinear Systems. Prentice Hall, Upper Saddle River (2002)

18. Kurzhanski, Valyi, I.: Ellipsoidal Calculus for Estimation and Control. Birkhäuser, Boston (1997)
19. Li, S., Yang, X., Yang, D.: Active disturbance rejection control for high pointing accuracy and rotation speed. Automatica **45**(8), 1854–1860 (2009)
20. Magni, L., Scattolini, R.: Model predictive control of continuous-time nonlinear systems with piecewise constant control. IEEE Trans. Autom. Control **49**(6), 900–906 (2004)
21. Mazo, M., Tabuada, P.: Decentralized event-triggered control over wireless sensor/actuator networks. IEEE Trans. Autom. Control **56**(10), 2456–2461 (2011)
22. Noack, B., Klumpp, V., Hanebeck, U.: State estimation with sets of densities considering stochastic and systematic errors. In: 12th International Conference on Information Fusion, 2009, pp. 1751–1758 (2009)
23. Postoyan, R., Nesic, D.: A framework for the observer design for networked control systems. IEEE Trans. Autom. Control **57**(5), 1309–1314 (2012)
24. Postoyan, R., van de Wouw, N., Nesic, D., Heemels, W.P.M.H.: Tracking control for nonlinear networked control systems. IEEE Trans. Autom. Control **59**(6), 1539–1554 (2014)
25. Ramirez-Neria, M., Garcia-Antonio, J., Sira-Ramirez, H., Velasco-Villa, M., Castro-Linares, R.: On the linear active rejection control of Thomson's jumping ring. In: American Control Conference (ACC) 2013, pp. 6643–6648 (2013)
26. Shi, D., Chen, T., Shi, L.: Event-triggered maximum likelihood state estimation. Automatica **50**(1), 247–254 (2014)
27. Shi, D., Chen, T., Shi, L.: On set-valued kalman filtering and its application to event-based state estimation. IEEE Trans. Autom. Control **60**(5), 1275–1290 (2015)
28. Sira-Ramirez, H., Linares-Flores, J., Garcia-Rodriguez, C., Contreras-Ordaz, M.: On the control of the permanent magnet synchronous motor: an active disturbance rejection control approach. IEEE Trans. Control Syst. Technol. **22**(5), 2056–2063 (2014)
29. Su, Y.X., Zheng, C.H., Duan, B.Y.: Automatic disturbances rejection controller for precise motion control of permanent-magnet synchronous motors. IEEE Trans. Indust. Electron. **52**(3), 814–823 (2005)
30. Tabuada, P.: Event-triggered real-time scheduling of stabilizing control tasks. IEEE Trans. Autom. Control **52**(9), 1680–1685 (2007)
31. Vincent, J., Morris, D., Usher, N., Gao, Z., Zhao, S., Nicoletti, A., Zheng, Q.: On active disturbance rejection based control design for superconducting RF cavities. Nuclear Instrum. Methods Phys. Res. Sect. A: Acceler., Spectrom., Detect. Assoc. Equip. **643**(1), 11–16 (2011)
32. Weerakkody, S., Mo, Y., Sinopoli, B., Han, D., Shi, L.: Multi-sensor scheduling for state estimation with event-based, stochastic triggers. IEEE Trans. Autom. Control **61**(9), 2695–2701 (2016)
33. Wu, D., Chen, K.: Design and analysis of precision active disturbance rejection control for noncircular turning process. IEEE Trans. Indust. Electron. **56**(7), 2746–2753 (2009)
34. Wu, J., Jia, Q.S., Johansson, K.H., Shi, L.: Event-based sensor data scheduling: trade-off between communication rate and estimation quality. IEEE Trans. Autom. Control **58**(4), 1041–1046 (2013)
35. Wu, Q.: Linear Algebra with Applications in Automatic Control. National Defense Industry Press (2011)
36. Wu, X., Liu, K., Bai, Y., Wang, J.: Towards event-triggered extended state observer for multi-agent systems. Neurocomputing **386**, 191–197 (2020)
37. Zhao, Z., Guo, B.: On active disturbance rejection control for nonlinear systems using time-varying gain. Eur. J. Control **23**, 100–108 (2015)
38. Zhao, Z., Guo, B.: On convergence of nonlinear active disturbance rejection control for siso nonlinear systems. J. Dyn. Control Syst. **22**, 385–413 (2016)
39. Zheng, Q., Gao, L., Gao, Z.: On stability analysis of active disturbance rejection control for nonlinear time-varying plants with unknown dynamics. In: 46th IEEE Conference on Decision and Control, pp. 3501–3506 (2007)
40. Zheng, Q., Gaol, L.Q., Gao, Z.: On stability analysis of active disturbance rejection control for nonlinear time-varying plants with unknown dynamics. In: 46th IEEE Conference on Decision and Control, pp. 3501–3506 (2007)

Chapter 4
Event-Triggered Active Disturbance Rejection Control

Based on our discussions on event-triggered ESO in Chap. 3, we further consider the problem of event-triggered ADRC design in this chapter. Specifically, we consider the problem of event-triggered sampled-data control for networked systems. To do this, a nonlinear continuous-time system with uncertainty and disturbance is introduced, for which an event-triggered control problem based on an ADRC approach is formulated. For this problem, we consider two different scenarios. First, we investigate the scenario that the control signal is transmitted at each event instant while the output measurements are transmitted in real time. For this scenario, an easy-to-implement event-triggering strategy based on the state observation information provided by an ESO is proposed, for which the asymptotic boundedness of the observation error is obtained and the stability of the closed-loop system can be guaranteed. By introducing some mild restrictions on the initial values of the observer and plant, the boundedness of observation error can be maintained at any time. Based on the results obtained for this scenario, we extend the obtained results to the more general scenario in which the transmission of the measurements and the control commands are determined by separate event-triggering mechanisms. We verify the performance of the proposed approach through experimental results on the direct current motor loading platform introduced earlier in this book.

4.1 Problem Description

In this chapter, we consider an n-dimensional nonlinear continuous-time system subject to uncertainty and disturbance

D. Shi et al., *Event-Triggered Active Disturbance Rejection Control*, Studies in Systems, Decision and Control 356,
https://doi.org/10.1007/978-981-16-0293-1_4

$$\begin{aligned} \dot{x}_1(t) &= x_2(t), x_1(t_0) = x_{10}, \\ &\vdots \\ \dot{x}_n(t) &= f(x(t), t) + u(t) + w(t), x_n(t_0) = x_{n0}, \\ y(t) &= x_1(t), \end{aligned} \tag{4.1}$$

where $x := [x_1, \ldots, x_n]^{\mathrm{T}} \in \mathbb{R}^n$ is the system state, $x(t_0) := [x_{10}, \ldots, x_{n0}]^{\mathrm{T}} \in \mathbb{R}^n$ is the initial state, $w \in \mathbb{R}$ is the disturbance, $f : \mathbb{R}^n \times \mathbb{R} \to \mathbb{R}$ is a function representing the nonlinear part with uncertainty, $u \in \mathbb{R}$ is the control input, and $y \in \mathbb{R}$ is the measurement output. For this system, we introduce the following assumptions, based on which the theoretic analysis in this chapter is built. Note that the assumptions here are consistent with those in Chap. 3.

Assumption 4.1 1. The nonlinear function $f(x, t)$ is unknown in the sense that neither its explicit expression nor an approximate estimate $\hat{f}(x, t)$ of $f(x, t)$ is available for controller design.
2. The function $f(x, t)$ and the external disturbance w are differentiable and satisfy

$$\begin{aligned} &|f(x, t)| + \left|\frac{\partial f}{\partial t}\right| \le c_0 + \sum_{j=1}^{n} c_j |x_j|, \text{ and } \left|\frac{\partial f}{\partial x_i}\right| \le c, \\ &|w(t)| + |\dot{w}(t)| \le b, \end{aligned} \tag{4.2}$$

for some positive constants c_j, b and c.

Note that extensive engineering practice (e.g., [1, 13]) indicates that the ADRC scheme does not rely on the specific process model. Thus, to investigate the theoretic performance of ADRC, one way of interpreting this "model-free" principle is to assume that the process and the disturbance are bounded but unknown [4]. Assumption 4.1 is a standard condition that is widely used in the developments of ADRC in the literature (e.g., [10, 15, 18]), which indicates the boundedness property of system uncertainty and disturbance. Although this assumption does not apply to all types of process dynamics, it can be satisfied by a large class of systems, due to the existence of hard constraints in engineering systems (e.g., operation restrictions and energy limitations [6, 11]).

For the system in (4.1), we consider the scenario that the measurement $y(t)$ and the control signal $u(t)$ are transmitted through a communication network (see, e.g., Figs. 4.1 and 4.2). To save communication resources, it is desirable to reduce the transmission rates while maintaining satisfactory control performance. This motivates the problem of simultaneously designing a suitable event-triggering mechanism that selectively controls the transmission and a controller ensuring closed-loop control performance. In this chapter, we show how to solve this problem using an ADRC approach based on our analysis on ET-ESO in the previous chapter. Specifically, we solve this problem through the following two directions:

1. As a first step, how to design ET-ADRC with asynchronous control update based on continuous output measurements such that the stability of the system can be guaranteed?
2. As a following-up step, how to jointly design ET-ADRC with event-triggered output measurement transmissions such that the stability of the system can be guaranteed?

These two problems will be discussed systematically in the next two sections.

4.2 Event-Triggered Control Based on Continuous-Time Observations

In this section, we present an event-triggered mechanism for ADRC based on continuous measurements (see Fig. 4.1). Since the full system states are not directly observed, we introduce an extended state observer of the form

$$
\begin{aligned}
\dot{\hat{x}}_1(t) &= \hat{x}_2(t) + \varepsilon^{n-1} g_1\left(\frac{y(t)-\hat{x}_1(t)}{\varepsilon^n}\right),\\
&\vdots\\
\dot{\hat{x}}_n(t) &= \hat{x}_{n+1}(t) + g_n\left(\frac{y(t)-\hat{x}_1(t)}{\varepsilon^n}\right) + u(t),\\
\dot{\hat{x}}_{n+1}(t) &= \frac{1}{\varepsilon} g_{n+1}\left(\frac{y(t)-\hat{x}_1(t)}{\varepsilon^n}\right),
\end{aligned}
\tag{4.3}
$$

where $\hat{x} := [\hat{x}_1, \ldots, \hat{x}_{n+1}]^{\mathrm{T}} \in \mathbb{R}^{n+1}$ is the observer state, $\hat{x}(t_0) := [\hat{x}_{10}, \ldots, \hat{x}_{(n+1)0}]^{\mathrm{T}} \in \mathbb{R}^{n+1}$ is the initial value of $\hat{x}$ for some constants $\hat{x}_{i0}$, $g_i(\cdot)$ are some pre-specified functions and $\hat{x}_{n+1}$ is the estimate of the extended state defined by

$$
x_{n+1} := f(x, t) + w. \tag{4.4}
$$

We emphasize that the uncertain part of system and the disturbance are treated as the extended state as shown in the above equation, the effect of which is observed by $\hat{x}_{n+1}$.

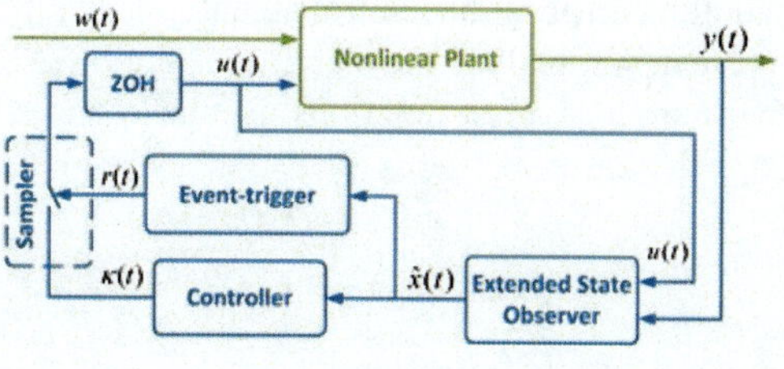

Fig. 4.1 The schematic of the event-triggered active disturbance rejection control

The parameter ε in Eq. (4.3) is a high-gain parameter, which is a crucial parameter determining the observer dynamics, and plays a key role in analyzing the control and the observation performance. To guarantee the observation performance, ε is set in $(0, 1)$. By setting this parameter to be suitably small, one can attenuate the effect of the disturbance in state observation while achieving desired control performance. For further understanding on high-gain parameter design, the interested authors can refer to [3, 4, 19].

Based on the above descriptions of the system in (4.1) and (4.3), we introduce the following event-triggered mechanism

$$r(t) = \begin{cases} 0, \text{ if } \sum_{i=1}^{n} k_i^2 \sigma_i^2(t) + \sigma_{n+1}^2(t) \le M_1 \varepsilon^{\frac{1}{2}}, \\ 1, \text{ otherwise,} \end{cases} \tag{4.5}$$

where $M_1 > 0$ and k_i are parameters, and σ_i is defined by

$$\sigma_i(t) := \hat{x}_i(t) - \hat{x}_i(t_k), \ i \in \{1, \ldots, n+1\}, \tag{4.6}$$

with t_k being the previous event-triggering time instant and k being the total number of triggered events during the time period $[0, t)$. The core idea of designing the above event-triggering condition is to restrict the upper bound of the sampling error [7, 12, 14]. To implement this event-triggering condition, the information of the "sampling error" of the observer state $\hat{x}$ is needed, which is defined by (4.6) and can be obtained based on the information of the sampler and extended state observer based on the scheme shown in Fig. 4.1.

Based on $\hat{x}$, a feedback control law $\kappa(\cdot)$ is designed as

$$\kappa(t) = -k_1 \hat{x}_1(t) \cdots - k_n \hat{x}_n(t) - \hat{x}_{n+1}(t). \tag{4.7}$$

From Fig. 4.1, the control signal $\kappa(t)$ is not continuously transmitted to the plant; in particular, the transmission is controlled by the event-triggering condition $r(t)$ such that the value of $\kappa(t)$ will be transmitted only when $r(t) = 1$. A zero-order hold (ZOH) module is used to hold the value of $\kappa(t)$ constant between two consecutive event instants and thus we have

$$u(t) = \kappa(t_k), \quad t \in [t_k, t_{k+1}). \tag{4.8}$$

In this way, the event-triggering condition $r(t)$ in (4.5) limits the upper bound of the sampling error of the control signal.

Next, we analyze the stability of the ET-ADRC scheme. Before continuing, we first introduce a few notations. Define $\tilde{x}_i$ and e_i as

$$\tilde{x}_i(t) := x_i(t) - \hat{x}_i(t), \tag{4.9}$$

$$e_i(t) := \tilde{x}_i(t)/\varepsilon^{n+1-i}, \ i \in \{1, \ldots, n+1\}. \tag{4.10}$$

From the definition of the sampling error in Eq. (4.6), the actual control signal $u(t)$ can be written as

$$u(t) = -\sum_{i=1}^{n} k_i x_i(t) + \sum_{i=1}^{n} k_i \tilde{x}_i(t) + \sum_{i=1}^{n} k_i \sigma_i(t) - \hat{x}_{n+1}(t) + \sigma_{n+1}(t). \tag{4.11}$$

Considering the system dynamics of x, we have

$$\begin{aligned}&\frac{d}{dt}(f(x,t)+w)\\ &= \frac{\partial f}{\partial t} + \frac{dw}{dt} + \sum_{i=1}^{n-1} \frac{\partial f}{\partial x_i} x_{i+1} + \frac{\partial f}{\partial x_n}\left(-\sum_{i=1}^{n} k_i x_i \right.\\ &\left. + \sum_{i=1}^{n} k_i \tilde{x}_i + \tilde{x}_{n+1} + \sum_{i=1}^{n} k_i \sigma_i + \sigma_{n+1}\right).\end{aligned} \tag{4.12}$$

According to Assumption 4.1 and the definition of $\tilde{x}_i$, we have

$$\begin{aligned}&\frac{d}{dt}(f(x,t)+w)\\ &\le \left|\frac{\partial f}{\partial t}\right| + \left|\frac{dw}{dt}\right| + \sum_{i=1}^{n-1}\left|\frac{\partial f}{\partial x_i}\right| |x_{i+1}| + \left|\frac{\partial f}{\partial x_n}\right|\left(\|x\|\sum_{i=1}^{n}|k_i|\right.\\ &\left. + \|e\|\sum_{i=1}^{n}|k_i|\varepsilon^{n+1-i} + \|e\| + \sum_{i=1}^{n}|k_i||\sigma_i| + |\sigma_{n+1}|\right)\\ &\le c_0 + \|x\|\sum_{j=1}^{n} c_j + b + \|x\|\sum_{j=1}^{n-1} c + c\left(\|x\|\sum_{i=1}^{n}|k_i|\right.\\ &\left. + \|e\|\left(\sum_{i=1}^{n}|k_i|\varepsilon^{n+1-i} + 1\right) + \sum_{i=1}^{n}|k_i||\sigma_i| + |\sigma_{n+1}|\right).\end{aligned}$$

For any scalars $a_1, a_2, \ldots, a_n$, the inequality $(a_1 + a_2 + \ldots + a_n)^2 \le n\sum_{i=1}^{n} a_i^2$ holds. Based on this relationship and the event-triggering condition $r(t)$ in (4.5), we have

$$\sum_{i=1}^{n}|k_i||\sigma_i| + |\sigma_{n+1}| \le \sqrt{n}\sqrt{\sum_{i=1}^{n} k_i^2\sigma_i^2 + \sigma_{n+1}^2} \le \sqrt{nM_1}\varepsilon^{\frac{1}{4}}.$$

This indicates

$$\frac{d}{dt}(f(x,t)+w) \le B_0 + B_1\|x\| + B_2\|e\|, \tag{4.13}$$

where $B_0 := c\sqrt{nM_1}\varepsilon^{\frac{1}{4}} + b + c_0$, $B_0 := c\sqrt{nM_1}\varepsilon^{\frac{1}{4}} + b + c_0$, $B_1 := c(n-1+\sum_{i=1}^{n}|k_i|) + \sum_{j=1}^{n} c_j$ and $B_2 := c(\sum_{i=1}^{n}|k_i|\varepsilon^{n+1-i}+1)$, respectively. Finally, since for any scalars a and b, $2ab \le a^2+b^2$ holds, there exist positive constants F_0, F_1 and F_2 such that

$$\left|\tfrac{d}{dt}(f(x(t),t)+w(t))\right|^2 \le F_0 + F_1\|x\|^2 + F_2\|e\|^2. \tag{4.14}$$

Before continuing, we introduce the following two conditions, which will play an important role in analyzing the asymptotic behavior of the ET-ADRC.

Condition **C1** *(Observer Design):* There exist positive constants $\beta_1, \lambda_1, \lambda_2, \lambda_3, \lambda_4$ and nonnegative-definite functions $\mathcal{P}(\cdot) : \mathbb{R}^{n+1} \to \mathbb{R}_0^+$ and $\mathcal{S}(\cdot) : \mathbb{R}^{n+1} \to \mathbb{R}_0^+$ which are continuous differentiable functions such that

$$\lambda_1\|z\|^2 \le \mathcal{P}(z) \le \lambda_2\|z\|^2,\ \lambda_3\|z\|^2 \le \mathcal{S}(z) \le \lambda_4\|z\|^2,$$
$$\sum_{i=1}^{n}\frac{\partial\mathcal{P}}{\partial z_i}(z_{i+1}-g_i(z_1)) - \frac{\partial\mathcal{P}}{\partial z_{n+1}}g_{n+1}(z_1) \le -\mathcal{S}(z),$$
$$\left\|\frac{\partial\mathcal{P}}{\partial z}\right\| \le \beta_1\|z\|.$$

Condition **C2** *(Controller Design):* k_i in (4.8) are some constants such that

$$K = \begin{bmatrix} 0 & 1 & 0 & \cdots & 0 \\ 0 & 0 & 1 & \cdots & 0 \\ \ddots & \ddots & \ddots & \ddots & \vdots \\ 0 & 0 & 0 & 0 & 1 \\ -k_1 & -k_2 & \cdots & -k_{n-1} & -k_n \end{bmatrix} \tag{4.15}$$

is Hurwitz.

Note that Condition **C1** is a class of typical assumptions in stability analysis of nonlinear systems [9], which potentially implies the Lyapunov properties of the nominal dynamics of the extended state observer. In particular, g_i can be chosen suitably (see Sect. 4.4 for an example) to fulfill Condition **C1**. In Eq. (4.15), $\{k_i\}$ are controller parameters designed to satisfy the requirement in Condition **C2**.

Based on the above discussions, we are ready to present the first theorem of this chapter. In the following result, we show that the event-triggering mechanism introduced in (4.5) can maintain the boundedness of the observation error and the stability of the system in (4.1).

Theorem 4.1 *Consider the closed-loop system* (4.1), (4.3), *and* (4.8) *under an event-triggering mechanism* (4.5). *Suppose Assumption 4.1 holds. Considering* $x(t_0) \in \mathcal{X}$

and $\hat{x}(t_0) \in \hat{\mathcal{X}}$, where $\hat{\mathcal{X}}, \mathcal{X}$ are any compact sets of $\mathbb{R}^{n+1}$, for any $M_1 > 0$, there exists a nonempty set $\mathfrak{E}_\varepsilon$ such that for any $\varepsilon \in \mathfrak{E}_\varepsilon$, x_i and $\hat{x}_i$ satisfy

$$\limsup_{t\to\infty} |x_i(t) - \hat{x}_i(t)| < O(\varepsilon^{n+\frac{13}{12}-i}),\ i \in \{1, \ldots, n+1\},$$
$$\limsup_{t\to\infty} \|x(t)\| \leq O(\varepsilon^{\frac{1}{12}}), \tag{4.16}$$

*if Conditions **C1–C2** are satisfied.*

Proof For the closed-loop system (4.1), (4.3) and (4.8) and any time $t \in [t_k, t_{k+1})$, we obtain that

$$\dot{e}_i(t) = \frac{1}{\varepsilon}e_{i+1}(t) - \frac{1}{\varepsilon}g_i(e_1(t)),\ i \in \{1, \ldots, n\},$$
$$\dot{e}_{n+1}(t) = h(t) - \frac{1}{\varepsilon}g_{n+1}(e_1(t)), \tag{4.17}$$

In the above equation,

$$h(t) := \frac{d}{dt}(f(x(t), t) + w(t)),$$

which describes the derivative of the uncertainty and disturbance. According to **C1**, the nominal part of the dynamics of e_i is stable while $h(t)$ is treated as the perturbation. From **C2**, it is obvious that there exists a positive semi-definite $Q > 0$ for the Hurwitz matrix K in (4.15) such that

$$QK + K^{\mathrm{T}}Q = -I. \tag{4.18}$$

Based on this observation, we can define a function $\mathcal{Q}(x)$ as

$$\mathcal{Q}(x) := x^{\mathrm{T}}Qx, \tag{4.19}$$

such that

$$\alpha_1\|x\|^2 \leq \mathcal{Q}(x) \leq \alpha_2\|x\|^2,\ \left\|\frac{\partial \mathcal{Q}}{\partial x}\right\| \leq \beta_2\|x\|, \tag{4.20}$$

where $\alpha_1 = \lambda_{\min}(Q)$, $\alpha_2 = \lambda_{\max}(Q)$ and $\beta_2 = 2\lambda_{\max}(Q)$. Consider a nonnegative-definite function $\mathcal{V}(x, e)$ defined by

$$\mathcal{V}(x, e) := \mathcal{Q}(x) + \mathcal{P}(e), \tag{4.21}$$

where $\mathcal{P}(\cdot)$ is provided in **C1**. From the definition of $\mathcal{V}(x, e)$ in (4.21), the system state x and the observation error e are combined to investigate the performance of this closed-loop system. By some straightforward computations, we obtain

$$\tfrac{d\mathcal{V}}{dt} := \mathfrak{T}_1 + \mathfrak{T}_2 + \mathfrak{T}_3 + \mathfrak{T}_4, \tag{4.22}$$

where

$$\mathfrak{T}_1 := \sum_{i=1}^{n-1} \tfrac{\partial \mathcal{Q}}{\partial x_i} x_{i+1} - \tfrac{\partial \mathcal{Q}}{\partial x_n} \sum_{i=1}^{n} k_i x_i, \tag{4.23}$$

$$\mathfrak{T}_2 := \tfrac{1}{\varepsilon}\left[\sum_{i=1}^{n} \tfrac{\partial \mathcal{P}}{\partial e_i}(e_{i+1} - g_i(e_1)) - \tfrac{\partial \mathcal{P}}{\partial e_{n+1}} g_{n+1}(e_1)\right], \tag{4.24}$$

$$\mathfrak{T}_3 := \tfrac{\partial \mathcal{P}}{\partial e_{n+1}} h(t), \tag{4.25}$$

$$\mathfrak{T}_4 := \tfrac{\partial \mathcal{Q}}{\partial x_n}\left(e_{n+1} + \sigma_{n+1} + \sum_{i=1}^{n} k_i \varepsilon^{n+1-i} e_i + k_i \sigma_i\right). \tag{4.26}$$

According to **C1** and (4.18), we have for $\mathfrak{T}_1$ and $\mathfrak{T}_2$

$$\mathfrak{T}_1 = x^{\mathrm{T}}(QK + K^{\mathrm{T}}Q)x = -\|x\|^2, \tag{4.27}$$

$$\mathfrak{T}_2 \leq -\frac{1}{\varepsilon}\mathcal{S}(e) \leq -\frac{\lambda_3}{\varepsilon}\|e\|^2. \tag{4.28}$$

In $\mathfrak{T}_3$, the total disturbance $f(x,t) + w$ is considered and we have

$$\begin{aligned}\mathfrak{T}_3 &\leq \left|\tfrac{\partial \mathcal{P}}{\partial e_{n+1}} \tfrac{d}{dt}(f(x(t),t) + w(t))\right| \\ &\leq \tfrac{1}{\varepsilon}\tfrac{\beta_1}{2}\varepsilon^{\frac{2}{3}}\|e\|^2 + \tfrac{\beta_1}{2}\varepsilon^{\frac{1}{3}}F_2\|e\|^2 + \tfrac{\beta_1}{2}\varepsilon^{\frac{1}{3}}F_1\|x\|^2 + \tfrac{\beta_1}{2}\varepsilon^{\frac{1}{3}}F_0.\end{aligned} \tag{4.29}$$

Similarly, for $\mathfrak{T}_4$, we have

$$\begin{aligned}\mathfrak{T}_4 &\leq \left|\tfrac{\partial \mathcal{Q}}{\partial x_n}\left(e_{n+1} + \sigma_{n+1} + \sum_{i=1}^{n} k_i(\varepsilon^{n+1-i}e_i + k_i\sigma_i)\right)\right| \\ &\leq \beta_2(n+1)\varepsilon^{\frac{1}{3}}\|x\|^2 + \tfrac{\beta_2}{2}\varepsilon^{-\frac{1}{3}}\left(\sum_{i=1}^{n} k_i^2\sigma_i^2 + \sigma_{n+1}^2\right) \\ &\quad + \tfrac{1}{\varepsilon}\tfrac{\beta_2}{2}\left(\sum_{i=1}^{n} k_i^2 \varepsilon^{2n-2i+\frac{8}{3}} + \varepsilon^{\frac{2}{3}}\right)\|e\|^2.\end{aligned} \tag{4.30}$$

From (4.27), (4.28), (4.29) and (4.30), the bound of $\frac{d\mathcal{V}}{dt}$ is given in terms of expressions of x and e. According to the above results, we have

$$\begin{aligned}\tfrac{d\mathcal{V}}{dt} &= \tfrac{d\mathcal{Q}}{dt} + \tfrac{d\mathcal{P}}{dt} = \sum_{i=1}^{n} \tfrac{\partial \mathcal{Q}}{\partial x_i}\tfrac{dx_i}{dt} + \sum_{i=1}^{n+1} \tfrac{\partial \mathcal{P}}{\partial e_i}\tfrac{de_i}{dt} \\ &\leq -\left[1 - \left(\beta_2(n+1)\varepsilon^{\frac{1}{3}} + \tfrac{\beta_1}{2}\varepsilon^{\frac{1}{3}}F_1\right)\right]\|x\|^2 \\ &\quad - \tfrac{1}{\varepsilon}\left[\lambda_3 - \tfrac{\beta_2}{2}\left(\sum_{i=1}^{n} k_i^2\varepsilon^{2n-2i+\frac{8}{3}} + \varepsilon^{\frac{2}{3}}\right) - \tfrac{\beta_1}{2}\varepsilon^{\frac{2}{3}}\right. \\ &\quad \left. - \tfrac{\beta_1}{2}\varepsilon^{\frac{4}{3}}F_2\right]\|e\|^2 + \tfrac{\beta_2}{2}\varepsilon^{-\frac{1}{3}}\left(\sum_{i=1}^{n} k_i^2\sigma_i^2 + \sigma_{n+1}^2\right) + \tfrac{\beta_1}{2}\varepsilon^{\frac{1}{3}}F_0.\end{aligned} \tag{4.31}$$

For notational brevity, we define $\gamma_1(\varepsilon)$, $\gamma_2(\varepsilon)$, $\gamma_3(\sigma)$, $\gamma(\varepsilon)$ as

$$\gamma_1(\varepsilon) := 1 - \left(\beta_2(n+1)\varepsilon^{\frac{1}{3}} + \tfrac{\beta_1}{2}\varepsilon^{\frac{1}{3}} F_1\right), \tag{4.32}$$
$$\gamma_2(\varepsilon) := \lambda_3 - \tfrac{\beta_2}{2}\left(\textstyle\sum_{i=1}^n k_i^2 \varepsilon^{2n-2i+\frac{8}{3}} + \varepsilon^{\frac{2}{3}}\right) - \tfrac{\beta_1}{2}\varepsilon^{\frac{2}{3}} - \tfrac{\beta_1 F_2}{2}\varepsilon^{\frac{4}{3}},$$
$$\gamma_3(\sigma) := \textstyle\sum_{i=1}^n k_i^2\sigma_i^2 + \sigma_{n+1}^2, \tag{4.33}$$
$$\gamma(\varepsilon) := \min\left\{\tfrac{\gamma_1(\varepsilon)}{\alpha_2}, \tfrac{\gamma_2(\varepsilon)}{\lambda_2\varepsilon}\right\}, \tag{4.34}$$

so that (4.31) can be written as

$$\dot{\mathcal{V}} \le -\gamma_1(\varepsilon)\|x\|^2 - \frac{1}{\varepsilon}\gamma_2(\varepsilon)\|e\|^2 + \frac{\beta_2}{2}\varepsilon^{-\frac{1}{3}}\gamma_3(\sigma) + \frac{\beta_1}{2}\varepsilon^{\frac{1}{3}}F_0.$$

Based on the expression of the event-triggering condition in (4.5), we have $\gamma_3(\sigma) \le M_1\varepsilon^{\frac{1}{2}}$. From **C1** and (4.20), the dynamics of $\mathcal{V}(x,e)$ can be bounded by

$$\dot{\mathcal{V}} \le -\frac{\gamma_1(\varepsilon)}{\alpha_2}Q(x) - \frac{\gamma_2(\varepsilon)}{\lambda_2\varepsilon}\mathcal{P}(e) + \frac{\beta_2}{2}M_1\varepsilon^{\frac{1}{6}} + \frac{\beta_1}{2}F_0\varepsilon^{\frac{1}{3}}.$$

Before continuing, we define $\mathfrak{E}_\varepsilon$ as

$$\mathfrak{E}_\varepsilon := \{\varepsilon|\gamma_1(\varepsilon) > 0, \gamma_2(\varepsilon) > 0, \varepsilon \in (0,1)\}. \tag{4.35}$$

From the definition of γ_1 and γ_2, we have $\gamma_1(\varepsilon) > 0$, $\gamma_2(\varepsilon) > 0$ for $\varepsilon = 0$ and $\gamma_1(\varepsilon) < 0$, $\gamma_2(\varepsilon) < 0$ for $\varepsilon \to \infty$, and therefore there exists $\Delta > 0$ such that $(0,\Delta) \subseteq \mathfrak{E}_\epsilon$, which implies that $\mathfrak{E}_\varepsilon$ is nonempty. For any $\varepsilon \in \mathfrak{E}_\varepsilon$, we have

$$\dot{\mathcal{V}}(x,e) \le -\gamma(\varepsilon)\mathcal{V}(x,e) + \frac{\beta_2}{2}M_1\varepsilon^{\frac{1}{6}} + \frac{\beta_1}{2}F_0\varepsilon^{\frac{1}{3}},$$

where $\gamma(\varepsilon)$ is defined in (4.34). By the comparison lemma in [9], we obtain that

$$\begin{aligned}
&\mathcal{V}(x(t), e(t)) \\
&\le \phi(t,t_0)\mathcal{V}(x(t_0), e(t_0)) \\
&+ \left(\frac{\beta_2}{2\gamma(\varepsilon)}M_1\varepsilon^{\frac{1}{6}} + \frac{\beta_1}{2\gamma(\varepsilon)}F_0\varepsilon^{\frac{1}{3}}\right)(1-\phi(t,t_0)),
\end{aligned} \tag{4.36}$$

where $\phi(t,t_0)$ is defined as

$$\phi(t,t_0) := \exp(-\gamma(\varepsilon)(t-t_0)).$$

When $t \to \infty$ such that $\phi(t,t_0) \to 0$, we have

$$\mathcal{V}(x(t), e(t)) \le \frac{\beta_2}{2\gamma(\varepsilon)}M_1\varepsilon^{\frac{1}{6}} + \frac{\beta_1}{2\gamma(\varepsilon)}F_0\varepsilon^{\frac{1}{3}}. \tag{4.37}$$

Considering **C1** and (4.20), we have

$$\mathcal{V}(x,e) = \mathcal{Q}(x) + \mathcal{P}(e) \geq \alpha_1 \|x\|^2 + \lambda_1 \|e\|^2$$

such that

$$\alpha_1 \|x\|^2 + \lambda_1 \|e\|^2 \leq \frac{\beta_2}{2\gamma(\varepsilon)} M_1 \varepsilon^{\frac{1}{6}} + \frac{\beta_1}{2\gamma(\varepsilon)} F_0 \varepsilon^{\frac{1}{3}}.$$

Hence, for the observation error e, we obtain

$$\limsup_{t\to\infty} \|e(t)\| \leq \sqrt{\frac{\beta_2 M_1 \varepsilon^{\frac{1}{6}}}{2\lambda_1 \gamma(\varepsilon)} + \frac{\beta_1 F_0 \varepsilon^{\frac{1}{3}}}{2\lambda_1 \gamma(\varepsilon)}}. \tag{4.38}$$

Then, from the definition of $\tilde{x}$ in (4.9), it is straightforward that

$$|x_i - \hat{x}_i| = |\tilde{x}_i| \leq \|e\| \varepsilon^{n+1-i}, \tag{4.39}$$

and then we have for $i \in \{1, \ldots, n+1\}$,

$$\limsup_{t\to\infty} |x_i(t) - \hat{x}_i(t)| \leq \varepsilon^{n+1-i} \sqrt{\frac{\beta_2 M_1 \varepsilon^{\frac{1}{6}}}{2\lambda_1 \gamma(\varepsilon)} + \frac{\beta_1 F_0 \varepsilon^{\frac{1}{3}}}{2\lambda_1 \gamma(\varepsilon)}}.$$

On the other hand, for system state x, we have

$$\limsup_{t\to\infty} \|x(t)\| \leq \sqrt{\frac{\beta_2 M_1 \varepsilon^{\frac{1}{6}}}{2\alpha_1 \gamma(\varepsilon)} + \frac{\beta_1 F_0 \varepsilon^{\frac{1}{3}}}{2\alpha_1 \gamma(\varepsilon)}}, \tag{4.40}$$

which completes the proof. □

From Theorem 4.1, we note that the precision of the ET-ADRC can be adjusted by appropriately selecting M_1. Specifically, from (4.37) and (4.40), for a smaller choice of parameter M_1, the event-triggering condition is tighter, which may lead to smaller bounds of observation error and state error but a larger number of measurement transmissions. Here, the restriction is that the gain parameter ε should be in the set $\mathfrak{E}_\varepsilon$ defined in (4.35). According to the structural properties of γ_1 and γ_2, it would be helpful to choose smaller values for the high-gain parameter to satisfy (4.35).

For many practical applications, it is possible to set the initial values in advance according to the specific operating conditions. Based on this observation, the following result can be obtained based on Theorem 4.1.

Corollary 4.1 *Consider the closed-loop system* (4.1), (4.3) *and* (4.8) *under an event-triggering mechanism* (4.5). *Suppose that Assumption* 1 *holds. For any given* $M_1 > 0$, *there exist a set* $\mathfrak{E}_\varepsilon$ *such that for any* $\varepsilon \in \mathfrak{E}_\varepsilon$, x_i *and* $\hat{x}_i$ *satisfy*

$$\max |x_i(t) - \hat{x}_i(t)| \le O(\varepsilon^{n+\frac{13}{12}-i}),\ i \in \{1, \ldots, n+1\},$$
$$\max \|x(t)\| \le O(\varepsilon^{\frac{1}{12}}), \forall t > t_0 \tag{4.41}$$

provided (1) ***C1*** *and* ***C2*** *are satisfied and (2)* $x(t_0)$ *and* $e(t_0)$ *satisfy*

$$\mathcal{Q}(x(t_0)) + \mathcal{P}(e(t_0)) \le \frac{\beta_2 M_1 \varepsilon^{\frac{1}{6}}}{2\gamma(\varepsilon)} + \frac{\beta_1 F_0 \varepsilon^{\frac{1}{3}}}{2\gamma(\varepsilon)},$$

where $\gamma(\varepsilon)$ *is defined in* (4.34).

Proof Since $\phi(t, t_0) \in (0, 1)$, we obtain that for any $L > 0$, $R > 0$,

$$L\phi(t, t_0) + R(1 - \phi(t, t_0)) \le \max\{L, R\}.$$

Thus, from (4.36), an upper bound of $\mathcal{V}(x, e)$ can be obtained as

$$\mathcal{V}(x(t), e(t)) \le \max\left\{\mathcal{V}(x(t_0), e(t_0)), \frac{\beta_2 M_1 \varepsilon^{\frac{1}{6}}}{2\gamma(\varepsilon)} + \frac{\beta_1 F_0 \varepsilon^{\frac{1}{3}}}{2\gamma(\varepsilon)}\right\}.$$

Since $\mathcal{Q}(x(t_0)) + \mathcal{P}(e(t_0)) \le \frac{\beta_2 M_1 \varepsilon^{\frac{1}{6}}}{2\gamma(\varepsilon)} + \frac{\beta_1 F_0 \varepsilon^{\frac{1}{3}}}{2\gamma(\varepsilon)}$, through a similar line of arguments as in those of (4.37)–(4.40), we have that for x_i and $\hat{x}_i$ and all $t > t_0$

$$\max |x_i(t) - \hat{x}_i(t)| < \varepsilon^{n+1-i} \sqrt{\frac{\beta_2 M_1 \varepsilon^{\frac{1}{6}}}{2\lambda_1 \gamma(\varepsilon)} + \frac{\beta_1 F_0 \varepsilon^{\frac{1}{3}}}{2\lambda_1 \gamma(\varepsilon)}},$$
$$\max \|x(t)\| \le \sqrt{\frac{\beta_2 M_1 \varepsilon^{\frac{1}{6}}}{2\alpha_1 \gamma(\varepsilon)} + \frac{\beta_1 F_0 \varepsilon^{\frac{1}{3}}}{2\alpha_1 \gamma(\varepsilon)}}, \tag{4.42}$$

where $1 \le i \le n+1$. This completes the proof. □

4.3 Separate Event-Triggered Observation and Control

In this section, we consider a more general scenario that ET-ADRC is designed with separate event-triggering conditions for measurement transmission and control command update (see Fig. 4.2). For this closed-loop scheme, the output measurement $y(t)$ is not continuously transmitted to ESO, and consequently the ESO is taken to have the form

$$\begin{aligned}
\dot{\hat{x}}_1(t) &= \hat{x}_2(t) + \varepsilon^{n-1} g_1\left(\frac{\mathcal{Y}(t) - \hat{x}_1(t)}{\varepsilon^n}\right), \\
&\vdots \\
\dot{\hat{x}}_n(t) &= \hat{x}_{n+1}(t) + g_n\left(\frac{\mathcal{Y}(t) - \hat{x}_1(t)}{\varepsilon^n}\right) + u(t), \\
\dot{\hat{x}}_{n+1}(t) &= \frac{1}{\varepsilon} g_{n+1}\left(\frac{\mathcal{Y}(t) - \hat{x}_1(t)}{\varepsilon^n}\right),
\end{aligned} \tag{4.43}$$

where $\mathcal{Y}(t) := y(t)$ for $t \in [t_k, t_{k+1})$ represents the previously transmitted output measurement y, and g_i are some chosen Lipschitz functions with $L_i > 0$. From Fig. 4.2, an additional event-triggering condition $r_2(t)$ determines the transmission of the value of $y(t)$ to ESO; only when $r_2(t) = 1$ will $y(t)$ be transmitted. We define sampling error of output as

$$\eta(t) := (\mathcal{Y}(t) - y(t))/\varepsilon^n, \tag{4.44}$$

and for notational brevity we write

$$\sum_{i=1}^{n+1} L_i := L, \tag{4.45}$$

$$\tilde{\gamma}_2(\varepsilon) := \gamma_2(\varepsilon) - \tfrac{\varepsilon^3}{2}, \tag{4.46}$$

$$\tilde{\gamma}(\varepsilon) := \min\left\{\tfrac{\gamma_1(\varepsilon)}{\alpha_2}, \tfrac{\tilde{\gamma}_2(\varepsilon)}{\lambda_2 \varepsilon}\right\}, \tag{4.47}$$

$$\delta_i(e_1(t), \eta(t)) := g_i(e_1(t) + \eta(t)) - g_i(e_1(t)). \tag{4.48}$$

Based on the results obtained in Theorem 4.1, we analyze the performance of the event-triggered state observation and control in the following theorem:

Theorem 4.2 *Consider the closed-loop system (4.1), (4.43), (4.8) and the following definition of the two event-triggering conditions:*

$$r_1(t) = \begin{cases} 0, \text{ if } \sum_{i=1}^n k_i^2 \sigma_i^2(t) + \sigma_{n+1}^2(t) \le M_1 \varepsilon^{\frac{1}{2}}, \\ 1, \text{ otherwise,} \end{cases} \tag{4.49}$$

and

$$r_2(t) = \begin{cases} 0, \text{ if } |\eta(t)|^2 \le M_2 \varepsilon^{\frac{13}{6}}, \\ 1, \text{ otherwise.} \end{cases} \tag{4.50}$$

Suppose that Assumption 1 holds. For $x(t_0) \in \mathcal{X}$ and $\hat{x}(t_0) \in \hat{\mathcal{X}}$, where $\hat{\mathcal{X}}, \mathcal{X}$ are any compact sets of $\mathbb{R}^{n+1}$ and for any $M_1 > 0$ and $M_2 > 0$, there exists a nonempty set $\tilde{\mathfrak{E}}_\varepsilon$ such that for any $\varepsilon \in \tilde{\mathfrak{E}}_\varepsilon$, x_i and $\hat{x}_i$ satisfy

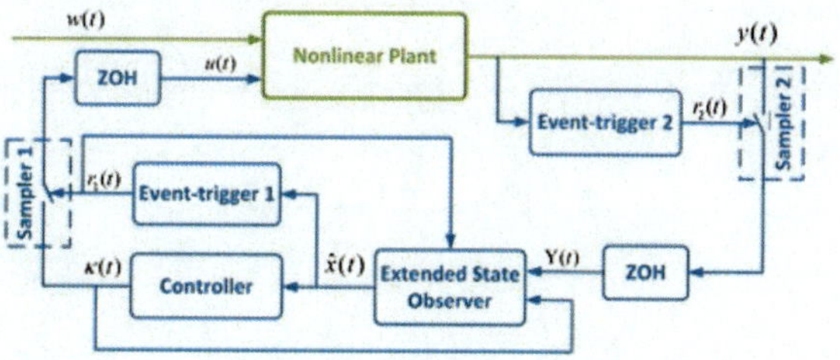

Fig. 4.2 The schematic of the event-triggered active disturbance rejection control with two triggers

$$\limsup_{t\to\infty} |x_i(t) - \hat{x}_i(t)| \le O(\varepsilon^{n+\frac{13}{12}-i}),\ i \in \{1, \dots, n+1\}$$
$$\limsup_{t\to\infty} \|x(t)\| \le O(\varepsilon^{\frac{1}{12}}), \tag{4.51}$$

provided the conditions in ***C1*** *and* ***C2*** *are satisfied.*

Proof For the closed-loop system (4.1), (4.43) and (4.8), we obtain that for e_i

$$\dot{e}_i(t) = \frac{1}{\varepsilon}e_{i+1}(t) - \frac{1}{\varepsilon}g_i(e_1(t)) - \frac{1}{\varepsilon}\delta_i(e_1(t), \eta(t)),$$
$$\dot{e}_{n+1}(t) = h(t) - \frac{1}{\varepsilon}g_{n+1}(e_1(t)) - \frac{1}{\varepsilon}\delta_{n+1}(e_1(t), \eta(t)).$$

Consider a nonnegative-definite function $\mathcal{V}(x, e) = \mathcal{Q}(x) + \mathcal{P}(e)$, where $\mathcal{P}(\cdot)$ and $\mathcal{Q}(\cdot)$ are provided in Condition C1 and (4.19), respectively. Then, for $\mathcal{V}(x, e)$ we have

$$\frac{d\mathcal{V}}{dt} := \mathfrak{T}_1 + \mathfrak{T}_2 + \mathfrak{T}_3 + \mathfrak{T}_4 + \mathfrak{T}_5, \tag{4.52}$$

where $\mathfrak{T}_1$–$\mathfrak{T}_4$ are defined in (4.23)–(4.26) and

$$\mathfrak{T}_5 := \frac{1}{\varepsilon}\left(\sum_{i=1}^{n+1} \frac{\partial \mathcal{P}}{\partial e_i}\delta_i(e_1, \eta)\right)$$

such that

$$\begin{aligned}\mathfrak{T}_5 &\le \tfrac{1}{\varepsilon}\beta_1 \|e\| \textstyle\sum_{i=1}^{n+1} |\delta_i(e_1, \eta)| \le \tfrac{1}{\varepsilon}\beta_1 \|e\| \sum_{i=1}^{n+1} L_i\, |\eta| \\ &\le \tfrac{1}{2}\varepsilon^2 \|e\|^2 + \tfrac{\beta_1^2}{2}\left(\textstyle\sum_{i=1}^{n+1} L_i\right)^2 |\eta|^2 \varepsilon^{-2}.\end{aligned} \tag{4.53}$$

Combining the results from the analysis of $\mathfrak{T}_1$–$\mathfrak{T}_4$ (*c.f.*, (4.29)–(4.28)), we have

$$\begin{aligned}\frac{d\mathcal{V}}{dt} &= \sum_{i=1}^{n} \frac{\partial Q}{\partial x_i}\frac{dx_i}{dt} + \sum_{i=1}^{n+1} \frac{\partial \mathcal{P}}{\partial e_i}\frac{de_i}{dt} \\ &\le -\left[1 - \left(\beta_2(n+1)\varepsilon^{\frac{1}{3}} + \frac{\beta_1}{2}\varepsilon^{\frac{1}{3}}F_1\right)\right]\|x\|^2 + \frac{\beta_1}{2}\varepsilon^{\frac{1}{3}}F_0 \\ &\quad - \frac{1}{\varepsilon}\left[\lambda_3 - \frac{\beta_2}{2}\left(\sum_{i=1}^{n} k_i^2\varepsilon^{2n-2i+\frac{8}{3}} + \varepsilon^{\frac{2}{3}}\right) - \frac{\beta_1}{2}\varepsilon^{\frac{2}{3}}\right. \\ &\quad \left. - \frac{\beta_1}{2}\varepsilon^{\frac{4}{3}}F_2 - \frac{1}{2}\varepsilon^{3}\right]\|e\|^2 + \frac{\beta_2}{2}\varepsilon^{-\frac{1}{3}}\left(\sum_{i=1}^{n} k_i^2\sigma_i^2 + \sigma_{n+1}^2\right) \\ &\quad + \frac{\beta_1^2}{2}\left(\sum_{i=1}^{n+1} L_i\right)^2 |\eta|^2\varepsilon^{-2}. \end{aligned} \tag{4.54}$$

Furthermore, for the event-triggering conditions in (4.49) and (4.50), we have $\gamma_3(\sigma) \le M_1\varepsilon^{\frac{1}{2}}$ and $|\eta(t)|^2 \le M_2\varepsilon^{\frac{13}{6}}$. From **C1** and (4.20), we further have

$$\begin{aligned}\dot{\mathcal{V}} &\le -\frac{\gamma_1(\varepsilon)}{\alpha_2}Q(x) - \frac{\tilde{\gamma}_2(\varepsilon)}{\lambda_2\varepsilon}\mathcal{P}(e) \\ &\quad + \frac{\beta_1^2}{2}L^2M_2\varepsilon^{\frac{1}{6}} + \frac{\beta_2}{2}M_1\varepsilon^{\frac{1}{6}} + \frac{\beta_1}{2}F_0\varepsilon^{\frac{1}{3}}.\end{aligned}$$

Next, we define $\tilde{\mathfrak{E}}_\varepsilon$ as

$$\tilde{\mathfrak{E}}_\varepsilon := \{\varepsilon | \gamma_1(\varepsilon) > 0, \tilde{\gamma}_2(\varepsilon) > 0, \varepsilon \in (0,1)\}, \tag{4.55}$$

such that for any $\varepsilon \in \tilde{\mathfrak{E}}_\varepsilon$, we have

$$\begin{aligned}\dot{\mathcal{V}}(x,e) &\le -\tilde{\gamma}(\varepsilon)\mathcal{V}(x,e) \\ &\quad + \frac{\beta_1^2}{2}L^2M_2\varepsilon^{\frac{1}{6}} + \frac{\beta_2}{2}M_1\varepsilon^{\frac{1}{6}} + \frac{\beta_1}{2}F_0\varepsilon^{\frac{1}{3}},\end{aligned}$$

where $\tilde{\gamma}(\varepsilon)$ is defined in (4.47). Through a similar line of arguments as those in (4.37)–(4.40), we observe that for x_i and $\hat{x}_i$,

$$\begin{aligned}&\limsup_{t\to\infty} |x_i(t) - \hat{x}_i(t)|,\ i \in \{1,\ldots,n+1\} \\ &< \varepsilon^{n+1-i}\sqrt{\frac{\beta_2 M_1\varepsilon^{\frac{1}{6}}}{2\lambda_1\tilde{\gamma}(\varepsilon)} + \frac{\beta_1 F_0\varepsilon^{\frac{1}{3}}}{2\lambda_1\tilde{\gamma}(\varepsilon)} + \frac{\beta_1^2 L^2 M_2\varepsilon^{\frac{1}{6}}}{2\lambda_1\tilde{\gamma}(\varepsilon)}}.\end{aligned} \tag{4.56}$$

and

$$\limsup_{t\to\infty} \|x(t)\| \le \sqrt{\frac{\beta_2 M_1\varepsilon^{\frac{1}{6}}}{2\alpha_1\tilde{\gamma}(\varepsilon)} + \frac{\beta_1 F_0\varepsilon^{\frac{1}{3}}}{2\alpha_1\tilde{\gamma}(\varepsilon)} + \frac{\beta_1^2 L^2 M_2\varepsilon^{\frac{1}{6}}}{2\alpha_1\tilde{\gamma}(\varepsilon)}}. \tag{4.57}$$

This completes the proof. □

Before proceeding, we show the nonexistence of Zeno phenomenon in the proposed event-triggered ADRC schemes. Consider the event-triggering condition $r_2(t)$ in (4.50). According to the definition of the sampling error $\eta(t)$, we know that

$$\varepsilon^n \eta(t) = \int_t^{t_k} \dot{x}_1(\tau) d\tau = \int_t^{t_k} x_2(\tau) d\tau.$$

From the boundedness of x obtained in Theorem 4.2, we have

$$|\eta(t)| \leq \frac{1}{\varepsilon^n} \int_t^{t_k} x_2(\tau) d\tau \leq (t - t_k) B_x(M_1, M_2, \varepsilon) \tag{4.58}$$

for some $B_x(M_1, M_2, \varepsilon) > 0$ which can be calculated from (4.57). Based on (4.58), we can obtain that at time instant $t \geq t_k$, the sampling error will not exceed $(t - t_k) B_x(M_1, M_2, \varepsilon)$. Thus, if we define the inter-triggering period limit $\tau_{miet,x} > 0$ by

$$\tau_{miet,x} := \sqrt{M_2 \varepsilon^{\frac{13}{12}}} / B_x(M_1, M_2, \varepsilon), \tag{4.59}$$

then from $r_2(t)$, it is obvious that

$$t_{k+1} - t_k \geq \tau_{miet,x}.$$

Next, for the event-triggering condition $r_1(t)$ in (4.49), we consider the definition of $\sigma_i(t)$ provided in (4.6), from which we can obtain

$$\begin{aligned} \sigma_i(t) &= \int_{t_k}^{t} \dot{\hat{x}}_i(\tau) d\tau = \int_{t_k}^{t} \hat{x}_{i+1}(\tau) + g_i \left(\frac{y(\tau) - \hat{x}_1(\tau)}{\varepsilon^n} \right) d\tau \\ &\leq \int_{t_k}^{t} |x_{i+1}(\tau)| + |(x_{i+1}(\tau) - \hat{x}_{i+1}(\tau))| + L_i \eta(\tau) \\ &\quad + \frac{L_i}{\varepsilon^n} |x_1(\tau) - \hat{x}_1(\tau)| d\tau, \quad i \in \{1, \ldots, n-1\}. \end{aligned} \tag{4.60}$$

According to the definition of $r_2(t)$ and the properties of x and $x_i - \hat{x}_i$ shown in Theorem 4.2, we can obtain

$$|\sigma_i(t)| \leq (t - t_k) B_{\sigma_i}(M_1, M_2, \varepsilon), \qquad i \in \{1, \ldots, n-1\},$$

for some $B_{\sigma_i}(M_1, M_2, \varepsilon) > 0$. This discussion can be easily extended to the case $i \in \{n, n+1\}$ such that $B_{\sigma_n}(M_1, M_2, \varepsilon)$ and $B_{\sigma_{n+1}}(M_1, M_2, \varepsilon)$ are also obtained. Recalling the event-triggering condition $r_1(t)$, we can define

$$B_\sigma(M_1, M_2, \varepsilon) := \sum_{i=1}^{n} k_i^2 (B_{\sigma_i}(M_1, M_2, \varepsilon))^2 + (B_{\sigma_{n+1}}(M_1, M_2, \varepsilon))^2,$$

such that we have

$$t_{k+1} - t_k \geq \tau_{miet,\sigma}$$

for the event-triggering condition $r_1(t)$, where the lower bound on the average sampling period $\tau_{miet,\sigma}$ is defined by

$$\tau_{miet,\sigma} := \sqrt{M_1/B_\sigma(M_1, M_2, \varepsilon)}\varepsilon^{\frac{1}{4}}. \tag{4.61}$$

This completes the discussion on the nonexistence of Zeno behavior for the case that both the transmissions of system output signal and control signal are event-triggered as shown in Fig. 4.2. Zeno-freeness can also be obtained for the simpler case that only the transmission of control signal is determined by an event-triggered strategy in Fig. 4.1 through a similar lines of arguments to those in (4.60)–(4.61) by considering $M_2 = 0$, since the output measurement is transmitted in real time. In this way, the non-existence of Zeno behavior for the proposed event-triggered ADRC schemes can be ensured.

4.4 Experimental Results

In this section, to validate the obtained theoretic results, we apply the ET-ADRC to the DC motor loading system that we introduced in Chap. 2. The specific structure of the motor loading platform is shown in Fig. 4.3, which is composed of a speed torque sensor, an optical encoder, a magnetic brake, an inertial loading mechanism, and a DC motor (Fig. 4.3). The DC motor is mounted on the bracket and connected with a speed torque sensor. To load inertia to this motor, a timing belt pulley via a torque coupling is set at the other end of the speed torque sensor. The belt pulley is connected with the magnetic brake to apply sliding friction load to this motor. As shown in Fig. 4.3, the platform can also load the motor by a magnetic brake with a maximum output torque of 20 N·m, a driving voltage of 24 VDC, and a maximum driving current of 1210 mA. The platform offers sensors to measure the speed of the DC motor. The platform converts the measurement signals to an analog signal within [−10, 10]V. To apply the stabilization results to a tracking problem, we construct the tracking error dynamics such that if the error system is stabilized, the goal of tracking can be achieved. The details are presented as follows:

The dynamics of DC motor is generally described as

$$M\ddot{y} = F_{ele}(u) - V_{frc}\dot{y} + F_{fri}(\dot{y}) + F_{cog}(y) + F_{dis}, \tag{4.62}$$

where y is the displacement, u denotes the input control signal, M is the inertia, V_{frc} is the viscous friction coefficient, $F_{ele}(u)$ is the electromagnetic driving force of the motor, $F_{fri}(\cdot)$ represents Coulomb friction, $F_{cog}(\cdot)$ represents the position dependent cogging force, and F_{dis} represents the lumped modeling errors and external

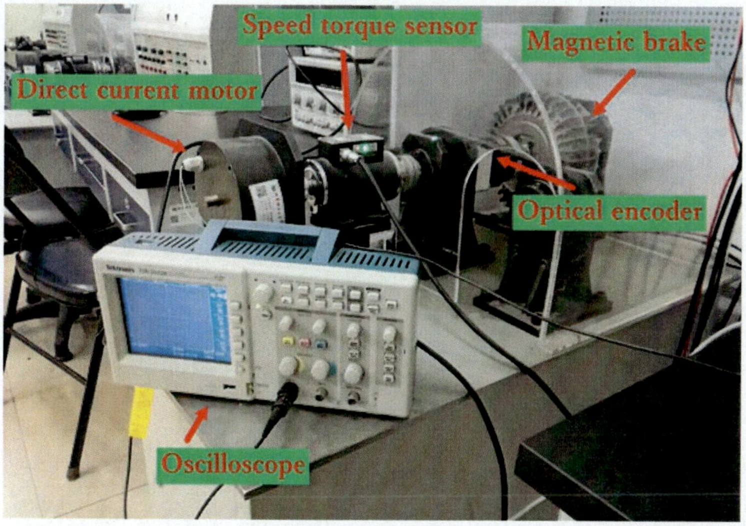

Fig. 4.3 The DC motor loading platform

disturbances. If we write

$$\zeta_1 := y, \zeta_2 := \dot{y},$$

then based on the technique introduced in [2, 16], the dynamics of DC motor in (4.62) can be expressed as

$$\begin{aligned} \dot{\zeta}_1 &= \zeta_2, \\ \dot{\zeta}_2 &= F(\zeta) + \frac{A}{M}u + w, \end{aligned} \tag{4.63}$$

where

$$F(\zeta) := -\frac{V_{frc}}{M}\zeta_2 + \sum_{i=1}^{n_f}\left(\frac{S_i}{M}\sin\left(\frac{2\pi i}{P}\zeta_1\right) + \frac{C_i}{M}\cos\left(\frac{2\pi i}{P}\zeta_1\right)\right) - \frac{A_f}{M}\arctan(\beta\zeta_2),$$

$w := \frac{F_{dis}}{M}$, and A, A_f, P, β, n_f, S_i, P_i are some constants parameterizing the dynamics of ζ. Considering the energy restrictions in practical application, we design the reference trajectory with the idea that the reference speed ζ_2^* is bounded with $|\zeta_2^*| \le C_\zeta$. The reference trajectory ζ^* of the DC motor is designed as $\zeta_1(t) \to \zeta_1^*(t)$, $\zeta_2(t) \to \zeta_2^*(t)$. To describe the tracking performance of the closed-loop system, we define the tracking error as

$$x(t) := \zeta(t) - \zeta^*(t), \tag{4.64}$$

$$\dot{\zeta}_1^*(t) = \zeta_2^*(t), \tag{4.65}$$

$$\dot{\zeta}_2^*(t) = D(t), \tag{4.66}$$

where $D(\cdot)$ is the designed trajectory function satisfying $|D(t)| + |\dot{D}(t)| \leq C_D$ for some $C_D > 0$. In this way, the tracking error dynamics for the DC torque motor can be represented by

$$\begin{aligned} \dot{x}_1 &= x_2, \\ \dot{x}_2 &= f(x,t) + \frac{A}{M}u + w, \end{aligned} \tag{4.67}$$

where $f(x,t) := F(x + \zeta^*) - D(t)$. To model the difference between the input channel parameter A/M and the nominal value, we pre-design a signal $U := \frac{A}{M}u$ as the input to the extended state observer, and the control signal u is constructed by $u = \frac{M}{A}(-k_1\hat{x}_1(t_k) \cdots - k_n\hat{x}_n(t_k) - \hat{x}_{n+1}(t_k))$. Thus, the model of DC motor can be modeled in the form of (4.1). Furthermore, if we define c, c_0, c_1 and c_2 as

$$\begin{aligned} c &:= \max\left\{\frac{2\pi}{P}\sum_{i=1}^{n_f}\left(\frac{|iS_i|}{M} + \frac{|iC_i|}{M}\right), \frac{V_{frc}}{M} + \frac{A_f}{M}\beta\right\}, \\ c_0 &:= \frac{A_f\pi}{2M} + \sum_{i=1}^{n_f}\left(\frac{|S_i|}{M} + \frac{|C_i|}{M}\right) + C_D + \frac{V_{frc}}{M}C_\zeta, \\ c_1 &:= 0, \quad c_2 := \frac{V_{frc}}{M}, \end{aligned} \tag{4.68}$$

then Assumption 4.1 can be satisfied while enabling the tracking goal for the DC torque motor. Finally, considering the energy restrictions in a practical application, the boundedness properties for the disturbance term is naturally expected.

We first consider the case in which the event-triggering mechanism is applied for control signal transmission. The extended state observer is of the form in (4.3) with $n = 2$, $\hat{x}(0) = [0, 0, 0]^{\mathrm{T}}$ and g_i are given as

$$g_1(z_1) = 3z_1 + \varphi(z_1), \quad g_2(z_1) = 3z_1, \quad g_3(z_1) = z_1,$$

where $\varphi(\theta)$ is defined by

$$\varphi(\theta) = \begin{cases} -\frac{1}{4}, & \theta < -\frac{\pi}{2}, \\ \frac{1}{4}\sin\theta, & -\frac{\pi}{2} \leq \theta \leq \frac{\pi}{2}, \\ \frac{1}{4}, & \theta > \frac{\pi}{2}. \end{cases} \tag{4.69}$$

The feedback control law κ is designed as

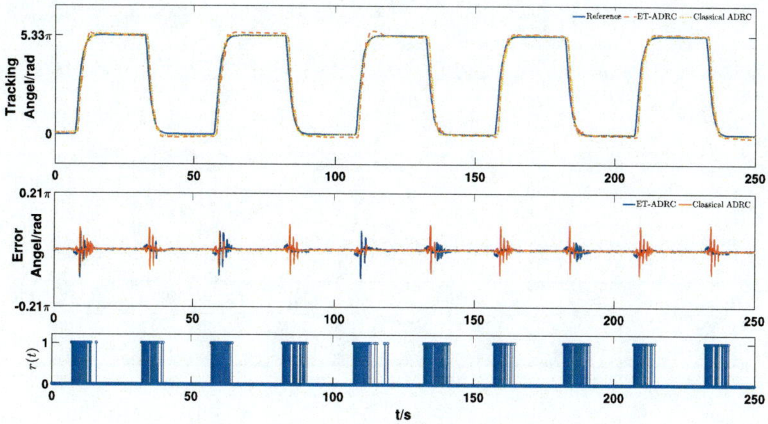

Fig. 4.4 Event-triggered ADRC with $M_1 = 0.022$ and $\varepsilon = 0.2$: Subfigure-1 shows the tracking performance; Subfigure-2 shows the tracking error of DC torque motor and Subfigure-3 illustrates the event-triggering condition

$$\kappa(t) = -6\hat{x}_1(t) - 9\hat{x}_2(t) - \hat{x}_3(t), \tag{4.70}$$

such that Condition **C2** holds.

Then, an event-triggering condition is introduced as

$$r(t) = \begin{cases} 0, \text{ if } 36\sigma_1^2(t) + 81\sigma_2^2(t) + \sigma_3^2(t) \le M_1\varepsilon^{\frac{1}{2}}, \\ 1, \text{ otherwise.} \end{cases}$$

To implement the ET-ADRC scheme, we first select $M_1 = 0.022$ and $\varepsilon = 0.2$. The performance of the controller is shown in Fig. 4.4. We observe that satisfactory performance can be maintained by ET-ADRC. Moreover, the control signal and the event-triggering signal $r(t)$ are illustrated in Fig. 4.5. It can be shown that the control signal updates only at the time instants when $r(t) = 1$.

Next, we consider the case in which the event-triggering mechanism is applied for transmission of both control signals and output measurements. The event-triggering condition $r_1(t)$ for control signal transmission is the same as (4.71), and the event-triggering condition $r_2(t)$ for output measurement transmission is of the form:

$$r_2(t) = \begin{cases} 0, \text{ if } |\eta(t)|^2 \le M_2\varepsilon^{\frac{13}{6}}, \\ 1, \text{ otherwise.} \end{cases} \tag{4.71}$$

To implement the ET-ADRC scheme, we select $M_1 = 0.044$, $M_2 = 4.086$ and $\varepsilon = 0.2$. The performance of this control scheme is shown in Fig. 4.6, and the control signal and sampled measurement are illustrated in Fig. 4.7, where we can observe that $r_1(t)$ and $r_2(t)$ are triggered separately.

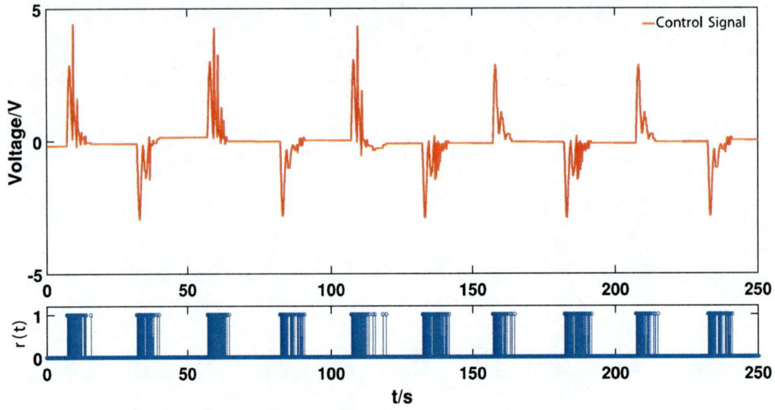

Fig. 4.5 Control signal of ET-ADRC with $M_1 = 0.022$ and $\varepsilon = 0.2$

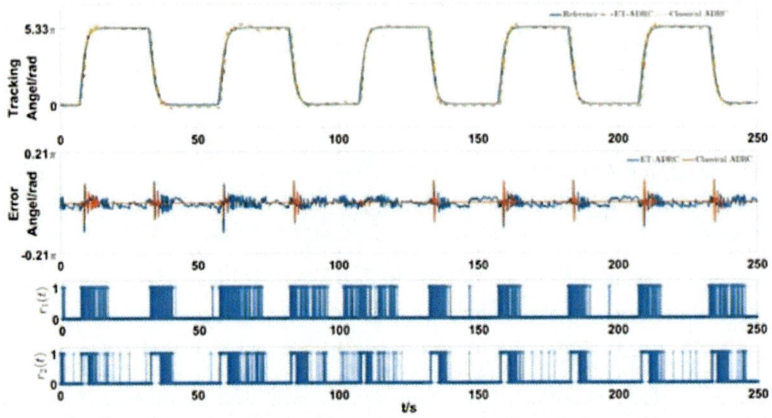

Fig. 4.6 Event-triggered ADRC with $M_1 = 0.044$, $M_2 = 4.086$ and $\varepsilon = 0.2$: the first plot shows the tracking performance; the second plot shows the tracking error of DC torque motor and the third plot illustrates the event-triggering condition

On the other hand, the trends of the control and observation performance with the increase of the event-triggering condition parameters M_1 and M_2 are provided in Tables 4.1 and 4.2. To enable the comparison, the ET-ADRC controllers are implemented by difference approximation with the sampling time $t_s = 0.025$ s and the "triggering ratio" is defined by

$$\text{Triggering ratio} = \frac{\text{Total trigger times}}{2 \times \text{Experiment time}/t_s}.$$

Based on the experimental results, the comparison of performance between ET-ADRC and standard ADRC is also obtained. Following the approach in [5, 17],

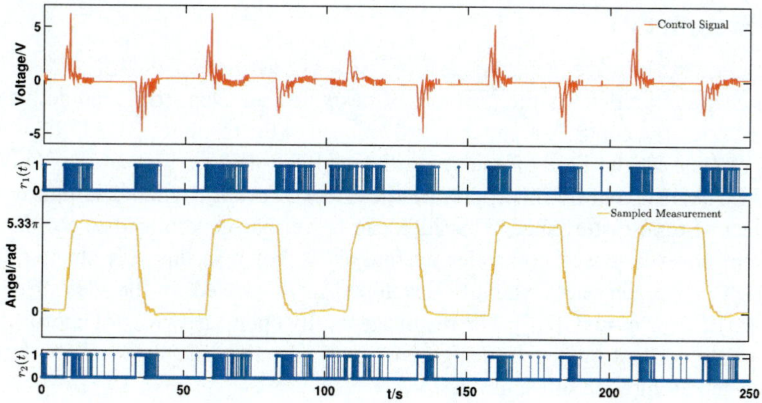

Fig. 4.7 Control signal and sampled measurement of event-triggered ADRC with $M_1 = 0.044$, $M_2 = 4.086$ and $\varepsilon = 0.2$

Table 4.1 Performance comparison for different M_1 with $M_2 = 0$

$M_1 \Downarrow$	Tracking error	Observation error	Triggering ratio
0.0894	0.6770/rad	0.0117/rad	0.5183
0.0447	0.5730/rad	0.0109/rad	0.5196
0.0224	0.5664/rad	0.0090/rad	0.5319
0	0.3855/rad	0.0079/rad	1

Table 4.2 Performance comparison for different M_2 with $M_1 = 0.022$

$M_2 \Downarrow$	Tracking error	Observation error	Triggering ratio
4.623	0.6167/rad	0.1458/rad	0.0613
3.269	0.5614/rad	0.0754/rad	0.1180
2.312	0.5564/rad	0.0352/rad	0.1135
0	0.5529/rad	0.0090/rad	0.5196

the measurement signal and control signal are transmitted in real time for standard ADRC. For standard ADRC, the average tracking error and observation error are 0.3836 rad and 0.0079 rad. Compared with standard ADRC, the performance of ET-ADRC is degraded within an acceptable level while the output measurements and control signals scheme are transmitted according to the event-triggered transmission schemes. Moreover, we observe that with the decrease of the event-triggering ratio through adjusting the event-triggering parameters, the performance of ET-ADRC approaches that of the standard ADRC (Tables 4.1 and 4.2).

4.5 Summary

In this chapter, we discuss the problem of designing an event-triggered active disturbance rejection controller, and the performance of the resultant ET-ADRC is analyzed. In general, the idea of analyzing the sampled-data performance is built on that of CT-ADRC, but by incorporating the effect of event-triggered sampling into the effect of total disturbance. This idea can be further generalized to the case of high-gain observer-based controller design, as is shown in the next chapter. This ET-ADRC approach was originally developed in [8], based on the idea of event-triggered ESO proposed in [7]. The results generally open a new way of approaching event-triggered control system design and many research questions can be further investigated along with this approach.

References

1. Castaneda, L., Luviano-Juarez, A., Chairez, I.: Robust trajectory tracking of a delta robot through adaptive active disturbance rejection control. IEEE Trans. Control Syst. Technol. **23**(4), 1387–1398 (2015)
2. Chen, Z., Yao, B., Wang, Q.: Adaptive robust precision motion control of linear motors with integrated compensation of nonlinearities and bearing flexible modes. IEEE Trans. Industr. Inf. **9**(2), 965–973 (2013). https://doi.org/10.1109/TII.2012.2225439
3. Guo, B., Zhao, Z.: On the convergence of an extended state observer for nonlinear systems with uncertainty. Syst. Control Lett. **60**(6), 420–430 (2011)
4. Guo, B., Zhao, Z.: On convergence of nonlinear active disturbance rejection for MIMO system. SIAM J. Control Optim. **51**(2), 1727–1757 (2013)
5. Guo, B., Zhao, Z.: Active Disturbance Rejeciton Control for Nonlinear Systems: An Introduction. Wiley, Hoboken (2016)
6. He, N., Shi, D.: Event-based robust sampled-data model predictive control: a non-monotonic lyapunov function approach. IEEE Trans. Circuits Syst. I Regul. Pap. **62**(10), 2555–2564 (2015)
7. Huang, Y., Wang, J., Shi, D., Shi, L.: Toward event-triggered extended state observer. IEEE Trans. Autom. Control **63**(6), 1842–1849 (2018)
8. Huang, Y., Wang, J., Shi, D., Wu, J., Shi, L.: Event-triggered sampled-data control: an active disturbance rejection approach. IEEE/ASME Trans. Mechatron. **24**(5), 2052–2063 (2019)
9. Khalil, H.K.: Nonlinear Systems. Prentice Hall, Upper Saddle River (2002)
10. Liu, F., Li, Y., Cao, Y., She, J., Wu, M.: A two-layer active disturbance rejection controller design for load frequency control of interconnected power system. IEEE Trans. Power Syst. **31**(4), 3320–3321 (2016)
11. Magni, L., Scattolini, R.: Model predictive control of continuous-time nonlinear systems with piecewise constant control. IEEE Trans. Autom. Control **49**(6), 900–906 (2004)
12. Mazo, M., Tabuada, P.: Decentralized event-triggered control over wireless sensor/actuator networks. IEEE Trans. Autom. Control **56**(10), 2456–2461 (2011)
13. Ramirez-Neria, M., Garcia-Antonio, J., Sira-Ramirez, H., Velasco-Villa, M., Castro-Linares, R.: On the linear active rejection control of thomson's jumping ring. In: American Control Conference (ACC) 2013, pp. 6643–6648 (2013)
14. Tabuada, P.: Event-triggered real-time scheduling of stabilizing control tasks. IEEE Trans. Autom. Control **52**(9), 1680–1685 (2007)
15. Tan, W., Fu, C.: Linear active disturbance-rejection control: analysis and tuning via imc. IEEE Trans. Industr. Electron. **63**(4), 2350–2359 (2016)

16. Xu, L., Yao, B.: Adaptive robust precision motion control of linear motors with negligible electrical dynamics: theory and experiments. IEEE/ASME Trans. Mechatron. **6**(4), 444–452 (2001). https://doi.org/10.1109/3516.974858
17. Zhao, Z., Guo, B.: On active disturbance rejection control for nonlinear systems using time-varying gain. Eur. J. Control **23**, 100–108 (2015)
18. Zhao, Z., Guo, B.: On convergence of nonlinear active disturbance rejection control for siso nonlinear systems. J. Dyn. Control Syst. **22**, 385–413 (2016)
19. Zheng, Q., Gaol, L.Q., Gao, Z.: On stability analysis of active disturbance rejection control for nonlinear time-varying plants with unknown dynamics. In: 46th IEEE Conference on Decision and Control, pp. 3501–3506 (2007)

Chapter 5
A High-Gain Approach to Event-Triggered Control

In the previous chapters, we consider the problem of event-triggered controller design in the standard framework of ADRC. A few limitations, however, exist in the formulation we considered. First, only the single-input-single-output case is investigated, and the more generic multiple-input-multiple output case remains to be explored. Second, the ADRC has the advantage of achieving closed-loop stability without requiring system information, but when certain prior model structure information is available, it is not clear how to exploit such information in controller design. To address these questions, we extend our previous analysis and design to the scenario of event-based control with high-gain observers in this chapter.

Specifically, an event-triggered high-gain observer-based output feedback control scheme is introduced by designing separate event-triggering conditions that only rely on the measurable signals of the controlled system for the system outputs and the control signal, and the performance of the resultant control scheme is analyzed. Explicit relationships between the asymptotic upper bounds of the observation error and the parameters of the event-based sampler are developed to quantify the effect of event-based sampling. The theoretic analysis shows that the observation error can be guaranteed to be bounded and the closed-loop system can be stabilized asymptotically. Based on these results, special cases in which only system outputs or control signals are event-triggered are discussed. Comparative experiments are carried out on the DC torque motor system that we used earlier to validate the theoretic results, indicating that the proposed control strategy can achieve satisfactory control performance.

D. Shi et al., *Event-Triggered Active Disturbance Rejection Control*,
Studies in Systems, Decision and Control 356,
https://doi.org/10.1007/978-981-16-0293-1_5

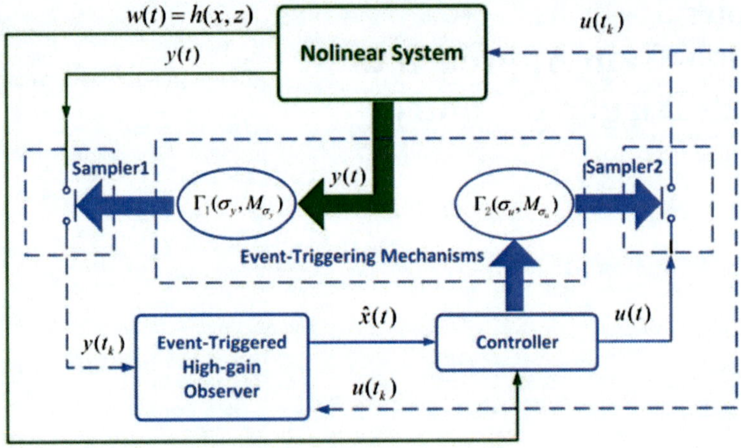

Fig. 5.1 Schematic of the event-triggered HGO-based control scheme, where the green arrows mean that the signals are transmitted in real time, the dotted lines mean that the signals are sampled, and the blue arrows represent the routes of event-triggering mechanisms

5.1 Problem Formulation

Consider the event-triggered high-gain observer (HGO) scheme in Fig. 5.1. In this chapter, our discussions focus on a multivariable nonlinear continuous-time control system of the following form [1]:

$$\dot{x} = Ax + B\phi(x, z, u), \tag{5.1}$$

$$\dot{z} = \psi(x, z, u), \tag{5.2}$$

$$y = Cx, \tag{5.3}$$

$$w = h(x, z), \tag{5.4}$$

where $x \in \mathcal{X} \subseteq \mathbb{R}^a$ and $z \in \mathcal{Z} \subseteq \mathbb{R}^l$ are the system states, $u \in \mathcal{U} \subseteq \mathbb{R}^m$ is the control signal, functions $\phi : \mathcal{X} \times \mathcal{Z} \times \mathcal{U} \rightarrow \mathbb{R}^b$ and $\psi : \mathcal{X} \times \mathcal{Z} \times \mathcal{U} \rightarrow \mathbb{R}^l$ are the nonlinear parts, $y \in \mathcal{Y} \subseteq \mathbb{R}^b$ and $w \in \mathcal{W} \subseteq \mathbb{R}^s$ are system outputs. The $a \times a$ matrix A, the $a \times b$ matrix B and the $b \times a$ matrix C are given by

$$A = \text{blockdiag}[A_1, A_2, \ldots, A_b],$$
$$B = \text{blockdiag}[B_1, B_2, \ldots, B_b],$$
$$C = \text{blockdiag}[C_1, C_2, \ldots, C_b],$$

where

$$A_i = \begin{bmatrix} 0 & 1 & \cdots & \cdots & 0 \\ 0 & 0 & 1 & \cdots & 0 \\ \vdots & & & & \vdots \\ 0 & \cdots & \cdots & 0 & 1 \\ 0 & \cdots & \cdots & \cdots & 0 \end{bmatrix}_{a_i \times a_i}, \quad B_i = \begin{bmatrix} 0 \\ 0 \\ \vdots \\ 0 \\ 1 \end{bmatrix}_{a_i \times 1},$$

$$C_i = \begin{bmatrix} 1 & 0 & \cdots & 0 \end{bmatrix}_{1 \times a_i},$$

and $1 \leq i \leq b$, $a = a_1 + \cdots + a_b$. For this system, we still take the emulation approach and assume that a stable state feedback control law u of the following form

$$\dot{v} = p(v, x, w), \tag{5.5}$$

$$u = q(v, x, w), \tag{5.6}$$

exists, such that the closed-loop system (5.1)–(5.6) satisfies the following assumption.

Assumption 5.1 $\phi,\ \psi,\ p,\ q$ are locally Lipschitz functions over the domain of interest, $p,\ q$ are globally bounded functions of x, and $\phi(0, 0, 0)$, $\psi(0, 0, 0)$, $p(0, 0, 0)$, $q(0, 0, 0)$, $h(0, 0)$ are all equal to 0. In addition, the origin of the closed-loop system is asymptotically stable.

In this chapter, we look into an output feedback problem for system (5.1). To do this, we consider a high-gain observer (HGO) of the following form:

$$\dot{\hat{x}} = A\hat{x} + B\phi_0(\hat{x}, w, u) + H(y - C\hat{x}), \tag{5.7}$$

where

$$H = \text{blockdiag}[H_1, \ldots, H_b] \tag{5.8}$$

$$H_i = \begin{bmatrix} \alpha_1^i/\varepsilon & \alpha_2^i/\varepsilon^2 & \ldots & \alpha_{a_i}^i/\varepsilon^{a_i} \end{bmatrix}^{\mathrm{T}},$$

ε is a high-gain parameter, and α_j^i are some chosen positive constants such that the roots of

$$s^{a_i} + \alpha_1^i s^{a_i - 1} + \cdots + \alpha_{a_i - 1}^i s + \alpha_{a_i}^i = 0 \tag{5.9}$$

are in the open left-half plane for all $i = 1, \cdots, b$. In (5.7), ϕ_0 is a nominal model of ϕ, which satisfies the following assumption.

Assumption 5.2 ϕ_0 is a locally Lipschitz function over the domain of interest, and is a globally bounded function of x. In addition, $\phi_0(0, 0, 0) = 0$.

With the above observer, the control signal (5.5)–(5.6) can be implemented as

$$\dot{v} = p(v, \hat{x}, w), \tag{5.10}$$

$$u = q(v, \hat{x}, w). \tag{5.11}$$

Note that the design of the controller may use different methods, but the key part is to choose the controller functions $p(\cdot)$ and $q(\cdot)$ such that Assumption 5.1 can be satisfied. For instance, if we choose $p = 0$, the static control signal $u = q(x, w)$ can be treated as a special case of (5.5)–(5.6). A commonly used controller design method is that to choose $u(t) = -\sum_{i=1}^{n} k_i x_i(t)$, where k_i are some parameters such that the matrix

$$K = \begin{bmatrix} 0 & 1 & 0 & \dots & 0 \\ 0 & 0 & 1 & \dots & 0 \\ \dots & \dots & \dots & \dots & \dots \\ 0 & 0 & 0 & \dots & 1 \\ -k_1 & -k_2 & -k_3 & \dots & -k_n \end{bmatrix}$$

is Hurwitz. This controller design satisfies Assumption 5.1 following the analysis in [2]. As for (5.9), since the chosen parameters α_j^i together with the high-gain parameter ε actually determine the observer gain, the choice of α_j^i should consider the requirements of the observer's pole assignment to determine the decaying speed of the observation error. In general, the response speed of the observer should be a little faster than that of the state feedback system.

As suggested in [1], the global boundedness requirement of the control signal in Assumption 5.1 can be typically achieved by saturating the functions $p(\cdot)$ and $q(\cdot)$. Indeed, in high-gain observer-based control, saturating the control signal is an effective way to avoid the influence of peaking phenomenon.

To characterize the observation error, we define e_{ij} as

$$e_{ij} := (x_{ij} - \hat{x_{ij}})/(\varepsilon^{a_i - j}),\ 1 \le i \le b,\ 1 \le j \le a_i \tag{5.12}$$

With this definition, we have $\hat{x} = x - D(\varepsilon)e$, where

$$\begin{aligned} e &= [e_{11}, \cdots, e_{1a_1}, \cdots, e_{b1}, \dots, e_{ba_b}]^{\mathrm{T}}, \\ D(\varepsilon) &= \mathrm{blockdiag}[D_1, \dots, D_b], \\ D_i &= \mathrm{diag}[\varepsilon^{a_i - 1}, \dots, 1]_{a_i \times a_i}. \end{aligned}$$

In this chapter, since the system states $x(t)$ are supposed to be unknown, an event-triggering scheme that the outputs of the system and the control signals are transmitted at asynchronous time instants determined by separate event-triggering conditions is considered. The schematic of the event-triggered HGO-based control scheme is provided in Fig. 6.1, where we observe that the system outputs and the control signal are only transmitted at some significant instants, which are determined by the event-triggering conditions Γ_1 and Γ_2, respectively. Based on this discussion, define the sampling error as

$$\sigma_y(t) := y(t_{k_y}) - y(t), \quad t \in [t_{k_y}, t_{k_y+1}), \tag{5.13}$$

$$\sigma_u(t) := u(t_{k_u}) - u(t), \quad t \in [t_{k_u}, t_{k_u+1}), \tag{5.14}$$

where t_{k_y} and t_{k_u} are the transmission instants determined by the corresponding event-triggering conditions. According to the general idea of the HGO, we mainly discuss the case that $\varepsilon < 1$ in our analysis. The event-triggered HGO has the following form:

$$\dot{\hat{x}} = A\hat{x} + B\phi_0(\hat{x}, w, u + \sigma_u) + H(y + \sigma_y - C\hat{x}). \tag{5.15}$$

In system (5.1)–(5.4), $w(t)$ could represent the system outputs which are complication of the system states $x(t)$ and $z(t)$, and it may not exist when all the system outputs can be expressed by $y(t)$. In this chapter, we consider the general case that not all system outputs are transmitted through an event-triggered fashion. In particular, since the system outputs $y(t)$ are used to estimate the system states $x(t)$ through HGO, $y(t)$ denote the event-triggered output signals and $w(t)$ represent the non-event-triggered outputs. Note that this treats the case when all the outputs are event-triggered (namely, $w(t) = 0$) as a special case. In the scenarios that the system outputs $w(t)$ are considered, both $w(t)$ and the estimated system states $\hat{x}$ are used to construct the control signal $u(t)$. In addition, since $w(t)$ are the functions of $x(t)$ and $z(t)$, the variation of $z(t)$ will affect the event-triggered sampling of $u(t)$.

Noticing that $D(\varepsilon)e = x - \hat{x}$, we can obtain that

$$\begin{aligned}\dot{e} &= D^{-1}(\varepsilon)[Ax + B\phi - A\hat{x} - B\phi_0 - H(y + \sigma_y - C\hat{x})] \\ &= (1/\varepsilon)A_0 e + Bg(x, z, v, D(\varepsilon)e) - U(\varepsilon)\sigma_y,\end{aligned}$$

where

$$\begin{aligned}&(1/\varepsilon)A_0 = D^{-1}(\varepsilon)(A - HC)D(\varepsilon), \\ &U(\varepsilon) = D^{-1}(\varepsilon)H = \text{diag}[U_1, \ldots, U_b], \\ &U_i = \left[\alpha_1^i/\varepsilon^{a_i}\ \alpha_2^i/\varepsilon^{a_i}\ \ldots\ \alpha_{a_i}^i/\varepsilon^{a_i}\right]^{\mathrm{T}},\end{aligned}$$

and $g(x, z, v, D(\varepsilon)e, \sigma_u) = \phi(x, z, q(v, \hat{x}, w) + \sigma_u) - \phi_0(\hat{x}, w, q(v, \hat{x}, w) + \sigma_u)$. Then the event-triggered closed-loop system can be formulated as

$$\dot{x} = Ax + B\phi(x, z, q(v, x - D(\varepsilon)e, w) + \sigma_u), \tag{5.16}$$
$$\dot{z} = \psi(x, z, q(v, x - D(\varepsilon)e, w) + \sigma_u), \tag{5.17}$$
$$\dot{v} = p(v, x - D(\varepsilon)e, w), \tag{5.18}$$
$$\dot{e} = (1/\varepsilon)A_0 e + Bg(x, z, v, D(\varepsilon)e, \sigma_u) - U(\varepsilon)\sigma_y. \tag{5.19}$$

Take $\mathcal{L} = [x^{\mathrm{T}}, z^{\mathrm{T}}, v^{\mathrm{T}}]^{\mathrm{T}}$, and (5.16)–(5.18) can be rewritten as

$$\dot{\mathcal{L}} = f(\mathcal{L}, D(\varepsilon)e, \sigma_u), \tag{5.20}$$

with the initial states $\mathcal{L}(0) = [x^{\mathrm{T}}(0), z^{\mathrm{T}}(0), v^{\mathrm{T}}(0)]^{\mathrm{T}} = [x_0^{\mathrm{T}}, z_0^{\mathrm{T}}, v_0^{\mathrm{T}}]^{\mathrm{T}} \in \mathbb{Q}_1$, $\hat{x}(0) = \hat{x}_0 \in \mathbb{Q}_2$, where $\mathbb{Q}_1$ is a compact set in the interior of $\mathbb{P}$ and $\mathbb{Q}_2$ is a compact set in the interior of $\mathbb{R}^a$, and

$$e(0) = D^{-1}(\varepsilon)(x_0 - \hat{x}_0) = e_0.$$

By substituting $e = 0$ and $\sigma_y = 0$ in (5.20), we can obtain the reduced system

$$\dot{\mathcal{L}} = f(\mathcal{L}, 0, 0), \tag{5.21}$$

which actually represents the closed-loop system under state feedback. Let $(\mathcal{L}(t, \varepsilon), e(t, \varepsilon))$ denote the trajectories of system (5.16)–(5.18) starting from $(\mathcal{L}(0), e(0))$. For notational brevity, define $\tilde{\gamma}_1(\varepsilon)$, $\tilde{\gamma}_2(\varepsilon)$, $\tilde{\gamma}(\varepsilon)$ and $\tilde{P}_\sigma(\varepsilon)$ as

$$\tilde{\gamma}_1(\varepsilon) := c_3 - \beta\lambda_2 L_5^* \varepsilon^{\frac{1}{3}}, \tag{5.22}$$

$$\tilde{\gamma}_2(\varepsilon) := \beta - 2\beta\lambda_2 L_6^* \varepsilon - \frac{1}{2}\varepsilon^{\frac{2}{3}}[M_2 L_{12}^* + 2\beta\lambda_2(L_5^* + L_7^* + 1)], \tag{5.23}$$

$$\tilde{\gamma}(\varepsilon) := \min\left\{\frac{1}{c_2}\tilde{\gamma}_1(\varepsilon), \frac{1}{\beta\varepsilon\lambda_2}\tilde{\gamma}_2(\varepsilon)\right\}, \tag{5.24}$$

$$\tilde{P}_\sigma(\varepsilon) := M_2\left(L_{11}^* M_{\sigma_u} + \frac{1}{2}L_{12}^* \varepsilon^{\frac{1}{3}}\right) + \beta\lambda_2\varepsilon^{\frac{1}{3}}(L_7^* M_{\sigma_u}^2 + M_{\sigma_y}^2). \tag{5.25}$$

In the remainder of this chapter, we mainly investigate the following problems:

1. Considering the high-gain observer and the system, can we design an event-triggering scheme to reduce the transmission rates of the system outputs and the control signal without deteriorating the control performance?
2. How will the proposed event-triggering schemes affect the control performance of the closed-loop system, in terms of observation error and tracking error?

5.2 Theoretic Results

In this section, the problems stated in the previous section are investigated. In particular, separate event-triggering conditions are proposed to control the sampling of the system outputs and the control signal. The theoretic analysis consists of two parts. First, we investigate the boundedness of the trajectories based on the event-triggering conditions. Then, we investigate the performance of the closed-loop system and quantify the effect of event-based sampling. For the first part, we present the event-triggering conditions in the following theorem.

Theorem 5.1 *Consider the system in* (5.19). *Suppose that Assumptions* 5.1–5.2 *hold, there exist* $\varepsilon_1^* > 0$ *and event-triggering conditions*

$$\Gamma_1 = \begin{cases} 0, \text{ if } \|U(\varepsilon)\sigma_y\| \leq M_{\sigma_y}, \\ 1, \text{ otherwise}, \end{cases}$$

$$\Gamma_2 = \begin{cases} 0, \text{ if } \|\sigma_u\| \leq M_{\sigma_u}, \\ 1, \text{ otherwise}, \end{cases} \tag{5.26}$$

where M_{σ_y}, $M_{\sigma_u} > 0$ are some chosen constants, such that for all $\varepsilon \in (0, \varepsilon_1^*)$, the trajectories $(\mathcal{L}(t,\varepsilon), e(t,\varepsilon))$ starting in $\mathbb{Q}_1 \times \mathbb{Q}_2$ are bounded for all $t \geq 0$.

Proof From Assumption 5.1, the origin of (5.21) is asymptotically stable with a region of attraction $\mathbb{P}$. Hence, there exists a Lyapunov function $V(\mathcal{L})$ such that

$$W_1(\mathcal{L}) \leq V(\mathcal{L}) \leq W_2(\mathcal{L}) \tag{5.27}$$

$$\lim_{\mathcal{L}\to\partial\mathbb{P}} W_1(\mathcal{L}) = \infty \tag{5.28}$$

$$\frac{\partial V}{\partial \mathcal{L}} f(\mathcal{L}, 0) \leq -W_3(\mathcal{L}), \tag{5.29}$$

where $W_1(\mathcal{L})$, $W_2(\mathcal{L})$, and $W_3(\mathcal{L})$ are positive definite functions all defined and continuous on $\mathbb{P}$. For any finite $c > \max_{\mathcal{L}\in Q_1} V(\mathcal{L})$, the set $\mathbb{S} = \{\mathcal{L} \in \mathbb{P} | V(\mathcal{L}) \leq c\}$ is a compact subset of $\mathbb{P}$. Define $P(e) := e^{\mathrm{T}} P_0 e$, where P_0 is a positive definite matrix which satisfies $P_0 A_0 + A_0^{\mathrm{T}} P_0 = -I$, and we can obtain

$$\lambda_1 \|e\|^2 \leq P(e) \leq \lambda_2 \|e\|^2, \tag{5.30}$$

$$\frac{\partial P}{\partial e} A_0 e \leq -\|e\|^2, \tag{5.31}$$

where $\lambda_1 = \lambda_{\min}(P_0)$, $\lambda_2 = \lambda_{\max}(P_0)$. Define $\Omega := \mathbb{S} \times \{P(e) \leq \rho\varepsilon^2\}$. From Assumptions 5.1–5.2, we can obtain that for all $(\mathcal{L}, e) \in \mathbb{S} \times \mathbb{R}^a$,

$$\|f(\mathcal{L}, D(\varepsilon)e, \sigma_u)\| \leq \tilde{k}_1, \tag{5.32}$$

$$\|g(\mathcal{L}, D(\varepsilon)e, \sigma_u)\| \leq \tilde{k}_2, \tag{5.33}$$

where $\tilde{k}_1, \tilde{k}_2$ are some positive constants. For all $(\mathcal{L}, e) \in \Omega$ and any $0 < \varepsilon < 1$, we have

$$\|f(\mathcal{L}, D(\varepsilon)e, \sigma_u) - f(\mathcal{L}, 0, 0)\| \leq \tilde{L}_{11}\|\sigma_u\| + \tilde{L}_{12}\|e\|. \tag{5.34}$$

Then, for all $(\mathcal{L}, e) \in \mathbb{S} \times \{P(e) = \rho\varepsilon^2\}$,

$$\begin{aligned}
\dot{P}(e) &= \frac{\partial P}{\partial e}\dot{e} \\
&= \frac{\partial P}{\partial e}\left(\frac{1}{\varepsilon}A_0 e + Bg(\mathcal{L}, D(\varepsilon)e, \sigma_u) - U(\varepsilon)\sigma_y\right) \\
&\leq -\frac{1}{\varepsilon}\|e\|^2 + 2\lambda_2\|e\|(\tilde{k}_2 + M_{\sigma_y}).
\end{aligned} \tag{5.35}$$

When $P(e) = \rho\varepsilon^2$, we have $\frac{1}{\varepsilon}\|e\| \geq \sqrt{\frac{\rho}{\lambda_2}}$, and then we can obtain

$$\dot{P}(e) \leq -\left[\sqrt{\frac{\rho}{\lambda_2}} - 2\lambda_2(\tilde{k}_2 + M_{\sigma_y})\right]\|e\|.$$

Take $\rho = 9(\tilde{k}_2 + M_{\sigma_y})^2\lambda_2^3$, we have

$$\dot{P}(e) \leq 0. \tag{5.36}$$

Consider system (5.20), for all $(\mathcal{L}, e) \in \{\mathcal{L} \in \mathbb{P} | V(\mathcal{L}) = c\} \times \{P(e) \leq \rho\varepsilon^2\}$, we have

$$\begin{aligned}\dot{V}(\mathcal{L}) &= \frac{\partial V}{\partial \mathcal{L}} f(\mathcal{L}, D(\varepsilon)e, \sigma_u) \\ &\leq -W_3(\mathcal{L}) + M_1(\tilde{L}_{11}\|\sigma_u\| + \tilde{L}_{12}\|e\|),\end{aligned} \tag{5.37}$$

where $M_1 = \max_{\mathcal{L}\in\mathbb{S}}(\|\frac{\partial V}{\partial \mathcal{L}}\|)$. Taking

$$\tilde{\kappa} = \min_{\mathcal{L}\in\partial\mathbb{S}} W_3(\mathcal{L}) - M_1\tilde{L}_{11}M_{\sigma_u} > 0$$

and

$$\tilde{\varepsilon}_1 = \tilde{\kappa}/(M_1\tilde{L}_{12}\sqrt{\rho/\lambda_1}),$$

we have for all $0 < \varepsilon \leq \tilde{\varepsilon}_1$,

$$\dot{V}(\mathcal{L}) \leq 0. \tag{5.38}$$

From (5.70) and (5.38), we can obtain that the set Ω is positively invariant. Note that $\mathbb{Q}_1$ is in the interior of $\mathbb{S}$, and

$$\|\mathcal{L}(t, \varepsilon) - \mathcal{L}(0)\| \leq \|\dot{\mathcal{L}}(t, \varepsilon)\| \leq \tilde{k}_1 t, \tag{5.39}$$

then there exists a finite time $\tilde{t}_1$ such that for all $t \in [0, \tilde{t}_1]$, we have $\mathcal{L}(t, \varepsilon) \in \mathbb{S}$. During the same time interval, for $P(e) \geq \rho\varepsilon^2$, we have

$$\dot{P}(e) \leq -\frac{1}{\varepsilon}\|e\|^2 + 2\lambda_2\|e\|(\tilde{k}_2 + M_{\sigma_y}) \leq -\frac{1}{3\varepsilon\lambda_2}P(e),$$

and then

$$P(e) \leq P(e_0)\exp\left(-\frac{t}{3\varepsilon\lambda_2}\right),$$

$$\|e_0\| \leq \|\frac{x_0 - \hat{x}_0}{\varepsilon^{a_m - 1}}\| \leq \|\frac{k}{\varepsilon^{a_m - 1}}\|,$$

where $k = x_0 - \hat{x}_0$, $a_m = \max\{a_1, \dots a_b\}$. Hence, we have

$$P(e) \leq \frac{k^2\lambda_2}{\varepsilon^{2(a_m-1)}} \exp\left(-\frac{t}{3\varepsilon\lambda_2}\right) = \frac{\tilde{m}_2}{\varepsilon^{2(a_m-1)}} \exp\left(-\frac{\tilde{m}_1 t}{\varepsilon}\right), \tag{5.40}$$

where $\tilde{m}_1 = 1/(3\lambda_2)$ and $\tilde{m}_2 = k^2\lambda_2$. Letting

$$\frac{\tilde{m}_2}{\varepsilon^{2(a_m-1)}} \exp\left(-\frac{\tilde{m}_1 t}{\varepsilon}\right) \leq \rho\varepsilon^2,$$

we have $t \geq \frac{\varepsilon}{\tilde{m}_1} \ln \frac{\tilde{m}_2}{\rho\varepsilon^{2a_m}}$. Define $\tilde{t}(\varepsilon) := \frac{\varepsilon}{\tilde{m}_1} \ln \frac{\tilde{m}_2}{\rho\varepsilon^{2a_m}}$, and choose $\tilde{\varepsilon}_2$ small enough such that $\tilde{t}(\varepsilon) \leq \frac{1}{k_t}\tilde{t}_1$ where $k_t > 1$, then we have for all $0 < \varepsilon \leq \tilde{\varepsilon}_2$, $P(e(\tilde{t}(\varepsilon), \varepsilon)) \leq \rho\varepsilon^2$. Taking $\varepsilon_1^* = \min(\tilde{\varepsilon}_1, \tilde{\varepsilon}_2)$, and then for all $0 < \varepsilon \leq \varepsilon_1^*$, we can obtain that for $t \in [0, \tilde{t}(\varepsilon)]$, the boundedness of the trajectories $(\mathcal{L}(t, \varepsilon), e(t, \varepsilon))$ can be guaranteed by (5.73) and (5.75), and for all $t \geq \tilde{t}(\varepsilon)$, the boundedness of $(\mathcal{L}(t, \varepsilon), e(t, \varepsilon))$ is also guaranteed since the trajectories enter Ω during $[0, \tilde{t}(\varepsilon)]$ and remains in Ω for all $t \geq \tilde{t}(\varepsilon)$. This completes the proof. □

Theorem 5.1 shows the boundedness of the trajectories $(\mathcal{L}(t, \varepsilon), e(t, \varepsilon))$. The theoretic analysis consists of two parts. First, Eqs. (5.70) and (5.38) indicate that the properly chosen set Ω is positively invariant. Then, the closed-loop trajectories are proved to enter the set Ω in finite time. The value of the constant ε_1^* is influenced by the initial values of $(\mathcal{L}(t, \varepsilon), e(t, \varepsilon))$, which are related to the values of $\mathcal{L}(0)$ and $\hat{x}(0)$.

Next, to investigate the performance of the closed-loop system and quantify the effect of event-based sampling, we develop a quantitative relationship between the asymptotic upper bound of the observation error and the parameters of the event-based sampler. To do this, the following additional assumption is needed.

Assumption 5.3 For system (5.21), there exists a C^1 Lyapunov function $V_2(\mathcal{L})$ defined over a ball $B(o, r_2) \subseteq \mathbb{P}$ with $r_2 > 0$, such that for all $\mathcal{L} \in B(o, r_2)$, we have

1. $c_1\|\mathcal{L}\|^2 \leq V_2(\mathcal{L}) \leq c_2\|\mathcal{L}\|^2$
2. $\frac{\partial V_2}{\partial \mathcal{L}} f(\mathcal{L}, 0, 0) \leq -c_3\|\mathcal{L}\|^2$
3. $\|\frac{\partial V_2}{\partial \mathcal{L}}\| \leq M_2$

where c_1, c_2, c_3 and M_2 are some positive constants.

With this assumption, we present the following theorem.

Theorem 5.2 *Consider system (5.16)–(5.19) and the event-triggering conditions in (5.26), suppose that Assumptions 5.1–5.3 hold, there exists $\varepsilon_2^* > 0$ such that for every $0 < \varepsilon \leq \varepsilon_2^*$, we have for all $(\mathcal{L}, e) \in B(0, \eta) \times \{\|e\| \leq \eta\} = \Omega_1$ with $\eta < r_2$,*

$$\min\{t_{k_u+1} - t_{k_u}\} \geq \tau_1, \tag{5.41}$$

$$\min\{t_{k_y+1} - t_{k_y}\} \geq \tau_2, \tag{5.42}$$

$$\lim_{t\to+\infty} |x_{ij} - \hat{x}_{ij}| \leq \varepsilon^{a_i - j}\sqrt{\frac{\tilde{P}_\sigma(\varepsilon)}{\beta\lambda_1\tilde{\gamma}(\varepsilon)}}, \tag{5.43}$$

$$\lim_{t\to+\infty} \|\mathcal{L}\| \leq \sqrt{\frac{\tilde{P}_\sigma(\varepsilon)}{c_1\tilde{\gamma}(\varepsilon)}}, \tag{5.44}$$

where τ_1, τ_2 and β are some positive constants, $\tilde{\gamma}(\varepsilon)$ and $\tilde{P}_\sigma(\varepsilon)$ are defined in (5.22)–(5.25).

Proof For the sampling error $\sigma_u(t) = u(t_{k_u}) - u(t)$, when $t \in [t_{k_u}, t_{k_u+1})$, we have

$$u(t) - u(t_{k_u}) = \int_{t_{k_u}}^{t} \dot{u}(\tau)d\tau = \int_{t_{k_u}}^{t} \dot{q}(v(\tau), x(\tau) - D(\varepsilon)e(\tau), w(\tau))d\tau,$$
$$\|u(t) - u(t_{k_u})\| \leq \int_{t_{k_u}}^{t} \|\dot{q}(v(\tau), x(\tau) - D(\varepsilon)e(\tau), w(\tau))\|d\tau.$$

From Assumptions 5.1–5.2 and Theorem 5.1, when $0 < \varepsilon \leq \varepsilon_1^*$, we have $\|\dot{q}(v, x - D(\varepsilon)e, w)\| \leq M_q$, where M_q is a positive constant. Hence, we have

$$\|u(t) - u(t_{k_u})\| \leq (t - t_{k_u})M_q. \tag{5.45}$$

Due to the design of the event-triggering condition, when $t = t_{k_u+1}$, we have $\|u(t) - u(t_{k_u})\| \geq M_{\sigma_u}$. Take $\tau_1 = M_{\sigma_u}/M_q$, we have

$$\min\{t_{k_u+1} - t_{k_u}\} \geq \tau_1. \tag{5.46}$$

As for the sampling intervals of the system outputs, for the sampling error $\sigma_y(t) = y(t_{k_y}) - y(t)$, when $t \in [t_{k_y}, t_{k_y+1})$, we have

$$y(t) - y(t_{k_y}) = \int_{t_{k_y}}^{t} C\dot{x}(\tau)d\tau$$
$$= \int_{t_{k_y}}^{t} [CAx(\tau) + CB\phi(x(\tau), z(\tau), u(\tau) + \sigma_u(\tau))]d\tau,$$
$$\|y(t) - y(t_{k_y})\| \leq \int_{t_{k_y}}^{t} (\|x(\tau)\| + \|\phi(x(\tau), z(\tau), u(\tau) + \sigma_u(\tau))\|)d\tau.$$

From Assumptions 5.1–5.2 and Theorem 5.1, when $0 < \varepsilon \leq \varepsilon_1^*$, we have $\|x(t)\| \leq M_x$ and $\|\phi\| \leq M_\phi$, where M_x and M_ϕ are some positive constants. Hence, we have

$$\|y(t) - y(t_{k_y})\| \leq (t - t_{k_y})(M_x + M_\phi). \tag{5.47}$$

According to the event-triggering condition, when $t = t_{k_y+1}$, we have $\|U(\varepsilon)[y(t) - y(t_{k_y})]\| \geq M_{\sigma_y}$. Taking $\tau_2 = M_{\sigma_y}/\|U(\varepsilon)\|(M_x + M_\phi)$, we have

$$\min\{t_{k_y+1} - t_{k_y}\} \geq \tau_2. \tag{5.48}$$

Next, from Assumptions 5.1–5.2, we have for all $(\mathcal{L}, e) \in \Omega_1$,

$$\|g(\mathcal{L}, D(\varepsilon)e, \sigma_u)\| \leq L_5^*\|\mathcal{L}\| + L_6^*\|e\| + L_7^*\|\sigma_u\|, \tag{5.49}$$

$$\|f(\mathcal{L}, D(\varepsilon)e, \sigma_u) - f(\mathcal{L}, 0, 0)\| \leq L_{11}^*\|\sigma_u\| + L_{12}^*\|e\|. \tag{5.50}$$

Since $\tilde{V}_2(\mathcal{L}, e) = V_2(\mathcal{L}) + \beta P(e)$, we have

$$\begin{aligned}
\dot{\tilde{V}}_2 &= \frac{\partial V_2}{\partial \mathcal{L}} f(\mathcal{L}, D(\varepsilon)e, \sigma_u) + \beta \frac{\partial P}{\partial e}\dot{e} \\
&\leq -(c_3 - \beta\lambda_2 L_5^* \varepsilon^{\frac{1}{3}})\|\mathcal{L}\|^2 \\
&\quad - \frac{1}{\varepsilon}\left[\beta - 2\beta\lambda_2 L_6^*\varepsilon - \varepsilon^{\frac{2}{3}}\left(\frac{M_2 L_{12}^*}{2} + \beta\lambda_2(L_5^* + L_7^* + 1)\right)\right]\|e\|^2 \\
&\quad + M_2(L_{11}^* M_{\sigma_u} + \frac{1}{2}L_{12}^*\varepsilon^{\frac{1}{3}}) + \beta\lambda_2\varepsilon^{\frac{1}{3}}(L_7^* M_{\sigma_u}^2 + M_{\sigma_y}^2).
\end{aligned}$$

Choose β small enough and $0 < \varepsilon_2^* \leq \varepsilon_1^*$ such that for every $0 < \varepsilon \leq \varepsilon_2^*$, it holds that $\tilde{\gamma}_1(\varepsilon) > 0$, $\tilde{\gamma}_2(\varepsilon) > 0$, where $\tilde{\gamma}_1(\varepsilon)$ and $\tilde{\gamma}_2(\varepsilon)$ are defined in (5.22)–(5.24). Then we have

$$\begin{aligned}
\dot{\tilde{V}}_2 &\leq -\tilde{\gamma}_1(\varepsilon)\|\mathcal{L}\|^2 - \frac{1}{\beta\varepsilon}\tilde{\gamma}_2(\varepsilon)\beta\|e\|^2 + \tilde{P}_\sigma(\varepsilon) \\
&\leq -\frac{1}{c_2}\tilde{\gamma}_1(\varepsilon)V_2(\mathcal{L}) - \frac{1}{\beta\varepsilon\lambda_2}\tilde{\gamma}_2(\varepsilon)\beta P(e) + \tilde{P}_\sigma(\varepsilon) \\
&\leq -\tilde{\gamma}(\varepsilon)\tilde{V}_2 + \tilde{P}_\sigma(\varepsilon),
\end{aligned} \tag{5.51}$$

where $\tilde{P}_\sigma(\varepsilon)$ is defined in (5.25). Hence, we can obtain that

$$\begin{aligned}
\tilde{V}_2 &\leq \tilde{\phi}_2(t, t_0)\tilde{V}_2(\mathcal{L}_0, e_0) + \tilde{P}_\sigma(\varepsilon)\textstyle\int_{t_0}^t \tilde{\phi}_2(t, \tau)d\tau \\
&\leq \tilde{\phi}_2(t, t_0)\tilde{V}_2(\mathcal{L}_0, e_0) + (\tilde{P}_\sigma(\varepsilon)\,\tilde{\gamma}(\varepsilon))(1 - \tilde{\phi}_2(t, t_0)),
\end{aligned}$$

where $\tilde{\phi}_2(t, t_0) := \exp(-\tilde{\gamma}(\varepsilon)(t - t_0))$. Noticing that when $t \to +\infty$, $\tilde{\phi}_2(t, t_0) \to 0$, we can obtain that when $t \to +\infty$, $\tilde{V}_2 \leq \tilde{P}_\sigma(\varepsilon)/\tilde{\gamma}(\varepsilon)$, hence $c_1\|\mathcal{L}\|^2 + \beta\lambda_1\|e\|^2 \leq \tilde{P}_\sigma(\varepsilon)/\tilde{\gamma}(\varepsilon)$. With this, we conclude that

$$\lim_{t\to+\infty} |x_{ij} - \hat{x}_{ij}| \leq \varepsilon^{a_i - j}\sqrt{\tfrac{\tilde{P}_\sigma(\varepsilon)}{\beta\lambda_1\tilde{\gamma}(\varepsilon)}}, \tag{5.52}$$

$$\lim_{t\to+\infty} \|\mathcal{L}\| \leq \sqrt{\tfrac{\tilde{P}_\sigma(\varepsilon)}{c_1\tilde{\gamma}(\varepsilon)}}, \tag{5.53}$$

This completes the proof. □

For the closed-loop system in (5.16)–(5.19), the asymptotic observation error can be maintained in a range defined in (5.43) with the event-triggering conditions introduced in (5.26), and the asymptotic stability of the closed-loop system can be guaranteed. For the boundedness of the observation error, a general understanding is that for output feedback control schemes, the designed observer dynamics need to be stable to ensure the internal stability of the closed-loop system, which means that the observation error should be bounded. With this principle, we need to obtain an observer with bounded observation errors, based on which the feedback control laws are further developed. The event-triggering conditions are designed based on the values of the sampling errors of the system outputs and the control signal, which are $\|\sigma_y(t)\|$ and $\|\sigma_u(t)\|$, respectively, and present upper bounds of them. The system performance can be adjusted by changing the values of the parameters M_{σ_y} or M_{σ_u}. More specifically, the value of M_{σ_y} mainly influences the communication rate between the controlled system and the observer, while the value of M_{σ_u} mainly affects the communication rate between the controller and the system. Moreover, since the two sampling schemes co-exist in the same closed-loop system, these two communication rates will also influence each other.

The control signal $u(t)$ is designed based on the values of the estimated states $\hat{x}(t)$ and the system outputs $w(t)$, which are available in the control scheme. Since the states of the controlled system are supposed to be unknown, we use $u(t)$, which is a function of $\hat{x}(t)$ and $w(t)$, to design the event-triggering condition Γ_2; here $\hat{x}(t)$ denotes the estimate of the unknown $x(t)$. Besides, as $x(t)$ is unavailable, we need to use the information of $y(t)$ to obtain the state estimate $\hat{x}(t)$, which is why $y(t)$ is used to design the event-triggering condition Γ_1. In the design of the event-triggering condition, the choice of the event-triggering parameters is the most crucial part. On one hand, Eqs. (5.43)–(5.44) provide quantitative relationships between the control performance and the event-triggering parameters. Therefore, the event-triggering parameters can be calculated according to the desired control performance when parameters in (5.43)–(5.44) can be obtained. On the other hand, since larger values of M_{σ_y} and M_{σ_u} will lead to larger bounds of control error but lower communication rates and vice versa, we can estimate the event-triggering parameters in a trial-and-error fashion to achieve satisfactory control performance at minimum communication rate.

The physical meaning of the event-triggering conditions is that when the errors between the real-time system outputs or the control signal and the previously transmitted values become sufficiently large, the system outputs or the control signal will be updated and transmitted. Since the system states are supposed to be unknown, the event-triggering conditions proposed in (5.26) do not rely on the values of the system states, and the proposed upper bounds of the sampling errors are determined by the pre-defined event-triggering parameters M_{σ_y} and M_{σ_u} instead of the system states. Due to this fact, these event-triggering conditions can be used when the system states are unavailable. To implement the proposed event-triggering conditions, the real-time signals $y(t)$ and $u(t)$ are needed. In practical applications, since these signals can only be measured in discrete time, a relatively high sampling rate is needed. More specifically, we need to monitor the system outputs $y(t)$ and the control signal $u(t)$ in discrete time with a relatively small sampling interval to guarantee an approximately accurate implementation of the event-triggering conditions.

An important issue regarding the event-triggered sampling is the Zeno behavior. The non-existence of Zeno behavior in the event-triggered control scheme can be guaranteed if the system could avoid updating signals infinite times in a finite time interval. In fact, (5.41) and (5.42) present a minimum nonzero sampling interval of $y(t)$ and $u(t)$, respectively. The minimum nonzero sampling intervals show that there is no Zeno behavior in this event-based control scheme, which guarantees the existence of the solution when $t \to \infty$.

5.3 Discussions on Special Sampling Schemes

The theoretic analysis in the previous section considers the scenario that both the system outputs and the control signals are sampled. With the proposed approach, special cases when only system outputs or controlled signals are sampled can also be considered, which are discussed in this section.

5.3.1 *Case 1: Only Sampling Output*

In this subsection, the case that the output of the system is not transmitted in real time, but only at certain instants determined by event-triggering condition is discussed in detail. To do this, define the sampling error $\sigma(t)$ as

$$\sigma(t) := y(t_k) - y(t), \tag{5.54}$$

where $t \in [t_k, t_{k+1})$, and t_k is the transmission instant determined by the triggering condition. Then the event-triggered closed-loop system can be represented by

$$\dot{x} = Ax + B\phi(x, z, q(v, x - D(\varepsilon)e, w)), \tag{5.55}$$

$$\dot{z} = \psi(x, z, q(v, x - D(\varepsilon)e, w)), \tag{5.56}$$

$$\dot{v} = p(v, x - D(\varepsilon)e, w), \tag{5.57}$$

$$\dot{e} = \frac{1}{\varepsilon}A_0 e + Bg_1(x, z, v, D(\varepsilon)e) - U(\varepsilon)\sigma, \tag{5.58}$$

where $g_1(x, z, v, D(\varepsilon)e) = \phi(x, z, q(v, \hat{x}, w)) - \phi_0(\hat{x}, w, q(v, \hat{x}, w))$. Take $\mathcal{L} = [x^{\mathrm{T}}, z^{\mathrm{T}}, v^{\mathrm{T}}]^{\mathrm{T}}$, then the system (5.55)–(5.58) can be rewritten as

$$\dot{\mathcal{L}} = f(\mathcal{L}, D(\varepsilon)e). \tag{5.59}$$

The state feedback system is

$$\dot{\mathcal{L}} = f(\mathcal{L}, 0), \tag{5.60}$$

Define $\gamma_1(\varepsilon)$, $\gamma_2(\varepsilon)$ and $\gamma(\varepsilon)$ as

$$\gamma_1(\varepsilon) := d_3 - \frac{1}{2}\varepsilon^{\frac{1}{3}}(L_4 d_4 + 2\beta\lambda_2 L_2), \tag{5.61}$$

$$\gamma_2(\varepsilon) := \beta - 2\lambda_2 L_3 \beta\varepsilon - \frac{1}{2}\varepsilon^{\frac{2}{3}}(L_4 d_4 + 2\beta\lambda_2(1 + L_2)), \tag{5.62}$$

$$\gamma(\varepsilon) := \min\left\{\frac{1}{d_2}\gamma_1(\varepsilon), \frac{1}{\beta\varepsilon\lambda_2}\gamma_2(\varepsilon)\right\}. \tag{5.63}$$

Then we present the event-triggering condition in the following theorem, the proof of which is similar to that of Theorem 5.2.

Theorem 5.3 *Consider the system in* (5.59). *Suppose that Assumptions* 5.1–5.2 *hold, there exist* $\varepsilon_3^* > 0$ *and an event-triggering condition*

$$\Gamma_y = \begin{cases} 0, \text{ if } \|U(\varepsilon)\sigma\| \le M_\sigma, \\ 1, \text{ otherwise}, \end{cases} \tag{5.64}$$

where $M_\sigma > 0$ *is a constant such that for all* $\varepsilon \in (0, \varepsilon_3^*)$, *the trajectories* $(\mathcal{L}(t, \varepsilon), e(t, \varepsilon))$ *starting in* $\mathbb{Q}_1 \times \mathbb{Q}_2$ *are bounded for all* $t \ge 0$.

Proof From Assumptions 5.1–5.2, we can obtain that for all $(\mathcal{L}, e) \in \mathbb{S} \times \mathbb{R}^a$,

$$\|f(\mathcal{L}, D(\varepsilon)e)\| \le k_1, \tag{5.65}$$

$$\|g_1(\mathcal{L}, D(\varepsilon)e)\| \le k_2, \tag{5.66}$$

where k_1, k_2 are some positive constants. For all $(\mathcal{L}, e) \in \Omega$ and any $0 < \varepsilon < 1$, we have

$$\|f(\mathcal{L}, D(\varepsilon)e) - f(\mathcal{L}, 0)\| \le L_1\|e\|. \tag{5.67}$$

Then, we can obtain that for all $(\mathcal{L}, e) \in \mathbb{S} \times \{P(e) = \rho\varepsilon^2\}$,

$$\begin{aligned} \dot{P}(e) &= \frac{\partial P}{\partial e}\dot{e} \\ &= \frac{\partial P}{\partial e}\frac{1}{\varepsilon}(-A_0 e + Bg_1(x, z, v, D(\varepsilon)e) - U(\varepsilon)\sigma) \\ &\le -\frac{1}{\varepsilon}\|e\|^2 + 2\lambda_2\|e\|(k_2 + \|U(\varepsilon)\sigma\|). \end{aligned} \tag{5.68}$$

From (5.64), we have

$$\dot{P}(e) \le -\frac{1}{\varepsilon}\|e\|^2 + 2\lambda_2\|e\|(k_2 + M_\sigma). \tag{5.69}$$

When $P(e) = \rho\varepsilon^2$, we have $\frac{1}{\varepsilon}\|e\| \geq \sqrt{\frac{\rho}{\lambda_2}}$, then we obtain $\dot{P}(e) \leq -[\sqrt{\frac{\rho}{\lambda_2}} - 2\lambda_2(k_2 + M_\sigma)]\|e\|$. Take $\rho = 9(k_2 + M_\sigma)^2\lambda_2^3$, then we have

$$\dot{P}(e) \leq 0. \tag{5.70}$$

For system (5.59) and all $(\mathcal{L}, e) \in \{\mathcal{L} \in \mathbb{R} | V(\mathcal{L}) = c\} \times \{P(e) \leq \rho\varepsilon^2\}$, we have

$$\begin{aligned}
\dot{V}(\mathcal{L}) &= \frac{\partial V}{\partial \mathcal{L}} f(\mathcal{L}, D(\varepsilon)e) \\
&= \frac{\partial V}{\partial \mathcal{L}}[f(\mathcal{L}, D(\varepsilon)e) - f(\mathcal{L}, 0)] + \frac{\partial V}{\partial \mathcal{L}} f(\mathcal{L}, 0) \\
&\leq -W_3(\mathcal{L}) + M_1 L_1 \|e\|,
\end{aligned} \tag{5.71}$$

where $M_1 = \max_{\mathcal{L}\in\mathbb{S}}(\|\frac{\partial V}{\partial \mathcal{L}}\|)$. Hence $P(e) \leq \rho\varepsilon^2$, we have $\frac{1}{\varepsilon}\|e\| \leq \sqrt{\frac{\rho}{\lambda_1}}$, take $\varepsilon_1 = \min_{\mathcal{L}\in\partial\mathbb{S}} W_3(\mathcal{L})/(M_1 L_1 \sqrt{\rho/\lambda_1})$, then we have for all $0 < \varepsilon \leq \varepsilon_1$,

$$\dot{V}(\mathcal{L}) \leq 0. \tag{5.72}$$

From (5.70) and (5.72), we can obtain that the set Ω is positively invariant. Note that $\mathbb{Q}_1$ is in the interior of $\mathbb{S}$, and

$$\|\mathcal{L}(t, \varepsilon) - \mathcal{L}(0)\| \leq \|\dot{\mathcal{L}}(t, \varepsilon)\| \leq k_1 t, \tag{5.73}$$

there exists a finite time t_1 such that for all $t \in [0, t_1]$, we have $\mathcal{L}(t, \varepsilon) \in \mathbb{S}$. During the same time interval, for $P(e) \geq \rho\varepsilon^2$, we have

$$\begin{aligned}
\dot{P}(e) &\leq -\frac{1}{\varepsilon}\|e\|^2 + 2\lambda_2\|e\|(k_2 + M_\sigma) \\
&\leq -\frac{1}{\varepsilon}\|e\|^2 + \frac{2}{3}\sqrt{\frac{\rho}{\lambda_2}}\|e\| \\
&\leq -\frac{1}{\varepsilon}\|e\|^2 + -\frac{2}{3\varepsilon}\|e\|^2 \\
&\leq -\frac{1}{3\varepsilon\lambda_2} P(e),
\end{aligned} \tag{5.74}$$

then $P(e) \leq P(e_0)\exp(-\frac{t}{3\varepsilon\lambda_2})$, $\|e_0\| \leq \|\frac{x_0 - \hat{x}_0}{\varepsilon^{a_m - 1}}\| \leq \|\frac{k}{\varepsilon^{a_m - 1}}\|$, where $k = x_0 - \hat{x}_0$, $a_m = \max\{a_1, \ldots a_b\}$. Hence, we have

$$\begin{aligned}
P(e) &\leq \frac{k^2\lambda_2}{\varepsilon^{2(a_m - 1)}} \exp(-\frac{t}{3\varepsilon\lambda_2}) \\
&= \frac{m_2}{\varepsilon^{2(a_m - 1)}} \exp(-\frac{m_1 t}{\varepsilon}),
\end{aligned} \tag{5.75}$$

where $m_1 = 1/(3\lambda_2)$ and $m_2 = k^2\lambda_2$. Let $\frac{m_2}{\varepsilon^{2(a_m-1)}}\exp(-\frac{m_1 t}{\varepsilon}) \le \rho\varepsilon^2$, we have $t \ge \frac{\varepsilon}{m_1}\ln\frac{m_2}{\rho\varepsilon^{2a_m}}$. Define $t(\varepsilon) = \frac{\varepsilon}{m_1}\ln\frac{m_2}{\rho\varepsilon^{2a_m}}$ and choose ε_2 small enough such that $t(\varepsilon) \le \frac{1}{k_t}t_1$ with $k_t > 1$; then we have for all $0 < \varepsilon \le \varepsilon_2$, $P(e(t(\varepsilon), \varepsilon)) \le \rho\varepsilon^2$. Take $\varepsilon_3^* = \min(\varepsilon_1, \varepsilon_2)$, and then for all $0 < \varepsilon \le \varepsilon_3^*$, we can obtain that for $t \in [0, t(\varepsilon)]$, the boundedness of the trajectory $(\mathcal{L}(t, \varepsilon), e(t, \varepsilon))$ can be guaranteed by (5.73) and (5.75), and for all $t \ge t(\varepsilon)$, the boundedness of $(\mathcal{L}(t, \varepsilon), e(t, \varepsilon))$ is guaranteed since the trajectory enters Ω during $[0, t(\varepsilon)]$ and remains in Ω for all $t \ge t(\varepsilon)$. This completes the proof. □

Then, similar to the discussions in the previous section, to investigate the performance of the closed-loop system and quantify the effect of event-based sampling, an additional assumption is needed.

Assumption 5.4 For system (5.60), there exists a C^1 Lyapunov function $V_1(\mathcal{L})$ defined over a ball $B(o, r_1) \subseteq \mathbb{R}$, where $r_1 > 0$, such that for all $\mathcal{L} \in B(o, r_1)$, we have

1. $d_1\|\mathcal{L}\|^2 \le V_1(\mathcal{L}) \le d_2\|\mathcal{L}\|^2$
2. $\frac{\partial V_1}{\partial \mathcal{L}} f(\mathcal{L}, 0) \le -d_3\|\mathcal{L}\|$
3. $\|\frac{\partial V_1}{\partial \mathcal{L}}\| \le d_4\|\mathcal{L}\|$

where d_1, d_2, d_3, and d_4 are some positive constants.

Then we present the following theorem.

Theorem 5.4 *Consider system (5.55)–(5.58) and the event-triggering condition (5.64). Suppose that Assumptions 5.1–5.2, 5.3 hold, then there exists $\varepsilon_4^* > 0$ such that for every $0 < \varepsilon \le \varepsilon_4^*$, we have for all $(\mathcal{L}, e) \in B(0, \eta_1) \times \{\|e\| \le \eta_1\} = \Omega_2$, where $\eta_1 < r_1$,*

$$\lim_{t\to+\infty} |x_{ij} - \hat{x}_{ij}| \le \varepsilon^{a_i - j}\sqrt{\frac{\varepsilon^{\frac{1}{3}}\lambda_2}{\lambda_1\gamma(\varepsilon)}}M_\sigma, \tag{5.76}$$

$$\lim_{t\to+\infty} \|\mathcal{L}\| \le \sqrt{\frac{\varepsilon^{\frac{1}{3}}\lambda_2\beta}{d_1\gamma(\varepsilon)}}M_\sigma, \tag{5.77}$$

β being a positive constant.

Proof From Assumptions 5.1–5.2, we have for all $(\mathcal{L}, e) \in \Omega_2$,

$$\|g(\mathcal{L}, D(\varepsilon)e)\| \le L_2\|\mathcal{L}\| + L_3\|e\|, \tag{5.78}$$
$$\|f(\mathcal{L}, D(\varepsilon)e) - f(\mathcal{L}, 0)\| \le L_4\|e\|. \tag{5.79}$$

Consider $\bar{V}_1(\mathcal{L}, e) = V_1(\mathcal{L}) + \beta P(e)$, then we have

$$\begin{aligned}\dot{\bar{V}}_1 &= \frac{\partial V_1}{\partial \mathcal{L}} f(\mathcal{L}, D(\varepsilon)e) + \beta \frac{\partial P}{\partial e}\dot{e} \\ &\le -[d_3 - \frac{1}{2}\varepsilon^{\frac{1}{3}}(L_4 d_4 + 2\beta\lambda_2 L_2)]\|\mathcal{L}\|^2 \\ &\quad - \frac{1}{\varepsilon}[\beta - 2\lambda_2 L_3 \beta\varepsilon - \frac{1}{2}\varepsilon^{\frac{2}{3}}(L_4 d_4 + 2\beta\lambda_2 L_2)]\|e\|^2 \\ &\quad + 2\beta\lambda_2\|e\|M_\sigma \\ &\le -[d_3 - \frac{1}{2}\varepsilon^{\frac{1}{3}}(L_4 d_4 + 2\beta\lambda_2 L_2)]\|\mathcal{L}\|^2 \\ &\quad - \frac{1}{\varepsilon}[\beta - 2\lambda_2 L_3 \beta\varepsilon - \frac{1}{2}\varepsilon^{\frac{2}{3}}(L_4 d_4 + 2\beta\lambda_2 (1 + L_2))]\|e\|^2 \\ &\quad + \varepsilon^{\frac{1}{3}}\beta\lambda_2 M_\sigma^2. \end{aligned} \tag{5.80}$$

Choosing β small enough, there exists $0 < \varepsilon_4^* \le \varepsilon_3^*$ such that for every $0 < \varepsilon \le \varepsilon_4^*$, $\gamma_1(\varepsilon) > 0$, $\gamma_2(\varepsilon) > 0$, where $\gamma_1(\varepsilon)$ and $\gamma_2(\varepsilon)$ are defined in (5.61)–(5.63). Then we have

$$\begin{aligned}\dot{\bar{V}}_1 &\le -\gamma_1(\varepsilon)\|\mathcal{L}\|^2 - \frac{1}{\beta\varepsilon}\gamma_2(\varepsilon)\beta\|e\|^2 + \varepsilon^{\frac{1}{3}}\beta\lambda_2 M_\sigma^2 \\ &\le -\frac{1}{d_2}\gamma_1(\varepsilon)V_1(\mathcal{L}) - \frac{1}{\beta\varepsilon\lambda_2}\gamma_2(\varepsilon)\beta P(e) + \varepsilon^{\frac{1}{3}}\beta\lambda_2 M_\sigma^2 \\ &\le -\gamma(\varepsilon)\bar{V}_1 + \varepsilon^{\frac{1}{3}}\beta\lambda_2 M_\sigma^2. \end{aligned} \tag{5.81}$$

Hence, we can obtain that

$$\begin{aligned}\bar{V}_1 &\le \phi_1(t, t_0)\bar{V}_1(\mathcal{L}_0, e_0) + \varepsilon^{\frac{1}{3}}\beta\lambda_2 M_\sigma^2 \int_{t_0}^{t} \phi_1(t, \tau)d\tau \\ &\le \phi_1(t, t_0)\bar{V}_1(\mathcal{L}_0, e_0) + \varepsilon^{\frac{1}{3}}\beta\lambda_2 M_\sigma^2 \frac{1}{\gamma(\varepsilon)}(1 - \phi_1(t, t_0)) \end{aligned} \tag{5.82}$$

where $\phi_1(t, t_0) := \exp(-\gamma(\varepsilon)(t - t_0))$. Note that when $t \to +\infty$, $\phi_1(t, t_0) \to 0$, then we can obtain that when $t \to +\infty$, $\bar{V}_1 \le \varepsilon^{\frac{1}{3}}\beta\lambda_2 M_\sigma^2 \frac{1}{\gamma(\varepsilon)}$, hence $d_1\|\mathcal{L}\|^2 + \beta\lambda_1\|e\|^2 \le \varepsilon^{\frac{1}{3}}\beta\lambda_2 M_\sigma^2 \frac{1}{\gamma(\varepsilon)}$. Then we conclude that

$$\lim_{t\to+\infty} \|e\| \le \sqrt{\frac{\varepsilon^{\frac{1}{3}}\lambda_2}{\lambda_1\gamma(\varepsilon)}} M_\sigma, \tag{5.83}$$

$$\lim_{t\to+\infty} |x_{ij} - \hat{x}_{ij}| \le \varepsilon^{a_i - j}\sqrt{\frac{\varepsilon^{\frac{1}{3}}\lambda_2}{\lambda_1\gamma(\varepsilon)}} M_\sigma, \tag{5.84}$$

$$\lim_{t\to+\infty} \|\mathcal{L}\| \le \sqrt{\frac{\varepsilon^{\frac{1}{3}}\lambda_2\beta}{d_1\gamma(\varepsilon)}} M_\sigma. \tag{5.85}$$

This completes the proof. □

Theorem 5.4 develops a quantitative relationship between the asymptotic upper bound of the observation error and the parameters of the event-based sampler in Case 1. The event-triggering condition proposed in (5.64) presents an upper bound of the sampling error of the system output. The system performance can be adjusted by changing the value of the triggering parameter M_σ.

5.3.2 *Case 2: Only Sampling Control Signal*

In this section, the case that the control signal is sampled and transmitted only at certain instants determined by event-triggering condition is discussed. To do this, define the sampling error $\sigma_u(t)$ as

$$\sigma_u(t) := u(t_k) - u(t), \tag{5.86}$$

where $t \in [t_k, t_{k+1})$, and t_k is the transmission instant determined by the triggering condition. Then the event-triggered closed-loop system can be represented by

$$\dot{x} = Ax + B\phi(x, z, q(v, x - D(\varepsilon)e, w) + \sigma_u), \tag{5.87}$$
$$\dot{z} = \psi(x, z, q(v, x - D(\varepsilon)e, w) + \sigma_u), \tag{5.88}$$
$$\dot{v} = p(v, x - D(\varepsilon)e, w), \tag{5.89}$$
$$\dot{e} = \frac{1}{\varepsilon}A_0 e + Bg_2(x, z, v, D(\varepsilon)e, \sigma_u), \tag{5.90}$$

where $g_2(x, z, v, D(\varepsilon)e, \sigma_u) = \phi(x, z, q(v, \hat{x}, w) + \sigma_u) - \phi_0(\hat{x}, w, q(v, \hat{x}, w) + \sigma_u)$. Take $\mathcal{L} = [x^{\mathrm{T}}, z^{\mathrm{T}}, v^{\mathrm{T}}]^{\mathrm{T}}$, then the system (5.87)–(5.90) can be rewritten as

$$\dot{\mathcal{L}} = f(\mathcal{L}, D(\varepsilon)e, \sigma_u), \tag{5.91}$$

with the state feedback system satisfying

$$\dot{\mathcal{L}} = f(\mathcal{L}, 0, 0). \tag{5.92}$$

Define $\bar{\gamma}_1(\varepsilon)$, $\bar{\gamma}_2(\varepsilon)$, $\bar{\gamma}(\varepsilon)$ and $P_\sigma(\varepsilon)$ as

$$\bar{\gamma}_1(\varepsilon) := c_3 - \beta\lambda_2 L_5 \varepsilon^{\frac{1}{3}}, \tag{5.93}$$
$$\bar{\gamma}_2(\varepsilon) := \beta - 2\beta\lambda_2 L_6 \varepsilon - \frac{1}{2}\varepsilon^{\frac{2}{3}}[M_2\bar{L}_{12} + 2\beta\lambda_2(L_5 + L_7)], \tag{5.94}$$
$$\bar{\gamma}(\varepsilon) := \min\{\frac{1}{c_2}\bar{\gamma}_1(\varepsilon), \frac{1}{\beta\varepsilon\lambda_2}\bar{\gamma}_2(\varepsilon)\}, \tag{5.95}$$

$$P_\sigma(\varepsilon) := M_2\bar{L}_{11}M_{\sigma_u} + \frac{1}{2}M_2\bar{L}_{12}\varepsilon^{\frac{1}{3}} + \beta\lambda_2 L_7 M_{\sigma_u}^2\varepsilon^{\frac{1}{3}}. \tag{5.96}$$

Then we present the event-triggering condition in the following theorem.

Theorem 5.5 *Consider the system in (5.91). Suppose that Assumptions 5.1–5.2 hold, there exist $\varepsilon_5^* > 0$ and an event-triggering condition*

$$\Gamma_u = \begin{cases} 0, \text{ if } \|\sigma_u\| \leq M_{\sigma_u}, \\ 1, \text{ otherwise,} \end{cases} \tag{5.97}$$

where $M_{\sigma_u} > 0$ is a constant such that for all $\varepsilon \in (0, \varepsilon_5^*)$, the trajectories $(\mathcal{L}(t, \varepsilon), e(t, \varepsilon))$ starting in $\mathbb{Q}_1 \times \mathbb{Q}_2$ are bounded for all $t \geq 0$.

Proof According to Assumptions 5.1–5.2, we can obtain that for all $(\mathcal{L}, e) \in \mathbb{S} \times \mathbb{R}^{a'}$,

$$\|f(\mathcal{L}, D(\varepsilon)e, \sigma_u)\| \leq \bar{k}_1, \tag{5.98}$$

$$\|g_2(\mathcal{L}, D(\varepsilon)e, \sigma_u)\| \leq \bar{k}_2, \tag{5.99}$$

where $\bar{k}_1, \bar{k}_2$ are some positive constants. For all $(\mathcal{L}, e) \in \Omega$ and any $0 < \varepsilon < 1$, we have

$$\|f(\mathcal{L}, D(\varepsilon)e, \sigma_u) - f(\mathcal{L}, 0, 0)\| \leq L_{11}\|\sigma_u\| + L_{12}\|e\|. \tag{5.100}$$

for all $(\mathcal{L}, e) \in \{\mathcal{L} \in \mathbb{R} | V(\mathcal{L}) = c\} \times \{P(e) \leq \rho\varepsilon^2\}$, we have

$$\begin{aligned} \dot{V}(\mathcal{L}) &= \frac{\partial V}{\partial \mathcal{L}} f(\mathcal{L}, D(\varepsilon)e, \sigma_u) \\ &\leq -W_3(\mathcal{L}) + M_1(L_{11}\|\sigma_u\| + L_{12}\|e\|), \end{aligned} \tag{5.101}$$

where $M_1 = \max_{\mathcal{L}\in\mathbb{S}}(\|\frac{\partial V}{\partial \mathcal{L}}\|)$. Hence $P(e) \leq \rho\varepsilon^2$, we have $\frac{1}{\varepsilon}\|e\| \leq \sqrt{\frac{\rho}{\lambda_1}}$, take $\kappa = \min_{\mathcal{L}\in\partial\mathbb{S}} W_3(\mathcal{L}) - M_1 L_{11} M_{\sigma_u} > 0$ and $\varepsilon_{11} = \kappa/(M_1 L_{12}\sqrt{\rho/\lambda_1})$, then we have for all $0 < \varepsilon \leq \varepsilon_{11}$,

$$\dot{V}(\mathcal{L}) \leq 0. \tag{5.102}$$

Then for all $(\mathcal{L}, e) \in \mathbb{S} \times \{P(e) = \rho\varepsilon^2\}$,

$$\dot{P}(e) = \frac{\partial P}{\partial e}\dot{e} \leq -\frac{1}{\varepsilon}\|e\|^2 + 2\lambda_2\bar{k}_2\|e\|. \tag{5.103}$$

Since $P(e) = \rho\varepsilon^2$, we have $\frac{1}{\varepsilon}\|e\| \geq \sqrt{\frac{\rho}{\lambda_2}}$, take $\rho = 25\bar{k}_2^2\lambda_2^3$, we can obtain $\dot{P}(e) \leq 0$. Then we conclude that the set Ω is positively invariant. Similar to the proof in Sect. 5.3, there exists a finite time t_2 such that for all $t \in [0, t_2]$, $\mathcal{L}(t, \varepsilon) \in \mathbb{S}$. During the same time interval, for $P(e) \geq \rho\varepsilon^2$, we have

$$\dot{P}(e) \le \frac{\bar{m}_2}{\varepsilon^{2(a_m-1)}} \exp(-\frac{\bar{m}_1 t}{\varepsilon}), \quad (5.104)$$

where $\bar{m}_1 = 3/(5\lambda_2)$ and $\bar{m}_2 = k^2\lambda_2$. Choose $\varepsilon_{21} > 0$ such that $\bar{t}(\varepsilon) = \frac{\varepsilon}{\bar{m}_1} \ln \frac{\bar{m}_2}{\rho\varepsilon^{2a_m}} \le \frac{1}{k_t} t_2$, where $k_t > 1$, then we have for all $0 < \varepsilon \le \varepsilon_{21}$, $P(e(\bar{t}(\varepsilon), \varepsilon)) \le \rho\varepsilon^2$. Taking $\varepsilon_5^* = \min(\varepsilon_{11}, \varepsilon_{21})$, then for all $0 < \varepsilon \le \varepsilon_5^*$, similar to the analysis in Case A, we can obtain that the boundedness of the trajectory $(\mathcal{L}(t, \varepsilon), e(t, \varepsilon))$ is guaranteed. This completes the proof. □

Theorem 5.5 shows the boundedness of the trajectories in Case B. The value of the constant ε_5^* is influenced by the initial values of the trajectories.

Next we discuss the performance of the closed-loop system and quantify the effect of event-based sampling. For convenience, we restate Assumption 5.3 proposed in Sect. 5.3 of the manuscript here as follows.

Assumption 5.5 Consider system (5.92), there exists a C^1 Lyapunov function $V_2(\mathcal{L})$ defined over a ball $B(o, r_2) \subseteq \mathbb{R}$, where $r_2 > 0$, such that for all $\mathcal{L} \in B(o, r_2)$, we have

1. $c_1\|\mathcal{L}\|^2 \le V_2(\mathcal{L}) \le c_2\|\mathcal{L}\|^2$
2. $\frac{\partial V_2}{\partial \mathcal{L}} f(\mathcal{L}, 0) \le -c_3\|\mathcal{L}\|$
3. $\|\frac{\partial V_2}{\partial \mathcal{L}}\| \le M_2$

where c_1, c_2, c_3 and M_2 are some positive constants.

Then we present the following theorem.

Theorem 5.6 *Consider system (5.87)–(5.90) and the event-triggering condition (5.97), suppose that Assumptions 5.1–5.2, 4 hold, there exists $\varepsilon_6^* > 0$ such that for every $0 < \varepsilon \le \varepsilon_6^*$, we have for all $(\mathcal{L}, e) \in B(0, \eta_2) \times \{\|e\| \le \eta_2\} = \Omega_3$, where $\eta_2 < r_2$,*

$$\lim_{t\to+\infty} |x_{ij} - \hat{x}_{ij}| \le \varepsilon^{a_i - j}\sqrt{\frac{P_\sigma(\varepsilon)}{\beta\lambda_1\bar{\gamma}(\varepsilon)}}, \quad (5.105)$$

$$\lim_{t\to+\infty} \|\mathcal{L}\| \le \sqrt{\frac{P_\sigma(\varepsilon)}{c_1\bar{\gamma}(\varepsilon)}}, \quad (5.106)$$

where β is a positive constant.

Proof From Assumptions 5.1–5.2, we have for all $(\mathcal{L}, e) \in \Omega_3$,

$$\|g_2(\mathcal{L}, D(\varepsilon)e, \sigma_u)\| \le L_5\|\mathcal{L}\| + L_6\|e\| + L_7\|\sigma_u\|, \quad (5.107)$$

$$\|f(\mathcal{L}, D(\varepsilon)e, \sigma_u) - f(\mathcal{L}, 0, 0)\| \le \bar{L}_{11}\|\sigma_u\| + \bar{L}_{12}\|e\|. \quad (5.108)$$

Consider $\bar{V}_2(\mathcal{L}, e) = V_2(\mathcal{L}) + \beta P(e)$, then we have

$$\begin{aligned}\dot{\bar{V}}_2 &= \frac{\partial V_2}{\partial \mathcal{L}} f(\mathcal{L}, D(\varepsilon)e, \sigma_u) + \beta \frac{\partial P}{\partial e}\dot{e} \\ &\leq -(c_3 - \beta\lambda_2 L_5 \varepsilon^{\frac{1}{3}})\|\mathcal{L}\|^2 \\ &\quad - \frac{1}{\varepsilon}[\beta - 2\beta\lambda_2 L_6 \varepsilon - \frac{1}{2}\varepsilon^{\frac{2}{3}}(M_2 \bar{L}12 + 2\beta\lambda_2(L_5 + L_7))]\|e\|^2 \\ &\quad + M_2 \bar{L}_{11} M_{\sigma_u} + \frac{1}{2} M_2 \bar{L}_{12} \varepsilon^{\frac{1}{3}} + \beta\lambda_2 L_7 M_{\sigma_u}^2 \varepsilon^{\frac{1}{3}}. \end{aligned} \tag{5.109}$$

Choose β small enough and $0 < \varepsilon_6^* \leq \varepsilon_5^*$ such that for every $0 < \varepsilon \leq \varepsilon_6^*$, $\bar{\gamma}_1(\varepsilon) > 0$, $\bar{\gamma}_2(\varepsilon) > 0$, where $\bar{\gamma}_1(\varepsilon)$ and $\bar{\gamma}_2(\varepsilon)$ are defined in (5.93)–(5.95). Then we have

$$\begin{aligned}\dot{\bar{V}}_2 &\leq -\bar{\gamma}_1(\varepsilon)\|\mathcal{L}\|^2 - \frac{1}{\beta\varepsilon}\bar{\gamma}_2(\varepsilon)\beta\|e\|^2 + P_\sigma(\varepsilon) \\ &\leq -\frac{1}{c_2}\bar{\gamma}_1(\varepsilon)V_2(\mathcal{L}) - \frac{1}{\beta\varepsilon\lambda_2}\bar{\gamma}_2(\varepsilon)\beta P(e) + P_\sigma(\varepsilon) \\ &\leq -\bar{\gamma}(\varepsilon)\bar{V}_2 + P_\sigma(\varepsilon), \end{aligned} \tag{5.110}$$

where $P_\sigma(\varepsilon)$ is defined in (5.96). Hence, we can obtain that

$$\bar{V}_2 \leq \phi_2(t, t_0)\bar{V}_2(\mathcal{L}_0, e_0) + \frac{P_\sigma(\varepsilon)}{\bar{\gamma}(\varepsilon)}(1 - \phi_2(t, t_0)) \tag{5.111}$$

where $\phi_2(t, t_0) := \exp(-\bar{\gamma}(\varepsilon)(t - t_0))$. Note that when $t \to +\infty$, $\phi_2(t, t_0) \to 0$, then we can obtain that when $t \to +\infty$, $\bar{V}_2 \leq P_\sigma(\varepsilon)/\bar{\gamma}(\varepsilon)$, hence $c_1\|\mathcal{L}\|^2 + \beta\lambda_1\|e\|^2 \leq P_\sigma(\varepsilon)/\bar{\gamma}(\varepsilon)$. Then we conclude that

$$\lim_{t\to+\infty} |x_{ij} - \hat{x}_{ij}| \leq \varepsilon^{a_i - j}\sqrt{\frac{P_\sigma(\varepsilon)}{\beta\lambda_1\bar{\gamma}(\varepsilon)}}, \tag{5.112}$$

$$\lim_{t\to+\infty} \|\mathcal{L}\| \leq \sqrt{\frac{P_\sigma(\varepsilon)}{c_1\bar{\gamma}(\varepsilon)}}, \tag{5.113}$$

This completes the proof. □

The above theorem develops a quantitative relationship between the asymptotic upper bound of the observation error and the parameters of the event-based sampler in Case 2. The event-triggering condition proposed in (5.97) presents an upper bound of the sampling error of the control signal. Note that the system performance can be adjusted by changing the value of the triggering parameter M_{σ_u}.

5.4 Numerical Example

In this section, a numerical example is utilized to validate the proposed upper bounds of the observation error. Consider a plant of the following form:

$$\begin{aligned}\dot{x}_1(t) &= x_2(t), x_1(0) = 1,\\ \dot{x}_2(t) &= -f(x_1(t)) - x_2(t) + u(t), x_2(0) = 1,\\ y(t) &= x_1(t),\end{aligned} \tag{5.114}$$

where $\phi(x) = -f(x_1) - x_2$, and $f(s)$ is defined by

$$f(s) := \begin{cases} s + \frac{\pi}{2} - 1, \; s < -\frac{\pi}{2}, \\ \sin(s), \qquad -\frac{\pi}{2} \le s \le \frac{\pi}{2}, \\ s - \frac{\pi}{2} + 1, \; s > \frac{\pi}{2}, \end{cases} \tag{5.115}$$

By choosing $\phi_0(\hat{x}) = -\hat{x}_1 - \hat{x}_2(t)$, the high-gain observer has the form

$$\begin{aligned}\dot{\hat{x}}_1(t) &= \hat{x}_2(t) + \frac{4}{\varepsilon} \cdot (\xi(t) - \hat{x}_1(t)),\\ \dot{\hat{x}}_2(t) &= -\hat{x}_1(t) - \hat{x}_2(t) + \frac{3}{\varepsilon^2} \cdot (\xi(t) - \hat{x}_1(t)) + \mu(t),\\ \xi(t) &= y(t_{k_y}), \; t \in [t_{k_y}, t_{k_y+1}),\\ \mu(t) &= u(t_{k_u}), \; t \in [t_{k_u}, t_{k_u+1}).\end{aligned} \tag{5.116}$$

The control signal $u(t)$ is designed as $u(t) = -4\hat{x}_1 - \hat{x}_2$. Based on the above analysis, Assumptions 5.1–5.2 can be satisfied. Specifically, write K as

$$K = \begin{bmatrix} 0 & 1 \\ -4 & -1 \end{bmatrix}, \tag{5.117}$$

and by solving the equation $QK + K^T Q = -I$, we can obtain

$$Q = \begin{bmatrix} 2.625 & 0.125 \\ 0.125 & 0.625 \end{bmatrix}. \tag{5.118}$$

Hence, choosing $V_2(x) = x^T Qx + \frac{5}{4}\int_0^{x_1} f(s)ds$ and taking $c_1 = 0.6172$, $c_2 = 3.2578$, $c_3 = 0.9876$, and $M_2 = 11.2854$, Assumption 5.3 is satisfied. On the other hand, since

$$A_0 = \begin{bmatrix} -4 & 1 \\ -3 & 0 \end{bmatrix}, \tag{5.119}$$

and by solving the function $P_0 A_0 + A_0^{\mathrm{T}} P_0 = -I$, we have

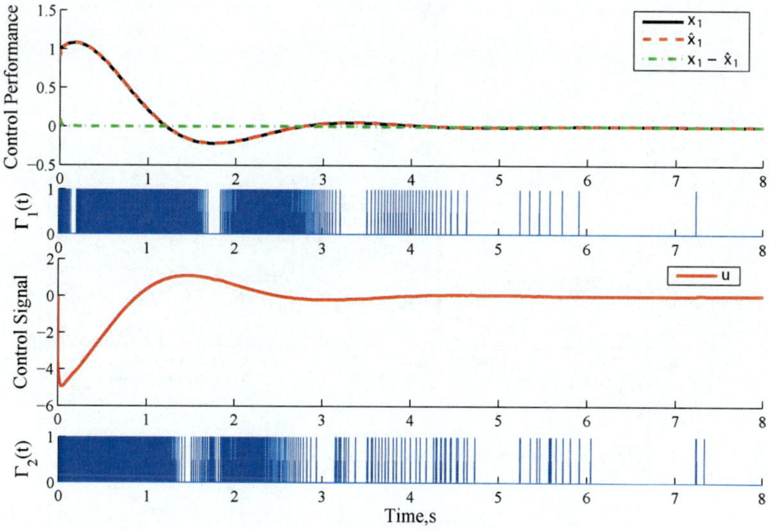

Fig. 5.2 Simulation results of the proposed event-triggered control scheme

$$P_0 = \begin{bmatrix} 0.5 & -0.5 \\ -0.5 & 0.8333 \end{bmatrix}, \tag{5.120}$$

then we can obtain $P(e)$ as $P(e) = \frac{1}{2}e_1^2 - e_1e_2 + \frac{5}{6}e_2^2$. Choosing $\lambda_1 = 0.1396$ and $\lambda_2 = 1.1937$, Eqs. (5.30)–(5.31) can be satisfied. As for $g(x, D(\varepsilon)e, \sigma_u) = \phi(x) - \phi_0(\hat{x})$, choosing $L_5^* = 1$, $L_6^* = 2$ and $L_7^* = 0$, (5.49) can be satisfied. Similarly, (5.50) can be satisfied by taking $L_{11}^* = 1$ and $L_{12}^* = 1.2$. Choosing $\varepsilon = 0.005$, $M_{\sigma_y} = 1$ and $M_{\sigma_u} = 0.01$, we can obtain that $\|x_1 - \hat{x}_1\| \leq 0.053$ at steady state according to Theorem 5.2. The simulation results are shown in Fig. 5.2, and the sampling intervals between two consecutive event-triggering instants are shown in Fig. 5.3.

From Fig. 5.2, we can observe that satisfactory control performance can be obtained in this HGO-based event-triggered control scheme. The control signal only updates when $\Gamma_2(t) = 1$. The maximum $\|x_1\|$ at steady state is 0.0347. The average observation error of x_1, which is $\|x_1 - \hat{x}_1\|$, at steady state is 0.0009, and the maximum $\|x_1 - \hat{x}_1\|$ at steady state is 0.0019. From Fig. 5.3, we can observe that nonzero sampling intervals can be guaranteed to avoid Zeno behavior. The simulation results show that the boundedness of the observation error can be guaranteed by Eq. (5.43).

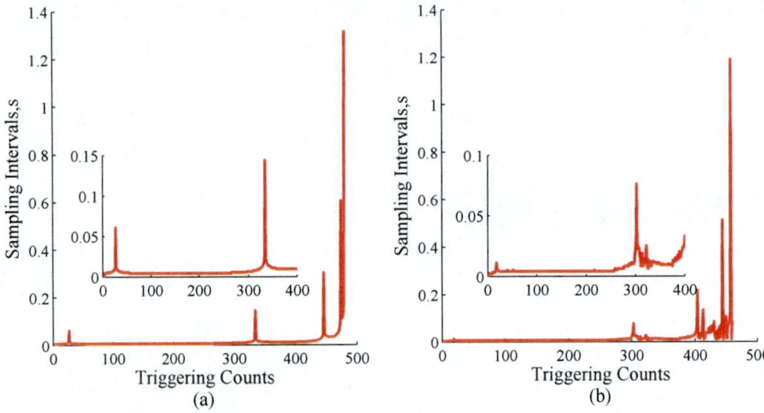

Fig. 5.3 Sampling intervals of **a** system output **b** control signal

5.5 Experimental Performance Evaluation

In this section, we further verify the obtained theoretical results through comparative experiments. The experiments are carried out on the DC torque motor platform that we introduced earlier (Fig. 5.4), which mainly consists of a motor load test bench and an embedded measurement and control test bench. The motor load test bench includes a permanent-magnet DC torque motor, a magnetic brake, a speed reducer, an inertial loading mechanism, and a speed sensor. The embedded measurement and control test bench can generate signals to control the motor and process the received signals, which can be considered as a controller. The goal of the experiments is to control the DC motor system to track a provided trajectory. For comparison purposes, a time-triggered continuous-time HGO (CT-HGO) controller is also implemented. In these experiments, the HGOs are implemented by difference approximation with the sampling time $t_s = 25$ ms. In order to investigate the tracking performance of the systems, we define the tracking error (E_T), the average sampling time (T_A) and $y(t)$ as follows:

$$E_T = \frac{1}{T}\int_0^T |y_r(t) - y_o(t)|dt, \tag{5.121}$$

$$T_A = \begin{cases} 25, & \text{for CT-HGO schemes,} \\ \frac{T}{N_s}, & \text{for ET-HGO schemes,} \end{cases} \tag{5.122}$$

$$y(t) = y_o(t) - y_r(t) \tag{5.123}$$

where $y_r(t)$ represents the reference trajectory, in this section that is a multitone sinusoid signal. $y_o(t)$ is the output of the controlled system, T represents the experimental time in milliseconds, and N_s is the total event-triggering counts in one experiment (Fig. 5.4).

By choosing $\phi_0 = 0$, the HGO is of the following form:

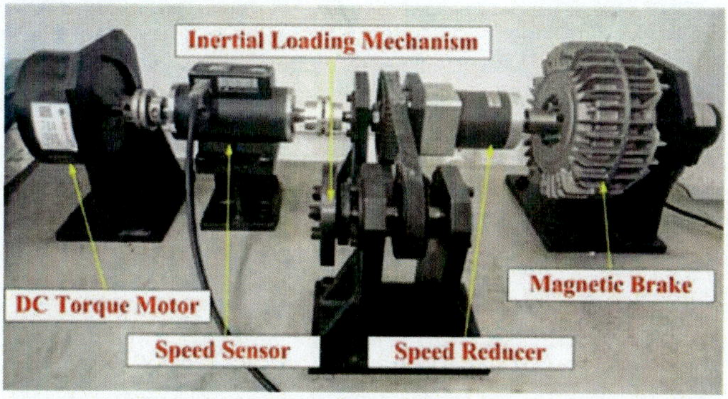

Fig. 5.4 The motor load test bench

$$\dot{\hat{x}}_1(t) = \hat{x}_2(t) + (2/\varepsilon) \cdot (y(t_{k_y}) - \hat{x}_1(t)),$$
$$\dot{\hat{x}}_2(t) = (1/\varepsilon^2) \cdot (y(t_{k_y}) - \hat{x}_1(t)) + u(t_{k_u}),$$

where the high-gain parameter $\varepsilon = 0.04$, and the control law $u(t)$ is designed as

$$u(t) = -13\hat{x}_1 - 3\hat{x}_2.$$

In the experiments, control signal $u(t)$ is saturated by $|u(t)| \leq 3.5$. According to the analysis in [4], the controlled system satisfies Assumption 5.1. Consider the closed-loop system and the event-triggering conditions in (5.26). For comparison purpose, two groups of experiments are carried out. One group evaluates the no-load tracking performance of the system, while another shows the performance with a load of 0.7 N·m. Besides, for each group, the tracking performance of the system without event-triggering conditions (namely, CT-HGO) is also analyzed. Due to the fact that the output and the control signal are both being sampled, their corresponding average sampling time (specifically, T_{A_y} and T_{A_u}, which are determined by Γ_1 and Γ_2, respectively) are calculated. The experimental results are shown in Figs. 5.5, 5.6, 5.7, 5.8, and the tracking error together with the average sampling time are summarized in Table 5.1 to evaluate the tracking performance of the systems. In each figure, the first plot shows the output and the reference signal, the second and the third plots represent the tracking error, and the event-triggering counts are shown in the fourth and the fifth plots. The control signals of the schemes are shown in Figs. 5.9, 5.10.

From the experimental results, we can observe that the output of the system can be controlled to achieve satisfactory tracking performance. The almost overlapping curves show that the ET-HGO schemes have similar tracking performances as CT-HGO schemes, and the observed state $\hat{x}_1$ can converge to the real physical state. From Figs. 5.9 and 5.10, we can observe that max $|u(t)| = 3.5$ due to the saturation effect. Compared with CT-HGO schemes, ET-HGO schemes can maintain a similar level of performance in terms of tracking error, but have an obvious decrease in

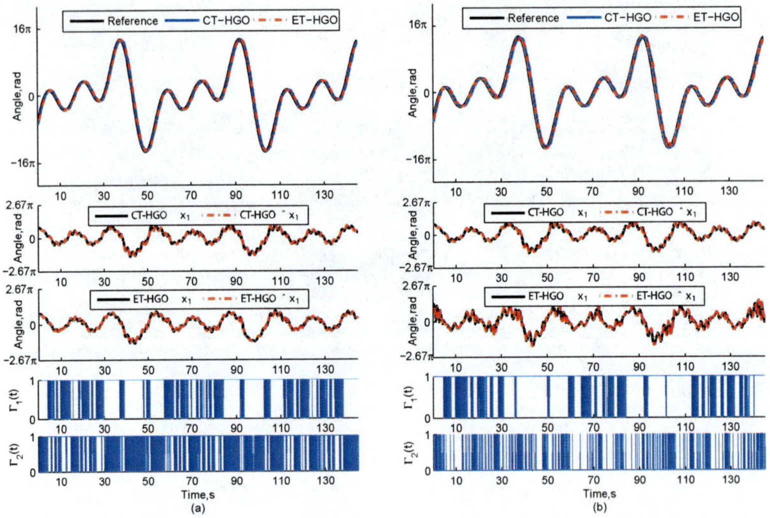

Fig. 5.5 Tracking performance (no-load) with $M_{\sigma_y} = 0.067$ and **a** $M_{\sigma_u} = 0.03$ **b** $M_{\sigma_u} = 0.7$

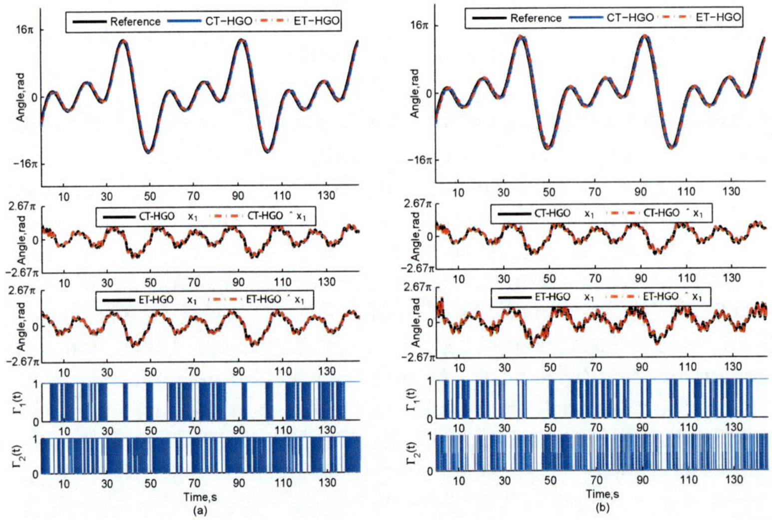

Fig. 5.6 Tracking performance with the load of 0.7 N·m, $M_{\sigma_y} = 0.067$ and **a** $M_{\sigma_u} = 0.03$ **b** $M_{\sigma_u} = 0.7$

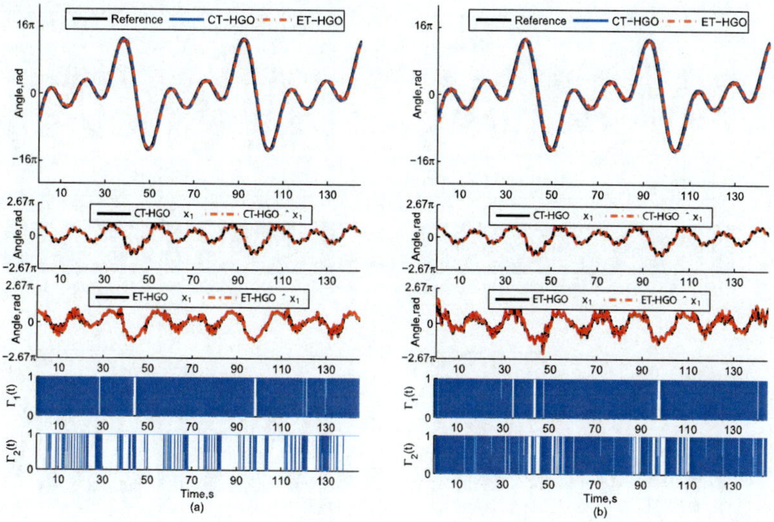

Fig. 5.7 Tracking performance (no-load) with $M_{\sigma_y} = 0.28$ and **a** $M_{\sigma_u} = 0.1$ **b** $M_{\sigma_u} = 0.7$

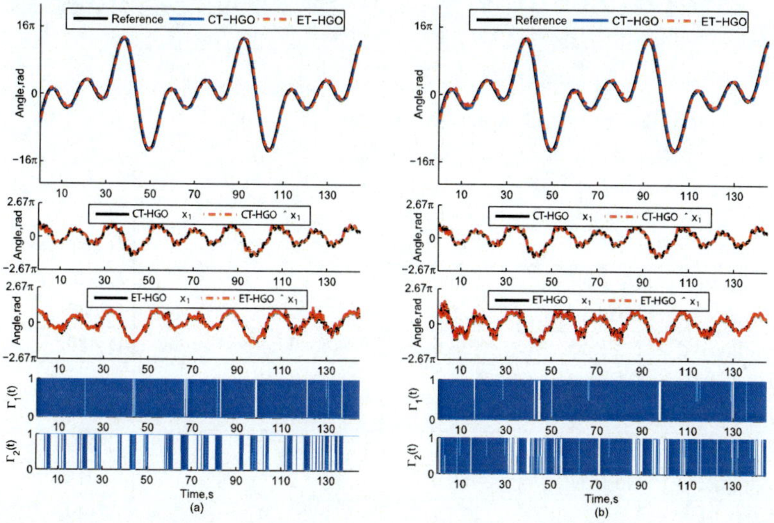

Fig. 5.8 Tracking performance with the load of 0.7 N·m, $M_{\sigma_y} = 0.28$ and **a** $M_{\sigma_u} = 0.1$ **b** $M_{\sigma_u} = 0.7$

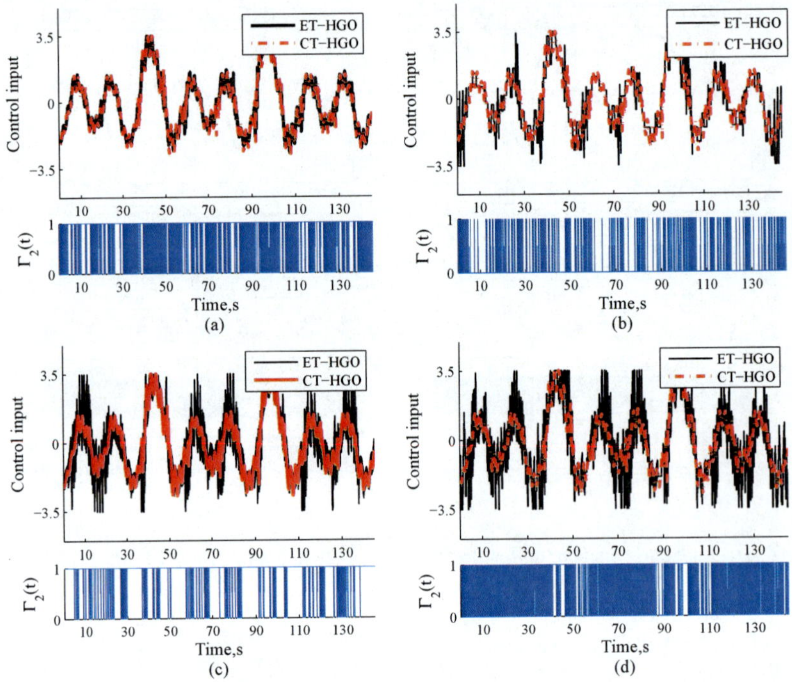

Fig. 5.9 Control signals (no-load) with **a** $M_{\sigma_y} = 0.067$, $M_{\sigma_u} = 0.03$ **b** $M_{\sigma_y} = 0.067$, $M_{\sigma_u} = 0.7$ (c) $M_{\sigma_y} = 0.28$, $M_{\sigma_u} = 0.1$ (d) $M_{\sigma_y} = 0.28$, $M_{\sigma_u} = 0.7$

Table 5.1 Performance of the proposed schemes

Experimental scenarios		T_{A_y} (ms)	T_{A_u} (ms)	Tracking error
Time-triggered	No-load	25	25	1.5926
	Load	25	25	1.7688
Event-triggered	No-load	33.6	37.4	1.5480
$\Gamma: 0.067,\ 0.03$	Load	34.5	35.9	1.7741
Event-triggered	No-load	32	327.4	1.6638
$\Gamma: 0.067,\ 0.7$	Load	33.7	303.1	1.8269
Event-triggered	No-load	69.9	27	1.5915
$\Gamma: 0.28,\ 0.1$	Load	75.7	27.8	1.7149
Event-triggered	No-load	67.1	69.5	1.8042
$\Gamma: 0.28,\ 0.7$	Load	72.2	73.2	1.8365

The notation "Event-triggered, $\Gamma: a,\ b$" in rows 3–6 of column 1 means that the experimental scenario is an event-triggering scheme with $M_{\sigma_y} = a$ and $M_{\sigma_u} = b$

average sampling rate. Note that the performance of event-triggering scheme ($\Gamma: 0.067,\ 0.03$) or ($\Gamma: 0.28,\ 0.7$) in row 3 or 6 of Table 5.1 is better than the time-triggering scheme when there is no load. Such a phenomenon is mainly caused by the uncertainties of the closed-loop system. More specifically, since the exact

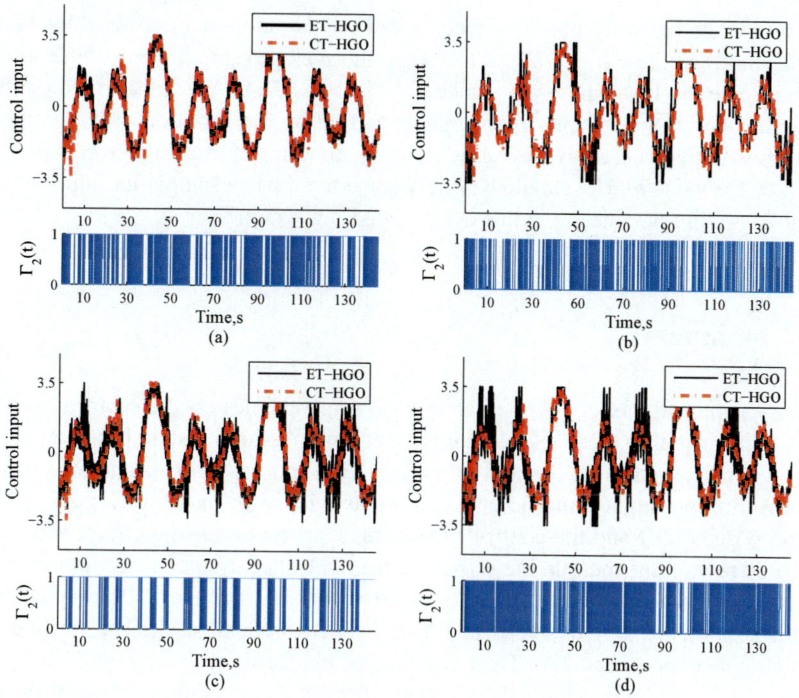

Fig. 5.10 Control signals with the load of 0.7 N·m and **a** $M_{\sigma_y} = 0.067$, $M_{\sigma_u} = 0.03$ **b** $M_{\sigma_y} = 0.067$, $M_{\sigma_u} = 0.7$ **c** $M_{\sigma_y} = 0.28$, $M_{\sigma_u} = 0.1$ **d** $M_{\sigma_y} = 0.28$, $M_{\sigma_u} = 0.7$

mathematical model of the DC torque motor is unknown, the function ϕ_0 in the high-gain observer, which is the nominal model of ϕ, is chosen as $\phi_0 = 0$. Due to this modeling uncertainty, transmitting signals more frequently may not lead to better control performance.

By tuning the values of the event-triggering parameters M_{σ_y} or M_{σ_u}, the average sampling time can be adjusted, which may influence the performance of the control system. In general, the tracking performance of the system may become worse with a larger average sampling time, but this phenomenon may not be very obvious when the average sampling time varies in a small range. Besides, we observe that the average sampling time T_{A_u} can be tuned a lot larger without significantly deteriorating the control performance compared with that of varying T_{A_y}. Based on this fact, the average sampling rate of the control signal can be adjusted a bit lower compared with that of the system output when choosing the value of M_{σ_u}. Finally, we also observe that the tracking errors of load schemes are larger than those of no-load schemes, and the average sampling time is also larger than no-load schemes with the same event-triggering parameter M_{σ_y} or M_{σ_u}.

Moreover, we can obtain that with the same M_{σ_y}, a larger T_{A_u} is likely to lead to a smaller T_{A_y}, and a larger T_{A_y} is likely to lead to smaller T_{A_u} with the same M_{σ_u}.

The difference between these two facts is that the variation of T_{A_y} caused by T_{A_u} is much smaller. If the value of M_{σ_y} is too large, the average sampling rate of the output may become too low, which may make T_{A_u} close to 25 ms. We can also obtain that the influence on the tracking performance caused by the variation of T_{A_y} is more obvious than that caused by the variation of T_{A_u} in terms of tracking error. Based on this fact, the value of M_{σ_y} should be firstly considered when setting the values of the event-triggering parameters in this event-triggering scheme.

5.6 Summary

In this chapter, the effect of the event-triggering schemes on the performance of the nonlinear output feedback control for a continuous-time system based on HGO has been discussed, with the aim of generalizing our previous discussions to the case of multiple-input-multiple-output systems with prior model information. For the proposed HGO and the controlled system, separate event-triggering conditions that only rely on the measurable signals of the controlled system are designed, with which the boundedness of the observation error can be maintained and the asymptotic stability of the system can be guaranteed. The theoretical results are verified by comparative experiments on a DC torque motor platform.

In Part II of this monograph, we shall provide more examples with different application backgrounds to further illustrate the effectiveness of the theoretic results introduced in the first part and to exemplify how the theoretic results can be applied to solve a particular problem.

References

1. Atassi, A.N., Khalil, H.K.: A separation principle for the stabilization of a class of nonlinear systems. IEEE Trans. Autom. Control **44**(9), 1672–1687 (1999)
2. Guo, B., Wu, Z., Zhou, H.: Active disturbance rejection control approach to output-feedback stabilization of a class of uncertain nonlinear systems subject to stochastic disturbance. IEEE Trans. Autom. Control **61**(6), 1613–1618 (2016)
3. Shi, D., Xue, J., Wang, J., Huang, Y.: Supplementary discussions on "A high-gain approach to event-triggered control with application to motor systems"(2018). www.escience.cn/people/dshi/
4. Shi, D., Xue, J., Zhao, L., Wang, J., Huang, Y.: Event-triggered active disturbance rejection control of DC torque motors. IEEE/ASME Trans. Mechatron. **22**(5), 2277–2287 (2017)
5. Tabuada, P.: Event-triggered real-time scheduling of stabilizing control tasks. IEEE Trans. Autom. Control **52**(9), 1680–1685 (2007)

Part II
Applications

Chapter 6
Event-Triggered Active Disturbance Rejection Control of DC Torque Motors

6.1 Background

With technological developments of power electronics and advanced control technology, servo systems have always been playing an increasingly important role in industrial engineering applications. For this reason, control algorithm design problems for servo systems received a lot of research attention in the recent years [1, 21, 24, 26, 29, 35]. In this chapter, we consider the problem of motion control of electrical motor systems and show how the proposed event-triggered ADRC in the earlier chapters of this book can be employed to achieve satisfactory results with a reduced average sampling rate. Through the results in this chapter, we also aim to exemplify how the developed theoretic results can be applied to solve a specific engineering design problem.

A number of control strategies have been investigated to ensure proper control performance for motor systems, e.g., PID control [18], fuzzy logic control (FLC) [7], ADRC [32], and adaptive robust control (ARC) [33]. For instance, Terzic and Jadric [28] introduced a method to estimate the speed and rotor position for a brushless DC motor (BLDCM) based on the application of an extended Kalman filter. Lim and Krishnan [20] introduced a novel current controller for a linear switched reluctance motor based on an ESO and a nonlinear proportional controller. Chen et al. [3] studied precision motion control of linear motors combined with parameter variations and disturbances, and proposed an ARC algorithm with simultaneous compensation for nonlinearities that affect control performance. A high-precision motion control approach for motors was introduced in [33], and a tracking control approach based on an output feedback robust controller with an ESO for motors was introduced in [34]. Gubara et al. [8] considered PID and FLC in motor speed control, and showed that FLC could achieve better performance than that of PID.

The motivating observation of considering event-triggered ADRC for motor systems, however, is that it is sometimes unnecessary to update the control signal when a motor system is working desirably in the absence of significant events (e.g., abrupt

D. Shi et al., *Event-Triggered Active Disturbance Rejection Control*,
Studies in Systems, Decision and Control 356,
https://doi.org/10.1007/978-981-16-0293-1_6

reference change, occurrence of disturbance). Moreover, in application scenarios with limited communication resources (e.g., when motors are controlled via wireless communication channels), transmitting the signal in real time may cause a huge amount of communication cost and computation power. In this chapter, an event-triggered position tracking problem for a DC torque motor system is considered. To deal with different kinds of disturbances and uncertainties in the system together with the additional disturbance induced by the intermittent event-triggering sampling scheme, an ADRC approach is employed in controller design, which has special advantages in dealing with internal uncertainties and external disturbances [6, 11, 13, 22, 25, 30, 36–40]. Noticeably, Liu et al. [22] applied the ADRC approach to a BLDCM system based on back-propagation neural network. Sira-Ramírez et al. [25] presented an ADRC scheme to achieve angular velocity trajectory tracking on a permanent-magnet synchronous motor combined with uncertainties and disturbances. Du et al. [6] designed an active disturbance rejection controller, and applied it to the sensorless control of internal permanent-magnet synchronous motors. Compared with these existing approaches, the key problems considered in this chapter are to figure out how the event-triggering condition can be designed and how the ADRC scheme can be adapted to achieve closed-loop position tracking and to implement the developed event-triggered system in an embedded microcontroller, so that we can experimentally evaluate whether the actual experimental performance is consistent with the theoretical analysis. The main contents of this chapter are summarized as follows:

1. Through the integration with an easy-to-implement event-triggering mechanism, an event-triggered ADRC scheme is proposed for the considered DC motor system. Since the states of the DC motor cannot be obtained directly, the event-triggering mechanism only needs the output of the system. With this condition, the output of the DC motor system is controlled to track the desired trajectory.
2. The effects that the event-triggered control mechanism brings to the closed-loop system are evaluated theoretically. The observation error is proved to be asymptotically bounded, and the asymptotic boundedness of the tracking error can be guaranteed. The upper bounds of the asymptotic observation error and the tracking error are provided and their relationship with the parameters in the event-triggering condition is explicitly characterized.
3. The experimental performance of the developed results is evaluated through a DC torque motor platform. For different reference signals and parameter choices in the event-triggering condition, satisfactory tracking performance (in comparison with time-triggered ADRC) is achieved at a reduced average sampling rate.

The remainder of this chapter is organized as follows: In Sect. 6.2, the mathematical model of the DC torque motor and the problem formulation are introduced. The performance of the event-triggered ADRC scheme is analyzed in Sect. 6.3. Comparative experimental results are presented to evaluate the actual performance of the system in Sect. 6.4. Some concluding remarks are summarized in Sect. 6.5.

6.2 Problem Description and System Dynamics

In this section, the mathematical model of the DC motor and the ET-ADRC scheme is introduced, and the tracking problem is formulated. To aid the description, the explicit form of the event-triggering condition is deferred to the next section.

Consider the event-triggered motor control scheme in Fig. 6.1. The dynamics of DC torque motor is described as

$$M\ddot{y} = F_m(u) - B\dot{y} + F_{\text{friction}}(\dot{y}) + F_{\text{cogging}}(y) + F_{\text{dis}}, \tag{6.1}$$

where y represents the displacement, u is the control signal, M denotes the inertia, B is the viscous friction coefficient, $F_m(u)$ is the electromagnetic driving force of the motor, $F_{\text{friction}}(\dot{y})$ represents the Coulomb friction, $F_{\text{cogging}}(y)$ denotes the position dependent cogging force, and F_{dis} models the lumped modeling errors and external disturbances.

According to [4], the Coulomb friction compensation can be expressed as

$$F_{\text{friction}}(\dot{y}) = -A_f S_f(\dot{y}), \tag{6.2}$$

where A_f is the Coulomb friction coefficient and $S_f(\dot{y})$ is the approximation of the sign function $\text{sgn}(\dot{y})$. In practice, we can define $S_f(\dot{y})$ as

$$S_f(\dot{y}) = \arctan(\beta\dot{y}) \tag{6.3}$$

for some sufficiently large constant β.

Considering the periodic properties of $F_{\text{cogging}}(y)$ with certain known constant $P > 0$, $F_{\text{cogging}}(y)$ can be approximated as

$$F_{\text{cogging}}(y) = \textstyle\sum_{i=1}^{n} \left(S_i \sin(\frac{2i\pi}{P}y) + C_i \cos(\frac{2i\pi}{P}y)\right) \tag{6.4}$$

through Fourier expansion for a sufficiently large n, where S_i and C_i are some constants.

According to [31], $F_m(u)$ is approximated with

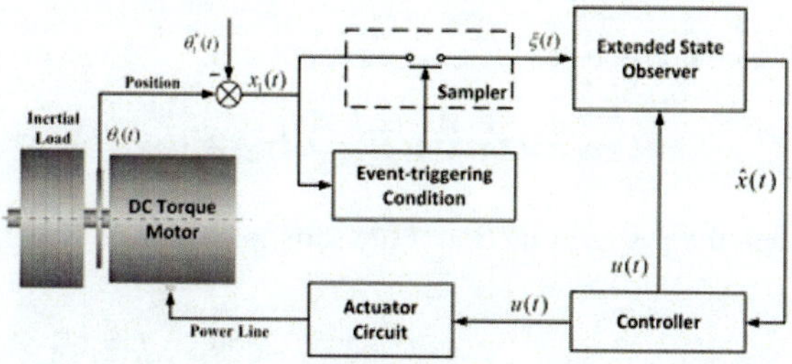

Fig. 6.1 Schematic of the event-triggered ADRC scheme

$$F_m(u) = A \cdot u \tag{6.5}$$

for some $A > 0$. Thus, combining the results in (6.2)–(6.5), if we write

$$\theta_1 := y, \quad \theta_2 := \dot{y}, \tag{6.6}$$

then the dynamics of DC motor in (6.1) can be expressed as

$$\begin{aligned}\dot{\theta}_1 &= \theta_2, \\ \dot{\theta}_2 &= -\tfrac{B}{M}\theta_2 + \textstyle\sum_{i=1}^n \left(\tfrac{S_i}{M}\sin(\tfrac{2i\pi}{P}\theta_1) + \tfrac{C_i}{M}\cos(\tfrac{2i\pi}{P}\theta_1)\right) \\ &\quad - \tfrac{A_f}{M}\arctan(\beta\theta_2) + \tfrac{F_{\text{dis}}}{M} + \tfrac{A}{M}u.\end{aligned} \tag{6.7}$$

For notional brevity, we define

$$\gamma_1 := \tfrac{B}{M}, \gamma_4 := \tfrac{A_f}{M}, \gamma_5 := \tfrac{1}{M}, \rho := \tfrac{A}{M}, \omega := \tfrac{2\pi}{P},$$

and write

$$\gamma_{2,i} := \tfrac{S_i}{M}, \gamma_{3,i} := \tfrac{C_i}{M}$$

for $i \in \{1, \ldots, n\}$, then the dynamics of the DC motor system can be rewritten as

$$\begin{aligned}\dot{\theta}_1(t) &= \theta_2(t), \\ \dot{\theta}_2(t) &= f(\theta(t)) + w(t) + \rho u(t),\end{aligned} \tag{6.8}$$

where the system function $f(\theta)$ and disturbance $w(t)$ are defined as

$$f(\theta) := -\gamma_1\theta_2 + \textstyle\sum_{i=1}^n \left(\gamma_{2,i}\sin(i\omega\theta_1) + \gamma_{3,i}\cos(i\omega\theta_1)\right) - \gamma_4\arctan(\beta\theta_2), \tag{6.9}$$

$$w(t) := \gamma_5 F_{\text{dis}}. \tag{6.10}$$

Through simple computation we have

$$\begin{aligned}\tfrac{\partial f}{\partial\theta_1} &= -\omega\textstyle\sum_{i=1}^n i\gamma_{3,i}\sin(i\omega\theta_1) + \omega\sum_{i=1}^n i\gamma_{2,i}\cos(i\omega\theta_1), \\ \tfrac{\partial f}{\partial\theta_2} &= -\gamma_1 - \tfrac{\gamma_4\beta}{1+\beta^2\theta_2^2}.\end{aligned} \tag{6.11}$$

If we choose b_0 as

$$b_0 := \max\{\omega\textstyle\sum_{i=1}^n(|i\gamma_2| + |i\gamma_3|), \gamma_1 + \gamma_4\beta\}, i \in \{1, 2\},$$

then the system function $f(\theta)$ satisfies

$$\left|\tfrac{\partial f}{\partial\theta_j}\right| \leq b_0, \quad j \in \{1, 2\}. \tag{6.12}$$

As the parameters of the system function $f(\theta)$ cannot be exactly known, we treat $f(\theta)$ as the uncertainties of the system. Equations (6.11)–(6.12) guarantee that the uncertainties of the motor system satisfy the boundedness assumption, i.e., $\left|\frac{\partial f}{\partial \theta_j}\right|$ is bounded. Moreover, considering the saturation phenomenon in practical applications, there exist some constants $C_w > 0$ and $C_\theta > 0$ such that

$$|w| + |\dot{w}| \leq C_w, \quad |\theta_2| \leq C_\theta. \tag{6.13}$$

In this chapter, we consider the scenario that the reference trajectory $\theta_1^*(t)$ of the DC torque motor satisfies

$$\begin{aligned} \dot{\theta}_1^*(t) &= \theta_2^*(t); \\ \dot{\theta}_2^*(t) &= r(t), \end{aligned} \tag{6.14}$$

where $|\dot{r}(t)| \leq C_r$ holds for some $C_r > 0$, and $|r(t)| \leq C_{r1}$ holds for some $C_{r1} > 0$. In order to investigate the tracking performance of the DC motor system in (6.8), we define the tracking error as

$$x(t) := \theta(t) - \theta^*(t), \tag{6.15}$$

where $x := [x_1, x_2]^T, \theta := [\theta_1, \theta_2]^T, \theta^* := [\theta_1^*, \theta_2^*]^T$, then the dynamics of the tracking error x_i has the following form:

$$\begin{aligned} \dot{x}_1(t) &= x_2(t); \\ \dot{x}_2(t) &= f(\theta(t)) - r(t) + w(t) + \rho u(t). \end{aligned} \tag{6.16}$$

In this chapter, our goal is to guarantee the boundedness of the tracking error. The external disturbances together with internal uncertainties, however, may deteriorate the tracking performance of the system. To estimate and compensate these disturbances, we define an extended state according to the standard ESO formulation [12, 13] as

$$x_3(t) := f(\theta(t)) - r(t) + w(t), \tag{6.17}$$

where x_3 represents the uncertainties and disturbances of the system. Based on this extended state, we introduce an event-triggered ESO of the following form:

$$\begin{aligned} \dot{\hat{x}}_1(t) &= \hat{x}_2(t) + \varepsilon g_1\left(\tfrac{\xi(t)-\hat{x}_1(t)}{\varepsilon^2}\right), \hat{x}_1(t_0) = \hat{x}_{10}, \\ \dot{\hat{x}}_2(t) &= \hat{x}_3(t) + g_2\left(\tfrac{\xi(t)-\hat{x}_1(t)}{\varepsilon^2}\right) + \rho u(t), \hat{x}_2(t_0) = \hat{x}_{20}, \\ \dot{\hat{x}}_3(t) &= \tfrac{1}{\varepsilon} g_3\left(\tfrac{\xi(t)-\hat{x}_1(t)}{\varepsilon^2}\right), \hat{x}_3(t_0) = \hat{x}_{30}, \end{aligned} \tag{6.18}$$

where $\hat{x} := [\hat{x}_1, \hat{x}_2, \hat{x}_3]^T \in \mathbb{R}^3$ is the observer state, ε is a high-gain parameter, $\hat{x}(t_0) := [\hat{x}_{10}, \hat{x}_{20}, \hat{x}_{30}]^T \in \mathbb{R}^3$ is the initial value and $\xi(t)$ is the output measurement given by

$$\xi(t) = \begin{cases} x_1(t_k), \text{ if } \Gamma(t) = 0, \\ x_1(t), \text{ otherwise,} \end{cases} \tag{6.19}$$

where $\Gamma(t)$ is the event-triggering condition which will be presented in the next section and t_k is the transmission instant determined by the event-triggering condition. Only when $\Gamma(t) = 1$ will the value of $\xi(t)$ be updated. In this observer scheme, $g_i(\cdot)$ are chosen as

$$g_1(z_1) = a_1 z_1 + \varphi(z_1), \quad g_2(z_1) = a_2 z_1, \quad g_3(z_1) = a_3 z_1,$$

where a_1, a_2, a_3 are some constants such that the matrix J

$$J := \begin{bmatrix} -a_1 & 1 & 0 \\ -a_2 & 0 & 1 \\ -a_3 & 0 & 0 \end{bmatrix} \tag{6.20}$$

is Hurwitz and $\varphi(\theta)$ is defined by

$$\varphi(\theta) = \begin{cases} -\frac{1}{4}, & \theta < -\frac{\pi}{2}, \\ \frac{1}{4}\sin\theta, & -\frac{\pi}{2} \le \theta \le \frac{\pi}{2}, \\ \frac{1}{4}, & \theta > \frac{\pi}{2}. \end{cases} \tag{6.21}$$

Define a function $P : \mathbb{R}^3 \to \mathbb{R}$ as

$$P(z) = \langle \tilde{P} z, z \rangle + \int_0^{z_1} \varphi(s) ds, \tag{6.22}$$

where the matrix $\tilde{P} > 0$ satisfies

$$\tilde{P} J + J^{\mathrm{T}} \tilde{P} = -I. \tag{6.23}$$

By choosing the functions g_1, g_2, g_3 and φ in the above form, we are able to identify a nonnegative-definite function $\Theta : \mathbb{R}^3 \to \mathbb{R}$ such that

$$\lambda_1 \|z\|^2 \le P(z) \le \lambda_2 \|z\|^2, \tag{6.24}$$

$$\lambda_3 \|z\|^2 \le \Theta(z) \le \lambda_4 \|z\|^2, \tag{6.25}$$

$$\sum_{j=1}^{2} \frac{\partial P}{\partial z_j}(z_{j+1} - g_j(z_1)) - \frac{\partial P}{\partial x_3} g_3(z_1) \le -\Theta(z), \tag{6.26}$$

$$\left\| \frac{\partial P}{\partial z} \right\| \le \tau_1 \|z\|, \tag{6.27}$$

where λ_1, λ_2, λ_3, λ_4, and τ_1 are some positive constants. The choice of $\varphi(\theta)$ is not unique, so long as it satisfies the requirements above. Note that the form of $\varphi(\theta)$ defined in (18) basically acts as a smooth approximation of the saturation nonlinearity. In this chapter, we design the control signal u as

$$u = -\tfrac{k_1}{\rho}\hat{x}_1 - \tfrac{k_2}{\rho}\hat{x}_2 - \tfrac{1}{\rho}\hat{x}_3, \tag{6.28}$$

where k_j are some constants such that the matrix H

$$H := \begin{bmatrix} 0 & 1 \\ -k_1 & -k_2 \end{bmatrix} \tag{6.29}$$

is Hurwitz. Obviously, there exists a matrix $\tilde{Q} > 0$ such that

$$\tilde{Q}H + H^T\tilde{Q} = -I. \tag{6.30}$$

Then we define $Q(z)$ as

$$Q(z) := z^T\tilde{Q}z, \tag{6.31}$$

and we have

$$\begin{aligned} &\varsigma_1\|z\|^2 \le Q(z) \le \varsigma_2\|z\|^2, \\ &\|\tfrac{\partial Q}{\partial z}\| \le \tau_2\|z\|, \end{aligned} \tag{6.32}$$

for some positive constants ς_1, ς_2 and τ_2 (e.g., we can choose $\varsigma_1 = \lambda_{\min}(\tilde{Q})$, $\varsigma_2 = \lambda_{\max}(\tilde{Q})$ and $\tau_2 = 2\lambda_{\max}(\tilde{Q})$). To describe the performance of the observer proposed in (6.18), we can define $\tilde{x}_j$ and e_j as

$$\begin{aligned} &\tilde{x}_j := x_j - \hat{x}_j, \\ &e_j := \tfrac{\tilde{x}_j}{\varepsilon^{3-j}}, \ \ j \in \{1, 2, 3\}. \end{aligned} \tag{6.33}$$

Then we define the sampling error $\sigma(t)$ as

$$\sigma(t) := \tfrac{x_1(t_k) - x_1(t)}{\varepsilon^2}, \tag{6.34}$$

where $t \in [t_k, t_{k+1})$, and for notional brevity, we define $\alpha_j(e_1, \sigma)$ as

$$\alpha_j(e_1, \sigma) := g_j(e_1 + \sigma) - g_j(e_1), \ \ j \in \{1, 2, 3\}. \tag{6.35}$$

In what follows, we mainly discuss the following problems:

1. Considering the event-triggered ESO and the DC motor system, can we propose an event-triggering condition to guarantee the boundedness of the observation error?
2. Based on the proposed event-triggering condition, can we control the system to track a desired trajectory, and guarantee the boundedness of the tracking error?

6.3 Event-Triggered ADRC Design

In this section, we show that the problems stated in the previous section can be solved by introducing an event-triggering condition that guarantees the performance of state observation and closed-loop control.

From (6.28), the control signal u defined above can be rewritten as

$$\begin{aligned} u &= -\tfrac{k_1}{\rho}x_1 - \tfrac{k_2}{\rho}x_2 - \tfrac{1}{\rho}\hat{x}_3 + \tfrac{k_1}{\rho}(x_1 - \hat{x}_1) + \tfrac{k_2}{\rho}(x_2 - \hat{x}_2) \\ &= -\textstyle\sum_{j=1}^{2} \tfrac{k_j}{\rho}x_j + \sum_{j=1}^{2} \tfrac{k_j}{\rho}\varepsilon^{3-j}e_j - \tfrac{1}{\rho}\hat{x}_3. \end{aligned} \tag{6.36}$$

For notational brevity, we define K as

$$K := \max\{k_j, k_j\varepsilon^{3-j}, 1\}, \ \ j \in \{1, 2\}. \tag{6.37}$$

Then we are ready to present the following theorem.

Theorem 6.1 *Consider the closed-loop system in (6.8), the ESO in (6.18) and the control law in (6.28). For a given $\Psi > 0$ and any initial values of x and $\hat{x}$, there exist $\varepsilon^* > 0$ and an event-triggering condition*

$$\Gamma(t) = \begin{cases} 0, \text{ if } \sum_{j=1}^{3} |\alpha_j(e_1, \sigma)| \le \Psi\varepsilon, \\ 1, \text{ otherwise}, \end{cases} \tag{6.38}$$

such that for any $\varepsilon \in (0, \varepsilon^*)$, it holds that

$$\limsup_{t\to\infty} |x_j - \hat{x}_j| \le \varepsilon^{3-j}\sqrt{\tfrac{\tau_1\varepsilon^{\frac{1}{3}}}{2\lambda_1\gamma}(F_0 + \Psi)}, \ \ j \in \{1, 2, 3\}, \tag{6.39}$$

$$\limsup_{t\to\infty} \|x\| \le \sqrt{\tfrac{\tau_1\varepsilon^{\frac{1}{3}}}{2\varsigma_1\gamma}(F_0 + \Psi)}. \tag{6.40}$$

Proof For the closed-loop system composed of (6.8), (6.18) and (6.28), considering the definition of the observation error in (6.33), we have

$$
\begin{aligned}
\dot{\tilde{x}}_j(t) &= \tilde{x}_{j+1}(t) - \varepsilon^{2-j} g_j\left(\tfrac{\xi(t)-\hat{x}_1(t)}{\varepsilon^2}\right) \qquad j \in \{1,2\}\\
&= \tilde{x}_{j+1}(t) - \varepsilon^{2-j} g_j\left(\tfrac{x_1(t)-\hat{x}_1(t)}{\varepsilon^2} + \tfrac{\xi(t)-x_1(t)}{\varepsilon^2}\right)\\
&= \tilde{x}_{j+1}(t) - \varepsilon^{2-j} g_j(e_1(t)) - \varepsilon^{2-j}\alpha_j(e_1(t),\sigma(t)),\\
\dot{\tilde{x}}_3(t) &= \tfrac{d}{dt}(f(\theta(t)) - r(t) + w(t)) - \tfrac{1}{\varepsilon} g_3\left(\tfrac{\xi(t)-\hat{x}_1(t)}{\varepsilon^2}\right)\\
&= \tfrac{d}{dt}(f(\theta(t)) - r(t) + w(t)) - \tfrac{g_3(e_1(t))}{\varepsilon} - \tfrac{\alpha_3(e_1(t),\sigma(t))}{\varepsilon}.
\end{aligned}
$$

Then, through simple computations, we obtain the dynamic equation of e_i in the following form:

$$
\begin{aligned}
\dot{e}_j(t) &= \tfrac{1}{\varepsilon}(e_{j+1}(t) - g_j(e_1(t)) - \alpha_j(e_1(t),\sigma(t))),\ j \in \{1,2\},\\
\dot{e}_3(t) &= \dot{x}_3(t) - \tfrac{1}{\varepsilon} g_3(e_1(t)) - \tfrac{1}{\varepsilon}\alpha_3(e_1(t),\sigma(t)).
\end{aligned} \tag{6.41}
$$

Define a nonnegative-definite function $V(x,e)$ as

$$
V(x,e) := Q(x) + P(e), \tag{6.42}
$$

where $Q(x)$ and $P(e)$ are provided in (6.22) and (6.31), respectively. For this function, we observe that

$$
\begin{aligned}
\tfrac{dV}{dt} &= \tfrac{dQ}{dt} + \tfrac{dP}{dt}\\
&= \tfrac{\partial Q}{\partial x_1} x_2 + \tfrac{\partial Q}{\partial x_2}(x_3 + \rho u) + \textstyle\sum_{j=1}^{3} \tfrac{\partial P}{\partial e_j}\dot{e}_j.
\end{aligned} \tag{6.43}
$$

From the dynamic of e_i in (6.41), we have

$$
\begin{aligned}
\tfrac{dV}{dt} &= \tfrac{\partial Q}{\partial x_1} x_2 + \tfrac{\partial Q}{\partial x_2}(x_3 - \textstyle\sum_{j=1}^{2} k_j x_j + \sum_{j=1}^{2} k_j \varepsilon^{3-j} e_j\\
&\quad - \hat{x}_3) + \textstyle\sum_{j=1}^{2} \tfrac{\partial P}{\partial e_j}[\tfrac{1}{\varepsilon}(e_{j+1} - g_j(e_1) - \alpha_j(e_1,\sigma))]\\
&\quad + \tfrac{\partial P}{\partial e_3}\left[\dot{x}_3 - \tfrac{1}{\varepsilon} g_3(e_1) - \tfrac{1}{\varepsilon}\alpha_3(e_1,\sigma)\right]\\
&\le \tfrac{\partial Q}{\partial x_1} x_2 - \tfrac{\partial Q}{\partial x_2}\textstyle\sum_{j=1}^{2} k_j x_j + \tfrac{\partial Q}{\partial x_2}(\sum_{j=1}^{2} k_j \varepsilon^{3-j} e_j + e_3)\\
&\quad + \tfrac{1}{\varepsilon}\left[\textstyle\sum_{j=1}^{2} \tfrac{\partial P}{\partial e_j}(e_{j+1} - g_j(e_1)) - \tfrac{\partial P}{\partial e_3} g_3(e_1)\right]\\
&\quad + \left|\tfrac{\partial P}{\partial e_3}\right| \cdot |\dot{x}_3| - \tfrac{1}{\varepsilon}\textstyle\sum_{j=1}^{3} \tfrac{\partial P}{\partial e_j}\alpha_j(e_1,\sigma).
\end{aligned} \tag{6.44}
$$

According to the definitions of $Q(\cdot)$ in (6.31), we observe

$$
\begin{aligned}
&\tfrac{\partial Q}{\partial x_1} x_2 - \tfrac{\partial Q}{\partial x_2}\textstyle\sum_{j=1}^{2} k_j x_j = 2x^T \tilde{Q} H x\\
&= x^T(\tilde{Q}H + H^T\tilde{Q})x = -\|x\|^2.
\end{aligned} \tag{6.45}
$$

Then, considering the function $P(\cdot)$, we have

$$\frac{1}{\varepsilon}[\sum_{j=1}^{2}\frac{\partial P}{\partial e_j}(e_{j+1}-g_j(e_1))-\frac{\partial P}{\partial e_3}g_3(e_1)]\leq-\frac{1}{\varepsilon}\Theta(e)\leq-\frac{\lambda_3}{\varepsilon}\|e\|^2. \tag{6.46}$$

According to inequality (6.32), we obtain that

$$\begin{aligned}
&\frac{\partial Q}{\partial x_2}(\sum_{j=1}^{2}k_j\varepsilon^{3-j}e_j+e_3)\leq\tau_2\|x\|\left|\sum_{j=1}^{2}k_j\varepsilon^{3-j}e_j+e_3\right|\\
&=\frac{1}{\varepsilon}\tau_2\left[\varepsilon^{\frac{2}{3}}\|x\|\varepsilon^{\frac{1}{3}}(\sum_{j=1}^{2}k_j\varepsilon^{3-j}|e_j|+|e_3|)\right]\\
&\leq\frac{3}{2}\varepsilon^{\frac{1}{3}}\tau_2\|x\|^2+\frac{\tau_2}{2\varepsilon}(\sum_{j=1}^{2}k_j^2\varepsilon^{\frac{20}{3}-2j}+\varepsilon^{\frac{2}{3}})\|e\|^2, \qquad (6.47)\\
&\left|\frac{\partial P}{\partial e_3}\right||\dot{x}_3|\leq\tau_1\|e\|\cdot|\dot{x}_3|\\
&=\frac{1}{\varepsilon}\tau_1\varepsilon^{\frac{1}{3}}\|e\|\varepsilon^{\frac{2}{3}}|\dot{x}_3|\leq\frac{1}{2\varepsilon}\tau_1(\varepsilon^{\frac{2}{3}}\|e\|^2+\varepsilon^{\frac{4}{3}}|\dot{x}_3|^2). \qquad (6.48)
\end{aligned}$$

Using the control signal u provided in (6.28), we further obtain

$$\begin{aligned}
&|\dot{x}_3|^2=|\dot{f}(\theta)-\dot{r}+\dot{w}|^2\\
&=|\frac{\partial f}{\partial\theta_1}\theta_2+\dot{w}-\dot{r}\\
&\quad+\frac{\partial f}{\partial\theta_2}(e_3-\sum_{j=1}^{2}k_jx_j+\sum_{j=1}^{2}k_j\varepsilon^{3-j}e_j+r)|^2\\
&\leq|b_0C_\theta+b_0(2K\|e\|+2K\|x\|+\|e\|+C_{r1})+C_w+C_r|^2\\
&\leq(A\|x\|+B\|e\|+C)^2, \qquad (6.49)
\end{aligned}$$

where A, B and C are defined as $A:=2b_0K$, $B:=b_0(2K+1)$, $C:=b_0(C_\theta+C_{r1})+C_r+C_w$. Then, define $F_0:=C^2+AC+BC$, $F_1:=A^2+AB+AC$, $F_2:=B^2+AB+BC$, such that

$$|\dot{x}_3|^2\leq F_0+F_1\|x\|^2+F_2\|e\|^2. \tag{6.50}$$

From the inequalities (6.48)–(6.50), we have

$$\begin{aligned}
\left|\frac{\partial P}{\partial e_3}\right||\dot{x}_3|&\leq\frac{\tau_1}{2\varepsilon}[\varepsilon^{\frac{2}{3}}\|e\|^2+\varepsilon^{\frac{4}{3}}(F_0+F_1\|x\|^2+F_2\|e\|^2)]\\
&\leq\frac{\tau_1\varepsilon^{-\frac{1}{3}}\|e\|^2}{2}+\frac{\tau_1\varepsilon^{\frac{1}{3}}F_0}{2}+\frac{\tau_1\varepsilon^{\frac{1}{3}}F_1\|x\|^2}{2}+\frac{\tau_1\varepsilon^{\frac{1}{3}}F_2\|e\|^2}{2}\\
&\leq\left(\frac{\tau_1\varepsilon^{-\frac{1}{3}}}{2}+\frac{\tau_1\varepsilon^{\frac{1}{3}}F_2}{2}\right)\|e\|^2+\frac{\tau_1\varepsilon^{\frac{1}{3}}F_1}{2}\|x\|^2+\frac{\tau_1\varepsilon^{\frac{1}{3}}F_0}{2}. \qquad (6.51)
\end{aligned}$$

According to $\|\frac{\partial P}{\partial z}\|\leq\tau_1\|z\|$ and the event-triggering condition (6.38), we obtain that

$$\begin{aligned}
&-\frac{1}{\varepsilon}\sum_{j=1}^{3}\frac{\partial P}{\partial e_j}\sigma_j(e_1,\sigma)\leq|\frac{1}{\varepsilon}\sum_{j=1}^{3}\frac{\partial P}{\partial e_j}\sigma_j(e_1,\sigma)|\\
&\leq\frac{1}{\varepsilon}\tau_1\|e\|\Psi\varepsilon\leq\left(\frac{\varepsilon^{-\frac{1}{3}}\|e\|^2}{2}+\frac{\varepsilon^{\frac{1}{3}}}{2}\right)\Psi\tau_1. \qquad (6.52)
\end{aligned}$$

Combining the above results and according to (6.44)–(6.47), (6.51) and (6.52), we have

$$
\begin{aligned}
\frac{dV}{dt} &\le -\|x\|^2 - \frac{\lambda_3}{\varepsilon}\|e\|^2 + \frac{3\tau_2}{2}\varepsilon^{\frac{1}{3}}\|x\|^2 + \Psi\tau_1\left(\frac{\varepsilon^{-\frac{1}{3}}\|e\|^2}{2} + \frac{\varepsilon^{\frac{1}{3}}}{2}\right) \\
&\quad + \frac{\tau_1\varepsilon^{\frac{1}{3}}F_0}{2} + \frac{\tau_1\varepsilon^{\frac{1}{3}}F_1}{2}\|x\|^2 + \left(\frac{\tau_1\varepsilon^{-\frac{1}{3}}}{2} + \frac{\tau_1\varepsilon^{\frac{1}{3}}F_2}{2}\right)\|e\|^2 \\
&\quad + \frac{\tau_2}{2\varepsilon}(\textstyle\sum_{j=1}^{2} k_j^2\varepsilon^{\frac{20}{3}-2j} + \varepsilon^{\frac{2}{3}})\|e\|^2 \\
&\le -\left(1 - \frac{3\tau_2}{2}\varepsilon^{\frac{1}{3}} - \frac{\tau_1}{2}\varepsilon^{\frac{1}{3}}F_1\right)\|x\|^2 + \frac{\tau_1\varepsilon^{\frac{1}{3}}F_0}{2} + \frac{\Psi\tau_1\varepsilon^{\frac{1}{3}}}{2} \\
&\quad - \frac{1}{\varepsilon}[\lambda_3 - \frac{\tau_2}{2}(\textstyle\sum_{j=1}^{2} k_j^2\varepsilon^{\frac{20}{3}-2j} + \varepsilon^{\frac{2}{3}}) - \left(\frac{\tau_1}{2}\varepsilon^{\frac{2}{3}}\right. \\
&\quad \left. + \frac{\tau_1}{2}\varepsilon^{\frac{4}{3}}F_2\right) - \frac{\varepsilon^{\frac{2}{3}}}{2}\Psi\tau_1]\|e\|^2.
\end{aligned} \tag{6.53}
$$

Then, write

$$
\begin{aligned}
\hbar_1 &:= \lambda_3 - \frac{\tau_2}{2}(\textstyle\sum_{j=1}^{2} k_j^2\varepsilon^{\frac{20}{3}-2j} + \varepsilon^{\frac{2}{3}}) \\
&\quad - \frac{1}{2}(\Psi\tau_1\varepsilon^{\frac{2}{3}} + \tau_1\varepsilon^{\frac{2}{3}} + \tau_1\varepsilon^{\frac{4}{3}}F_2), \\
\hbar_2 &:= 1 - \frac{3\tau_2}{2}\varepsilon^{\frac{1}{3}} - \frac{\tau_1}{2}\varepsilon^{\frac{1}{3}}F_1,
\end{aligned} \tag{6.54}
$$

and define ε^* as

$$
\varepsilon^* = \max\{\varepsilon \in \mathbb{R} | \hbar_1(\varepsilon) \ge 0, \hbar_2(\varepsilon) \ge 0, \text{ and } \varepsilon > 0\}. \tag{6.55}
$$

Note that $\hbar_1(\varepsilon)$ and $\hbar_2(\varepsilon)$ are monotonically decreasing functions of ε, that $\lim_{\varepsilon\to 0}\hbar_1(\varepsilon) > 0$, $\lim_{\varepsilon\to 0}\hbar_2(\varepsilon) > 0$, and that $\hbar_1(\varepsilon) < 0$, $\hbar_2(\varepsilon) < 0$ when ε becomes sufficiently large. Thus ε^* is well-defined. For $\varepsilon \in (0, \varepsilon^*)$, from inequality (6.53), we obtain

$$
\begin{aligned}
\frac{dV}{dt} &\le -\hbar_2\|x\|^2 - \frac{\hbar_1}{\varepsilon}\|e\|^2 + \frac{\tau_1}{2}\varepsilon^{\frac{1}{3}}F_0 + \frac{\Psi\tau_1}{2}\varepsilon^{\frac{1}{3}} \\
&\le -\frac{\hbar_2}{\varsigma_2}Q(x) - \frac{\hbar_1}{\varepsilon\lambda_2}P(e) + \frac{\tau_1}{2}\varepsilon^{\frac{1}{3}}F_0 + \frac{\Psi\tau_1}{2}\varepsilon^{\frac{1}{3}}.
\end{aligned} \tag{6.56}
$$

Then, we further define γ and $\delta(t, t_0)$ as

$$
\gamma = \min\left\{\frac{\hbar_2}{\varsigma_2}, \frac{\hbar_1}{\varepsilon\lambda_2}\right\}, \quad \delta(t, t_0) = \exp(-\gamma(t - t_0)).
$$

With these definitions, $\frac{dV(x,e)}{dt}$ can be rewritten as

$$
\frac{dV(x,e)}{dt} \le -\gamma V(x, e) + \frac{\tau_1}{2}\varepsilon^{\frac{1}{3}}F_0 + \frac{\Psi\tau_1}{2}\varepsilon^{\frac{1}{3}}. \tag{6.57}
$$

Then, we obtain that

$$
\begin{aligned}
& V(x, e) \\
\leq & V(x(t_0), e(t_0))\delta(t, t_0) + \left(\frac{\tau_1 \varepsilon^{\frac{1}{3}} F_0}{2} + \frac{\Psi \tau_1 \varepsilon^{\frac{1}{3}}}{2}\right) \int_{t_0}^{t} \delta(t, \vartheta) d\vartheta \\
\leq & V(x(t_0), e(t_0))\delta(t, t_0) + \left(\frac{\tau_1 \varepsilon^{\frac{1}{3}} F_0}{2\gamma} + \frac{\Psi \tau_1 \varepsilon^{\frac{1}{3}}}{2\gamma}\right) (1 - \delta(t, t_0)).
\end{aligned}
$$

When $t \to \infty$, $\delta(t, t_0) \to 0$, and we have

$$
V(x, e) \leq \frac{\tau_1}{2\gamma} \varepsilon^{\frac{1}{3}} F_0 + \frac{\Psi \tau_1 \varepsilon^{\frac{1}{3}}}{2\gamma}, \tag{6.58}
$$

hence

$$
P(e) + Q(x) \leq \frac{\tau_1}{2\gamma} \varepsilon^{\frac{1}{3}} F_0 + \frac{\Psi \tau_1 \varepsilon^{\frac{1}{3}}}{2\gamma}. \tag{6.59}
$$

From (6.32) and (6.59), we obtain that when $t \to \infty$

$$
\varsigma_1 \|x\|^2 + \lambda_1 \|e\|^2 \leq V \leq \frac{\tau_1}{2\gamma} \varepsilon^{\frac{1}{3}} F_0 + \frac{\Psi \tau_1 \varepsilon^{\frac{1}{3}}}{2\gamma}, \tag{6.60}
$$

then we have

$$
\|x\|^2 \leq \frac{\tau_1 \varepsilon^{\frac{1}{3}}}{2\varsigma_1 \gamma} F_0 + \frac{\Psi \tau_1 \varepsilon^{\frac{1}{3}}}{2\varsigma_1 \gamma}, \tag{6.61}
$$

and obtain

$$
\limsup_{t \to \infty} \|x\| \leq \sqrt{\frac{\tau_1 \varepsilon^{\frac{1}{3}}}{2\varsigma_1 \gamma} (F_0 + \Psi)}. \tag{6.62}
$$

Considering (6.60), we obtain that when $t \to \infty$

$$
\lambda_1 \|e\|^2 \leq \frac{\tau_1}{2\gamma} \varepsilon^{\frac{1}{3}} F_0 + \frac{\Psi \tau_1 \varepsilon^{\frac{1}{3}}}{2\gamma}. \tag{6.63}
$$

Therefore, for the observation error e, we are able to guarantee that

$$
\limsup_{t \to \infty} \|e\| \leq \sqrt{\frac{\tau_1 \varepsilon^{\frac{1}{3}}}{2\lambda_1 \gamma} (F_0 + \Psi)}, \tag{6.64}
$$

and from the definition of e in (6.33), we obtain for $j \in \{1, 2, 3\}$,

$$
\limsup_{t \to \infty} |x_j - \hat{x}_j| \leq \varepsilon^{3-j} \sqrt{\frac{\tau_1 \varepsilon^{\frac{1}{3}}}{2\lambda_1 \gamma} (F_0 + \Psi)}. \tag{6.65}
$$

This completes the proof. □

Based on this result, for the torque motor system in (6.8), the ESO in (6.18) and the control signal in (6.28), the event-triggering mechanism proposed in (6.38) guarantees the asymptotic boundedness of the observation error and the tracking error. Specifically, Eq. (6.62) guarantees the boundedness of the steady-state tracking error, and Eq. (6.65) guarantees the boundedness of the steady-state observation error. The parameterization of the functions g_i, $i = 1, 2, 3$ and $\varphi(\theta)$ chosen in Sect. 6.2 satisfy the requirements in (6.23)–(6.27), which helps ensure the input-to-state stability of the ESO. The function $\varphi(\theta)$ is the nonlinear part of the proposed ESO. If we choose $\varphi(\theta) = 0$, the ESO reduces to a linear one. With the proposed functions, we can infer from (6.65) that the estimated state variables will converge to the real physical states at steady state for properly chosen values of ε and Ψ. The proposed event-triggering condition is based on the value of the sampling error $\sigma(t)$, and presents an upper bound of $|\sigma(t)|$. Once $|\sigma(t)|$ reaches the predefined bound, the value of $\xi(t)$ will be updated. From the event-triggering condition, the system performance can be adjusted by changing the value of the parameter Ψ; this point is further analytically explained in (6.40), where quantitative relationships of the observation and tracking performance with the tuning parameter Ψ in the event-triggering condition are provided.

By assumption, the extended state x_3 in Eq. (6.17) is not directly known; however, it evolves in a unique (but unknown) way determined by the realization of the external disturbance signal and the values of the unknown parameters. From Eqs. (6.16)–(6.17) and the definition of $\xi(t)$ in (6.19), the information of $x_3(t)$ is reflected in the event-triggered output measurement $\xi(t)$, and $\hat{x}_3(t)$ in the extended state observer tries to learn the fixed but unknown behavior of $x_3(t)$ by exploring this information. In other words, taking advantage of the knowledge of $\xi(t)$, the "specified way" in which $\hat{x}_3$ evolves mimics that of $x_3(t)$. In particular, Eq. (6.39) of Theorem 6.1 indicates that the difference between $\hat{x}_3(t)$ and $x_3(t)$ can be sufficiently small as $t \to \infty$ for $\varepsilon \in (0, \varepsilon^*)$.

6.4 Experimental Performance Evaluation

In this section, the introduced event-triggered control scheme is evaluated on a DC torque motor platform (Fig. 6.2). This platform is used for illustration purposes in earlier chapters but is also briefly introduced here to make the chapter self-sustained. Specifically, the experimental platform mainly consists of a motor load test bench and an embedded measurement and control test bench. The motor load test bench contains a permanent-magnet DC torque motor, a magnetic brake, a speed reducer, an inertial loading mechanism, and a speed sensor. By adjusting the input current, the magnetic brake can simulate the adjustable sliding friction load. The embedded measurement and control test bench works as a controller, since it can generate signals to control the motor and process the received signals. In this section, we choose two different kinds of reference trajectory inputs, namely, a squarewave input and a multitone sinusoid input, to perform the experiments. Apart from the ET-ADRC

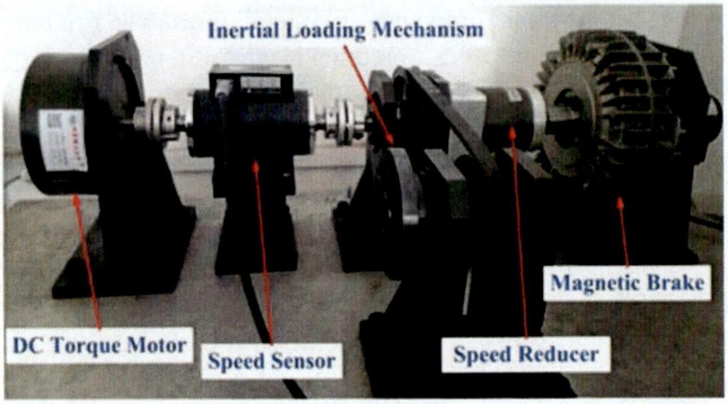

Fig. 6.2 The motor load test bench

controller, a time-triggered continuous-time ADRC (CT-ADRC) controller is also implemented for comparison purposes. In these experiments, the ADRC controllers are implemented by Euler approximation with the sampling time $t_s = 0.025$s. In order to evaluate the tracking performance of the system, we define the tracking error (E_T) and the average sampling time (T_A) as follows:

$$E_T = \frac{1}{T}\int_0^T |\theta_1^*(t) - \theta_1(t)|dt, \tag{6.66}$$

$$T_A = \begin{cases} 25, & \text{for CT-ADRC schemes,} \\ \frac{T}{N_s}, & \text{for ET-ADRC schemes,} \end{cases} \tag{6.67}$$

where we recall $\theta_1^*(t)$ represents the reference trajectory, $\theta_1(t)$ is the output of the controlled system, T represents the experimental time in milliseconds, and N_s is the total triggering counts in one experiment.

6.4.1 Experimental Results with Squarewave-Type Input

In this subsection, the tracking performance of the system with squarewave-type reference input is investigated. For the ESO in (6.18), g_i is designed as

$$g_1(z_1) = 3z_1 + \varphi(z_1), \quad g_2(z_1) = 3z_1, \quad g_3(z_1) = z_1, \tag{6.68}$$

and the high-gain parameter is chosen as $\varepsilon = 0.2$. The control law $\rho u_1(t)$ is designed as

$$\rho u_1(t) = -14\hat{x}_1 - \hat{x}_2 - \hat{x}_3, \tag{6.69}$$

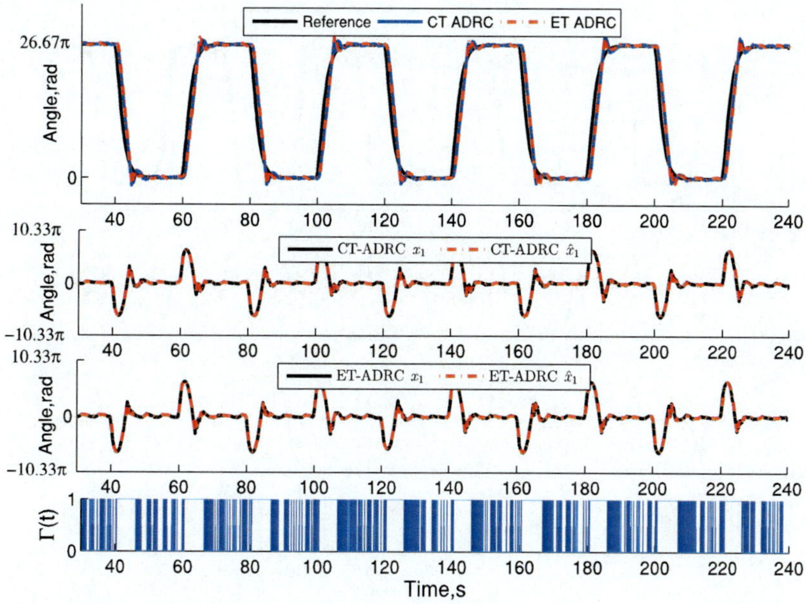

Fig. 6.3 No-load tracking performance with $\Psi = 0.35$

and the reference trajectory $\theta_1^*(t)$ is a squarewave signal processed by a low pass filter, so that $|r(t)|$ and $|\dot{r}(t)|$ are guaranteed to be bounded. Since the output of the system is controlled to track the processed signal, adding a low pass filter will not affect the accuracy of the estimated state variables. In addition, low pass filters are often adopted to process signals in many practical applications [15], since it can make the response smoother and can thus reduce overshoot.

Consider the closed-loop system and the event-triggering condition proposed in (6.38). We choose different values of Ψ so that we can obtain experimental results with different average communication rates. For comparison purpose, two groups of experiments are carried out. One group shows the no-load performance of the system, and the other group shows the performance of the system with the load of 0.7 Nm. The tracking performance of the system with no event-triggering condition (namely, CT-ADRC) is also evaluated for each group. These experimental results are shown in Figs. 6.3, 6.4, 6.5, 6.6. The tracking error (E_T) and the average sampling time (T_A) are summarized in Table 6.1.

From the experimental results, we observe that with the event-triggering control mechanism, the system can be controlled to track the desired trajectory with tracking error around 4. Based on the obtained tracking performance, we can infer that the uncertainties x_3 can be properly compensated in the feedback loop. Since the tracking performances of the ET-ADRC schemes are almost the same as the results of the time-triggered schemes, the CT-ADRC curves in the figures cannot be seen very clearly. The curves of x_1 and $\hat{x}_1$ partially show that the estimated states can converge

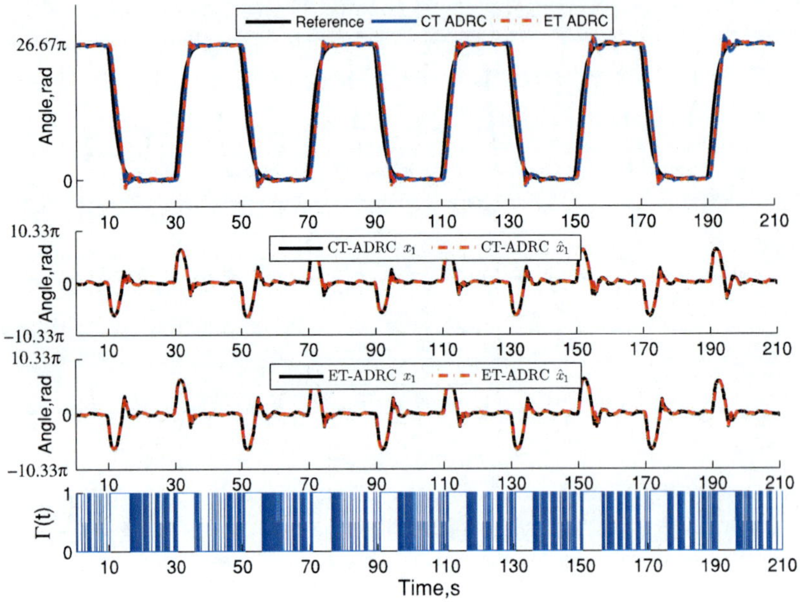

Fig. 6.4 Tracking performance with the load of 0.7 Nm and $\Psi = 0.35$

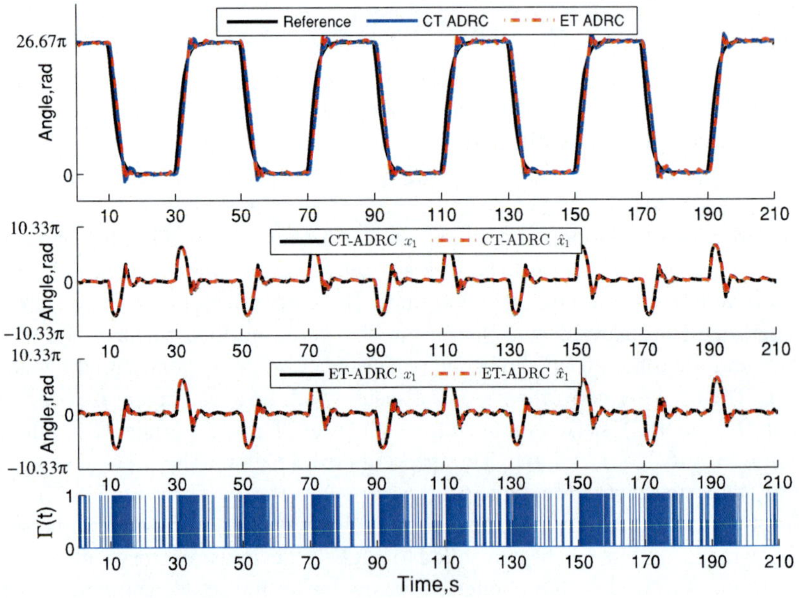

Fig. 6.5 No-load tracking performance with $\Psi = 1.4$

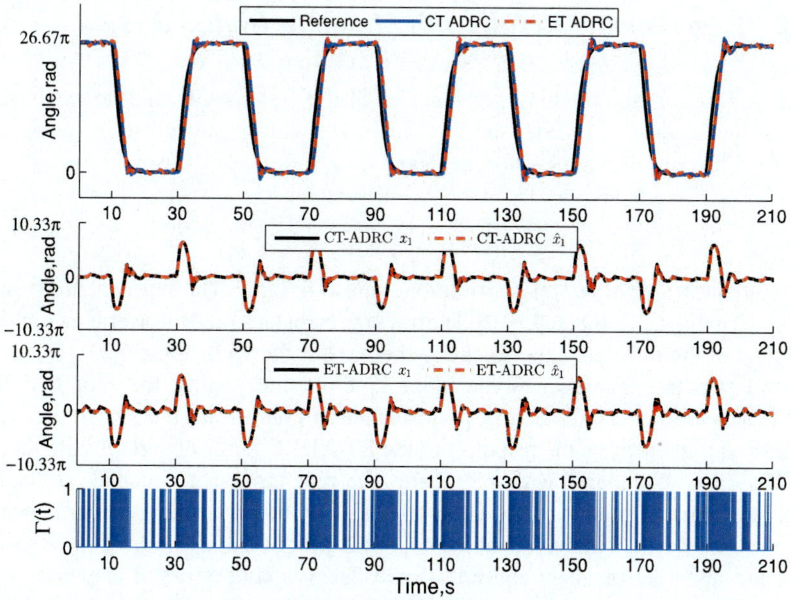

Fig. 6.6 Tracking performance with the load of 0.7 Nm and $\Psi = 1.4$

Table 6.1 Performance for squarewave-type input

Experimental schemes		Average sampling time (ms)	Tracking error
Time-triggered	No-load	25	4.0701
	Load	25	4.0400
Event-triggered	No-load	69	4.0607
$\Psi = 0.35$	Load	71.4	4.1641
Event-triggered	No-load	167.8	4.1756
$\Psi = 1.4$	Load	166.7	4.4246

to the real physical states. Compared with the performance of CT-ADRC schemes, the event-triggering approach has an obvious decrease in average sampling rate, and the proposed event-triggering schemes maintain a similar level of performance in terms of tracking error. Note that the tracking performance can be adjusted by changing the average sampling rate, which is implemented by tuning the triggering parameter Ψ. When Ψ is set to be larger, the average sampling rate will obviously become lower, but the tracking error will be more likely to be increased, which can be seen in Table 6.1. Although a lower average sampling rate is preferred in reducing resource consumption, it may cause negative effects on control performance.

6.4.2 Experimental Results with Multitone Sinusoid Input

In this subsection, the tracking performance of the system with multitone sinusoid input is investigated. Consider the ESO in (6.18) with g_i still given in (6.68) and $\varepsilon = 0.1$. The control law $\rho u_2(t)$ is designed as

$$\rho u_2(t) = -4\hat{x}_1 - \hat{x}_2 - \hat{x}_3. \tag{6.70}$$

The reference trajectory $\theta_1^*(t)$ is a multitone sinusoid signal. The experimental results are shown in Figs. 6.7, 6.8, 6.9, 6.10. The tracking error (E_T) and the average sampling time (T_A) of the different experimental schemes are shown in Table 6.2.

Similar to the squarewave-type input experimental results, the proposed ET-ADRC scheme has a satisfactory performance in controlling the system tracking the desired trajectory, as the response of the ET-ADRC almost overlaps with that of the CT-ADRC. Compared with the squarewave-type input experimental results, the multitone sinusoid input experimental results show better tracking performance in terms of smaller tracking errors at the cost of higher average sampling rates, as the reference signal keeps changing in this case. The tracking errors of responses with load are larger than that without load, and the effect of the load is more obvious compared with that for the squarewave-type input. This indicates that the event-triggering approach is more effective when the reference trajectory is smoother and steadier. Finally, we note that compared with the CT-ADRC schemes, the event-triggered

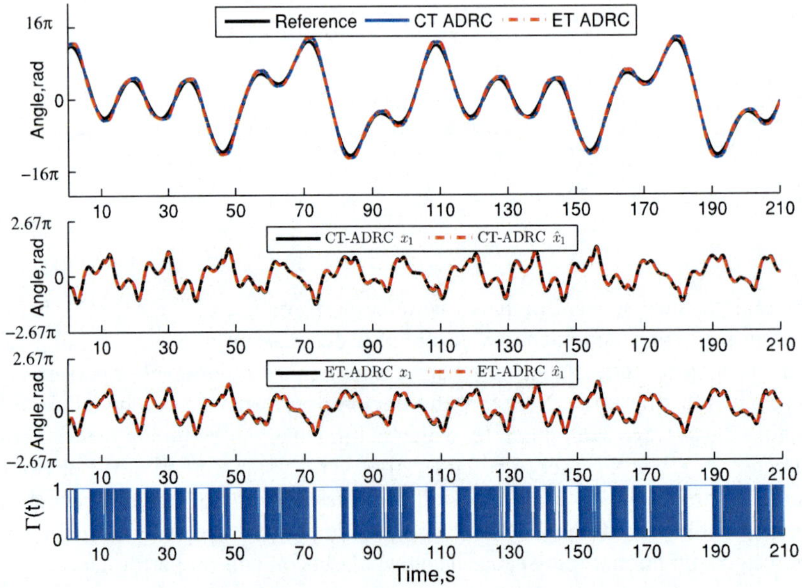

Fig. 6.7 No-load tracking performance with $\Psi = 0.35$

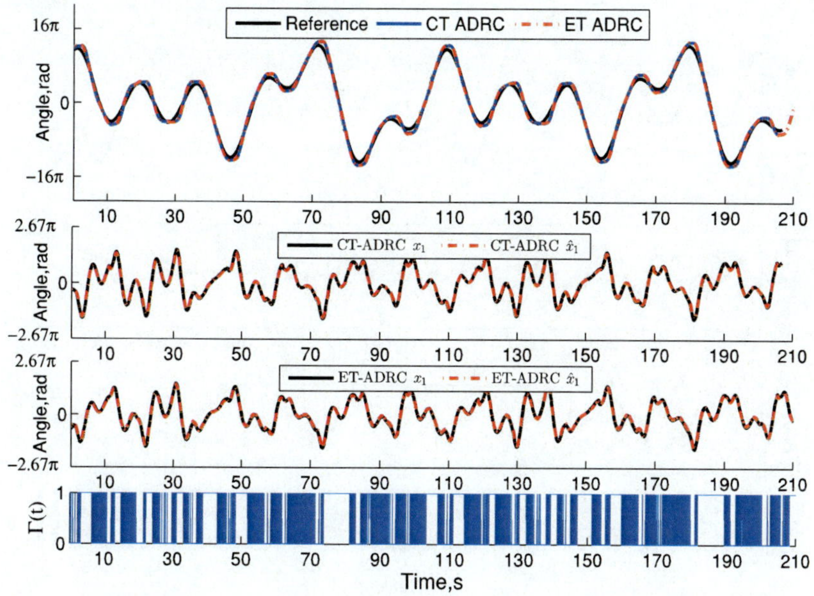

Fig. 6.8 Tracking performance with the load of 0.7 Nm and $\Psi = 0.35$

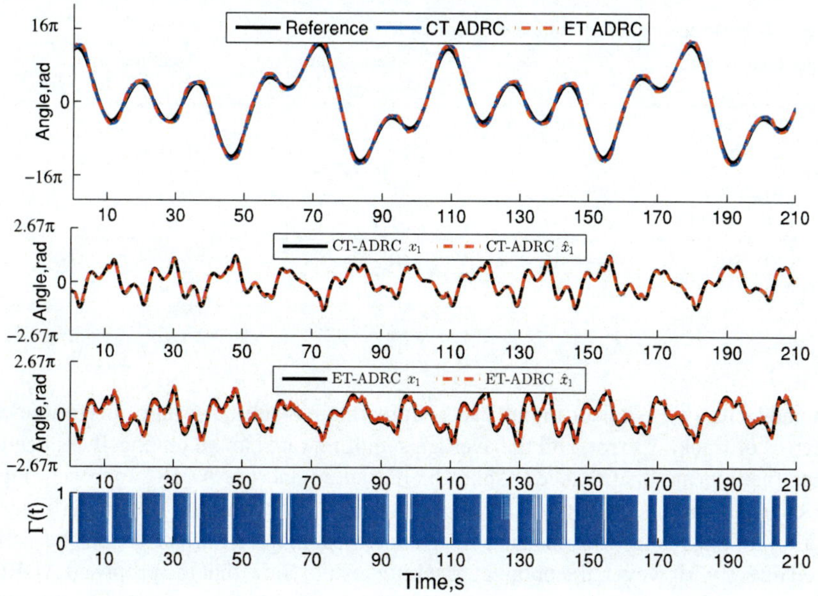

Fig. 6.9 No-load tracking performance with $\Psi = 1.4$

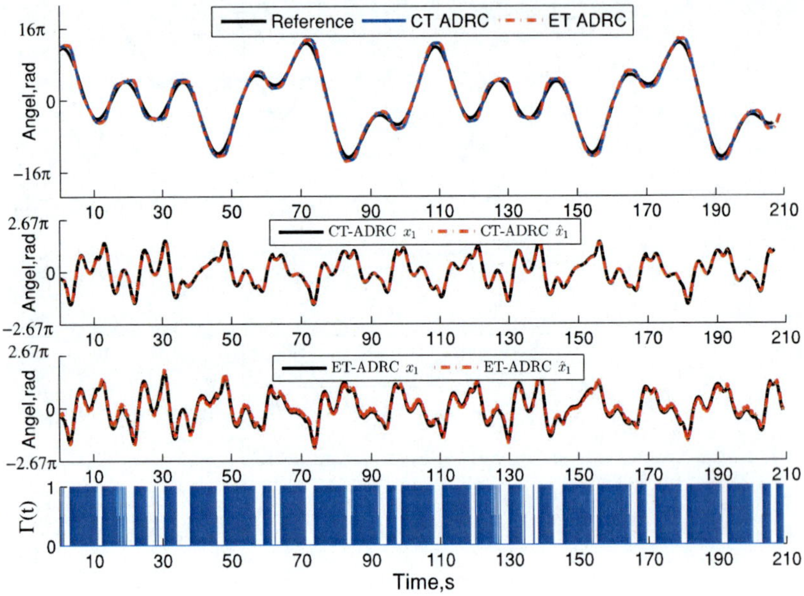

Fig. 6.10 Tracking performance with the load of 0.7 Nm and $\Psi = 1.4$

Table 6.2 Performance for multitone sinusoid input

Experimental schemes		Average sampling time (ms)	Tracking error
Time-triggered	No-load	25	1.5833
	Load	25	1.9406
Event-triggered	No-load	51.5	1.5820
$\Psi = 0.35$	Load	51.3	1.8471
Event-triggered	No-load	167.8	1.7291
$\Psi = 1.4$	Load	169.5	2.1466

schemes with two kinds of inputs can both maintain a similar level of performance in terms of tracking error, and the average sampling rate can be obviously reduced. Since the internal disturbances cannot be measured and the ADRC approach does not consider the uncertainty explicitly but takes all uncertainties and disturbances as a total disturbance, the internal interference suppression effect cannot be validated directly. However, the obtained tracking results show that the proposed ADRC schemes can achieve satisfactory performance in compensating for total disturbances, including internal uncertainty and external disturbances. Moreover, the CT-ADRC schemes are relatively less affected by external disturbances (in terms of sampling errors induced by event trigger) than ET-ADRC schemes, and therefore the fact that the ET-ADRC schemes can maintain a similar level of control performance (in terms of tracking error) to that of CT-ADRC schemes helps us indirectly validate the effect of the internal interference suppression for ET-ADRC schemes.

6.5 Summary

In this chapter, we show that the proposed event-triggered ADRC approach can be applied to control a DC torque motor. Through designing an event-triggering mechanism that only relies on the output of the controlled system, the effects of the event-triggering scheme are investigated both theoretically and experimentally. In particular, with the proposed ET-ADRC scheme, the observation error of the ESO and the tracking error of the system can be guaranteed to be asymptotically bounded. The actual performance of the ET-ADRC scheme is extensively evaluated through comparative experiments on a DC torque motor platform for different load conditions and reference signals.

References

1. An, H., Liu, J., Wang, C., Wu, L.: Approximate back-stepping fault-tolerant control of the flexible air-breathing hypersonic vehicle. IEEE/ASME Trans. Mechatron. **21**(3), 1680–1691 (2016)
2. Åström, K.J., Bo, B.: Comparison of periodic and event based sampling for first-order stochastic systems. In: Proceedings of IFAC World Congress (1999)
3. Chen, Z., Yao, B., Wang, Q.: Adaptive robust precision motion control of linear motors with integrated compensation of nonlinearities and bearing flexible modes. IEEE Trans. In. Inf. **9**(2), 965–973 (2013)
4. Chen, Z., Yao, B., Wang, Q.: Adaptive robust precision motion control of linear motors with integrated compensation of nonlinearities and bearing flexible modes. IEEE Trans. Ind. Inf. **9**(2), 965–973 (2013). https://doi.org/10.1109/TII.2012.2225439
5. Dong, H., Wang, Z., Shen, B., Ding, D.: Variance-constrained control for a class of nonlinear stochastic discrete time-varying systems: the event-triggered design. Automatica **72**, 28–36 (2016)
6. Du, B., Wu, S., Han, S., Cui, S.: Application of linear active disturbance rejection controller for sensorless control of internal permanent-magnet synchronous motor. IEEE Trans. Ind. Electron. **63**(5), 3019–3027 (2016)
7. Ghalehpardaz, S.L., Shafiee, M.: Speed control of DC motor using imperialist competitive algorithm based on PI-Like FLC. In: 2011 Third International Conference on Computational Intelligence, Modelling Simulation, pp. 28–33 (2011)
8. Gubara, W., Elnaim, M., Babiker, S.F.: Comparative study on the speed of DC motor using PID and FLC. In: 2016 Conference of Basic Sciences and Engineering Studies (SGCAC), pp. 24–29 (2016)
9. Guo, B., Zhao, Z.: On the convergence of an extended state observer for nonlinear systems with uncertainty. Syst. Control Lett. **60**(6), 420–430 (2011)
10. Guo, B., Zhao, Z.: On convergence of non-linear extended state observer for multi-input multi-output systems with uncertainty. IET Control Theory Appl. **6**(15), 2375–2386 (2012)
11. Guo, B.Z., Wu, Z.H., Zhou, H.C.: Active disturbance rejection control approach to output-feedback stabilization of a class of uncertain nonlinear systems subject to stochastic disturbance. IEEE Trans. Autom. Control **61**(6), 1613–1618 (2016)
12. Han, J.: The "extended state observer" of a class of uncertain systems. In: Control and Decision (1995)
13. Han, J.: From PID to active disturbance rejection control. IEEE Trans. Ind. Electron. **56**(3), 900–906 (2009)

14. He, N., Shi, D.: Event-based robust sampled-data model predictive control: A non-monotonic lyapunov function approach. IEEE Trans. Circuits Syst. I: Regul. Papers **62**(10), 2555–2564 (2015)
15. He, N., Shi, D., Forbes, M., Backström, J., Chen, T.: Robust tuning for machine-directional predictive control of MIMO paper-making processes. Control Eng. Pract. **55**, 1–12 (2016)
16. Heemels, W.P.M.H., Donkers, M.C.F.: Model-based periodic event-triggered control for linear systems. Automatica **49**(3), 698–711 (2013)
17. Jing, Z., Su, L., Liu, J.: Economic model predictive control with triggered evaluations: state and output feedback. J. Process Control **24**(8), 1197–1206 (2014)
18. Khubalkar, S.W., Chopade, A.S., Junghare, A.S., Aware, M.V.: Design and tuning of fractional order PID controller for speed control of permanent magnet brushless DC motor. In: 2016 IEEE First International Conference on Control, Measurement and Instrumentation (CMI), pp. 326–320 (2016)
19. Li, H., Shi, Y.: Event-triggered robust model predictive control of continuous-time nonlinear systems. Automatica **50**(5), 1507–1513 (2014)
20. Lim, H.S., Krishnan, R.: Novel measurement disturbance rejection current control for linear switched reluctance motor drives. In: 2007 IEEE Industry Applications Annual Meeting, pp. 2226–2233 (2007)
21. Lin, Y., Shi, Y., Burton, R.: Modeling and robust discrete-time sliding-mode control design for a fluid power electrohydraulic actuator (EHA) system. IEEE/ASME Trans. Mechatron. **18**(1), 1–10 (2013)
22. Liu, Z., Guo, H., Wang, D., Wu, Z., Xu, J.: Active-disturbance rejection control of brushless DC motor based on BP neural network. In: 2010 International Conference on Electrical and Control Engineering, pp. 3253–3256 (2010)
23. Meng, X., Chen, T.: Event based agreement protocols for multi-agent networks. Automatica **49**(7), 2125–2132 (2013)
24. Shi, Y., Shen, C., Buckham, B.: Integrated path planning and tracking control of an AUV: a unified receding horizon optimization approach. IEEE/ASME Trans. Mechatron. https://doi.org/10.1109/TMECH.2016.2612689
25. Sira-Ramírez, H., Linares-Flores, J., García-Rodríguez, C., Contreras-Ordaz, M.A.: On the control of the permanent magnet synchronous motor: an active disturbance rejection control approach. IEEE Trans. Control Syst. Technol. **22**(5), 2056–2063 (2014)
26. Sun, W., Gao, H., Kaynak, O.: Vibration isolation for active suspensions with performance constraints and actuator saturation. IEEE/ASME Trans. Mechatron. **20**(2), 675–683 (2015)
27. Tabuada, P.: Event-triggered real-time scheduling of stabilizing control tasks. IEEE Trans. Autom. Control **52**(9), 1680–1685 (2007)
28. Terzic, B., Jadric, M.: Design and implementation of the extended kalman filter for the speed and rotor position estimation of brushless DC motor. IEEE Trans. Ind. Electron. **48**(6), 1065–1073 (2001)
29. Wang, X., Gao, H., Kaynak, O., Sun, W.: Online deflection estimation of X-axis beam on positioning machine. IEEE/ASME Trans. Mechatron. **21**(1), 339–350 (2016)
30. Wu, C., Qi, R.: The simplified active disturbance rejection control for permanent magnet synchronous motor drive system. In: Proceedings of the 32nd Chinese Control Conference, pp. 4172–4176 (2013)
31. Xu, L., Yao, B.: Adaptive robust precision motion control of linear motors with negligible electrical dynamics: theory and experiments. IEEE/ASME Trans. Mechatron. **6**(4), 444–452 (2001). https://doi.org/10.1109/3516.974858
32. Yang, J., Dong, L., Liao, X.: Fractional order PD controller based on ADRC algorithm for DC motor. In: 2014 IEEE Conference and Expo Transportation Electrification Asia-Pacific (ITEC Asia-Pacific), pp. 1–6 (2014)
33. Yao, J., Jiao, Z., Ma, D.: Adaptive robust control of dc motors with extended state observer. IEEE Trans. Ind. Appl. **61**(7), 3630–3637 (2014)
34. Yao, J., Jiao, Z., Ma, D.: Output feedback robust control of direct current motors with nonlinear friction compensation and disturbance rejection. J. Dyn. Syst. Meas. Control **137**(4) (2015)

35. Zhang, L., Wang, S., Karimi, H.R., Jasra, A.: Robust finite-time control of switched linear systems and application to a class of servomechanism systems. IEEE/ASME Trans. Mechatron. **20**(5), 2476–2485 (2015)
36. Zhang, Y., Zhang, Y., Wang, J., Ma, R.: An active disturbance rejection control of induction motor using DSP+FPGA. In: 2013 25th Chinese Control and Decision Conference (CCDC), pp. 4047–4052 (2013)
37. Zhao, Z.L., Guo, B.Z.: On convergence of nonlinear active disturbance rejection control for MIMO systems. In: Proceedings of the 31st Chinese Control Conference, pp. 434–441 (2012)
38. Zhao, Z.L., Guo, B.Z.: Active disturbance rejection control approach to stabilization of lower triangular systems with uncertainty. Int. J. Robust Nonlinear Control **26**(11), 2314–2337 (2015)
39. Zheng, Q., Gao, Z.: On practical applications of active disturbance rejection control. In: Proceedings of the 29th Chinese Control Conference, pp. 6095–6100 (2010)
40. Zou, L., Wang, Z., Gao, H., Liu, X.: Event-triggered state estimation for complex networks with mixed time delays via sampled data information: the continuous-time case. IEEE Trans. Cybern. **45**(12), 2804–2815 (2015)

Chapter 7
Event-Triggered ADRC for Electric Cylinders with PD-Type Event-Triggering Conditions

7.1 Background

In this chapter, we illustrate the application of the proposed ET-ADRC approach through a different type of servo systems. Specifically, we consider a position tracking problem for electric cylinders used in a recently developed wheel-legged robotic system. Each leg of this robotic system is composed of six electric cylinders, which are controlled based on data from more than 12 sensors transmitted through a controller area network (CAN) and a user datagram protocol (UDP) network. The requirement of achieving satisfactory performance under limited communication resources motivates us to adopt an event-triggered control approach for the electric cylinders.

Several practical challenges are considered and overcome in designing the ET-ADRC for electric cylinders. On one hand, the stable operation of the robotic system requires the reference signal of electric cylinders to change over a wide range and a high-frequency band, which makes it difficult to design an event-triggering condition that can reduce triggering rate without compromising tracking performance. To deal with this challenge, a proportional-derivative-type (PD-type) event-triggering condition is designed. As the absolute value and the trend of the tracking error are both taken into account, the event-triggering condition is predictive, which leads to effective improvement in tracking accuracy and response speed of the controller and the decrease of event-triggering rate. On the other hand, although event-triggered PID algorithms were introduced in the literature [1, 8], the stability of the resultant closed-loop control system was not addressed systematically. Meanwhile, as the electric cylinders use roller-screws to realize the linear motion of inertia loads when they drive the robotic system, the sliding friction [31], which is nonlinear related to the sliding velocity, and torque disturbance [30] are nonnegligible in such systems. These observations motivate us to explore the convergence properties of event-triggered ADRC proposed for this application, which can be viewed as a nonlinear PID controller [33] and has the advantages of dealing with internal uncertainties [6] and external disturbances [3]. Since the Coulomb friction has defects in sliding friction analysis, the results in the previous chapter for the DC motor control system

D. Shi et al., *Event-Triggered Active Disturbance Rejection Control*, Studies in Systems, Decision and Control 356,
https://doi.org/10.1007/978-981-16-0293-1_7

Table 7.1 Nomenclature

Symbols	Description	Units
T_e	Electromagnetic torque of the electric cylinder	N·m
T_f	Friction torque on the electric cylinder	N·m
T_r	Radial torque of the electric cylinder	N·m
F_L	Load force	N
J	Inertia of the screw	kg·m^2
P	Lead of the screw	mm·r^{-1}
ω	Rotation velocity of the screw	rad·s^{-1}
z	Immeasurable internal friction state	rad
ϑ_0	Stiffness coefficient of bristles	N·m·rad^{-1}
ϑ_1	Damping coefficient of bristles	N·m·s·rad^{-1}
ϵ_2	Viscous friction coefficient	N·m·s·rad^{-1}
v_s	Stribeck velocity	m·s^{-1}
K_p	Input-torque ratio of the electric cylinder	N·m·V^{-1}
θ_1	Screw angular position	rad
θ_2	Screw angular velocity	rad·s^{-1}
u	Torque command	V
θ_1^*	Screw angular position reference	rad
θ_2^*	Screw angular velocity reference	rad·s^{-1}
b_0	Boundary of $\frac{\partial \eta(\theta)}{\partial \theta_i}$	–
b_1	Boundary of $\dot{\lambda}(t)$	–
b_2	Boundary of $u(t)$ and $\theta(t)$	–
B	Boundary of $x_i(t)$ and $\lambda(t)$	–
Ψ	Boundary of $\frac{1}{\varepsilon}\sum_{i=1}^{3} L_i\sigma(t)$	–

are not applicable to the electric cylinder control systems. As the LuGre friction model [19] takes viscous friction, Coulomb friction, static friction, and Stribeck friction into account, it can represent most friction behaviors of the electric cylinders and is considered in the convergence analysis in this chapter. The developed results are evaluated on an electric cylinder control system that can verify the control performance for the actuators of our quadruped robotic system. Position tracking of the system under different load conditions and reference inputs is obtained, which verifies the effectiveness of the introduced ET-ADRC.

Table 7.1 provides a necessary explanation of the notations and alphabetic symbols used in the rest sections. The rest of this chapter is organized as follows. The dynamics analysis of an electric cylinder is proposed in Sect. 7.2. The event-triggered ADRC of the electric cylinders is shown in detail in Sect. 7.3. Experiment results are presented to evaluate the actual performance of the system in Sect. 7.4. A brief summary of this chapter is provided in Sect. 7.5.

7.2 Nonlinear Modeling and Problem Formulation

The basic structure of electric cylinders is provided that is shown in Fig. 7.1, where T_e, T_f, T_r, and F_L represent the electromagnetic torque, friction torque, radial torque, and load force on the cylinder, respectively. As P represents the lead of the screw, we have

$$T_r = \frac{PF_L}{2\pi}.$$

For this system, the torque equation holds:

$$T_e - T_r - T_f = J\dot{\omega}, \tag{7.1}$$

where J and ω are the inertia and rotation velocity of the screw. The model of PMSM can be decoupled by using a vector control algorithm, which makes it easy to design the controller using traditional approaches. With this simplification, the model of the electric cylinder can be described as

$$\begin{cases} \dot{\theta}_1(t) = \theta_2(t), \\ J\dot{\theta}_2(t) = K_p u(t) - T_f(t) - K_h F_L(t), \end{cases} \tag{7.2}$$

where $\theta(t) = [\theta_1(t), \theta_2(t)]^T$ represents the state vector consisting of the angle and rotation speed of the screw, K_p denotes the input-torque ratio, $u(t)$ is the control input, $T_f(t)$ denotes the friction torque, and $K_h = \frac{P}{2\pi}$. To focus on the friction compensation, the effect of dead zone is ignored in our analysis.

With the LuGre model, the friction torque in (7.2) can be written as

$$\begin{cases} T_f = \vartheta_0 z + \vartheta_1 \dot{z} + \epsilon_2 \theta_2, \\ \dot{z} = \theta_2 - \dfrac{|\theta_2|}{g(\theta_2)} z, \\ g(\theta_2) = \epsilon_0 + \epsilon_1 e^{-(\theta_2/v_s)^2}, \end{cases} \tag{7.3}$$

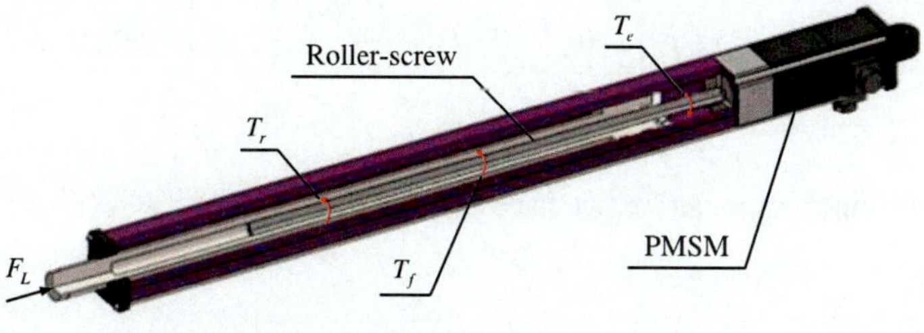

Fig. 7.1 Basic structure of electric cylinders

where z represents the immeasurable internal friction state, and the friction force parameters ϑ_0, ϑ_1, and ϵ_2 can be physically explained as the stiffness, the damping coefficient of bristles and the viscous friction coefficient. The function $g(\theta_2)$ is positive and describes the Stribeck effect with v_s being the Stribeck velocity. Due to the non-differentiable term $|\theta_2|$ in (7.3), a continuous approximation of the LuGre model introduced in [23] is employed, which has the form

$$\dot{z} = S_1(\theta_2)\theta_2 - \frac{S_2(\theta_2)}{g(\theta_2)}z, \tag{7.4}$$

with $S_1(\theta_2)$ and $S_2(\theta_2)$ being two approximation functions defined as

$$\begin{aligned} S_1(\theta_2) &= \left(\frac{2}{\pi}\arctan(k_v, \theta_2)\right)^2, \\ S_2(\theta_2) &= \frac{2\theta_2}{\pi}\arctan(k_v, \theta_2), \end{aligned} \tag{7.5}$$

where k_v is a positive constant. Even though this version of the LuGre model differs from other ones in minor details, it presents an important advantage since it can be proven to maintain the properties of boundedness and passivity. For the complete proof of these two properties, the interested readers can refer to [24]. For notional brevity, we define

$$\rho_1 = -\frac{\vartheta_0}{J},\ \rho_2 = -\frac{\vartheta_1}{J},\ \rho_3 = k_v,\ \rho_4 = -\frac{\epsilon_2}{J},\ \rho_5 = \frac{K_p}{J},\ \rho_6 = -\frac{K_h}{J}.$$

With the modified LuGre model for the friction torque, the electric cylinder system dynamics to be controlled can be described as

$$\begin{cases} \dot{\theta}_1(t) = \theta_2(t), \\ \dot{\theta}_2(t) = \eta(\theta(t)) + \lambda(t) + \rho_5 u(t), \end{cases} \tag{7.6}$$

where system function $\eta(\theta(t))$ and disturbance $\lambda(t)$ satisfy

$$\begin{cases} \eta(\theta(t)) = \rho_1 z(t) + \rho_2\left[S_1(\theta_2(t))\theta_2(t) - \dfrac{S_2(\theta_2(t))}{g(\theta_2(t))}z(t)\right] + \rho_4\theta_2(t), \\ \lambda(t) = \rho_6 F_L(t). \end{cases} \tag{7.7}$$

Through a simple computation, we have

$$\begin{cases} \dfrac{\partial\eta(\theta)}{\partial\theta_1} = 0, \\ \dfrac{\partial\eta(\theta)}{\partial\theta_2} = \rho_1\dfrac{dz}{d\theta_2} + \rho_2\left[\dfrac{dS_1}{d\theta_2}\theta_2 + S_1 - \dfrac{\frac{dS_2}{d\theta_2}g - S_2\frac{dg}{d\theta_2}}{g^2}z - \dfrac{S_2}{g}\dfrac{dz}{d\theta_2}\right] + \rho_4. \end{cases} \tag{7.8}$$

From (7.3), (7.5), and (7.8), there exists a positive constant b_0 satisfying

$$\left|\frac{\partial \eta(\theta)}{\partial \theta_{\rm i}}\right| \le b_0, \quad i \in \{1,2\}. \tag{7.9}$$

As the system function $\eta(\theta)$ cannot be known exactly, we treat $\eta(\theta)$ as the uncertainties of the system. Equations (7.8) and (7.9) guarantee that $\frac{\partial \eta(\theta)}{\partial \theta_{\rm i}}$ is bounded.

As the disturbances of the electric cylinders in our application are mainly from inertial loads, there exists a positive constant b_1 such that $\left|\dot{\lambda}(t)\right| \le b_1$.

Next, define the reference trajectory $\theta^*(t)$ as

$$\begin{cases} \theta^*(t) = [\theta_1^*(t), \theta_2^*(t)]^T, \\ \dot{\theta}_1^*(t) = \theta_2^*(t), \\ \dot{\theta}_2^*(t) = \upsilon(t). \end{cases} \tag{7.10}$$

Suppose that $|\upsilon(t)| \le C_{\upsilon 1}$ for some $C_{\upsilon 1} > 0$, and $|\dot{\upsilon}(t)| \le C_{\upsilon 2}$ holds for some $C_{\upsilon 2} > 0$. With the aim of investigating the tracking performance of the electric cylinder in (7.7), the tracking error is defined as

$$x(t) = \theta(t) - \theta^*(t), \tag{7.11}$$

where $x = [x_1, x_2]^T$. Then, the error equation is as shown

$$\begin{cases} \dot{x}_1(t) = x_2(t), \\ \dot{x}_2(t) = \eta(\theta(t)) - \upsilon(t) + \lambda(t) + \rho_5 u(t). \end{cases} \tag{7.12}$$

The external disturbances together with internal disturbances may deteriorate the tracking performance of the system. To estimate and compensate these disturbances, an extended state is introduced as

$$x_3(t) = \eta(\theta(t)) - \upsilon(t) + \lambda(t) \tag{7.13}$$

where the x_3 represents the total effect of the uncertainties and disturbances. An ET-ESO is then introduced as

$$\begin{cases} \dot{\hat{x}}_1(t) = \hat{x}_2(t) + \varepsilon h_1\left(\dfrac{\xi(t) - \hat{x}_1(t)}{\varepsilon^2}\right), \\ \dot{\hat{x}}_2(t) = \hat{x}_3(t) + h_2\left(\dfrac{\xi(t) - \hat{x}_1(t)}{\varepsilon^2}\right) + \rho_5 u(t), \\ \dot{\hat{x}}_3(t) = \dfrac{1}{\varepsilon} h_3\left(\dfrac{\xi(t) - \hat{x}_1(t)}{\varepsilon^2}\right), \\ \hat{x}_1(t_0) = \hat{x}_{10},\ \hat{x}_2(t_0) = \hat{x}_{20},\ \hat{x}_3(t_0) = \hat{x}_{30}, \end{cases} \tag{7.14}$$

where $\hat{x} = [\hat{x}_1, \hat{x}_2, \hat{x}_3]^T \in R^3$ is the observer state, ε is a high-gain parameter, $\hat{x}(t_0) = [\hat{x}_{10}, \hat{x}_{20}, \hat{x}_{30}]^T \in R^3$ is the initial value of the observer state, and $\xi(t)$ is given by

$$\xi(t) = \begin{cases} x_1(t_k), & \text{if } \Gamma(t) = 0, \\ x_1(t), & \text{otherwise.} \end{cases} \tag{7.15}$$

In the above equation, $\Gamma(t)$ is the event-trigger to be designed in this chapter, t_k denotes the previous event-triggering time instant, and k denotes the total number of triggered events during the time period $[t_0, t)$. The functions $h_i(d)$ are Lipschitz with $L_i > 0$. In our design, the $h_i(d)$'s are chosen as

$$\begin{cases} h_1(d) = a_1 d + \varphi(d), \\ h_2(d) = a_2 d, \\ h_3(d) = a_3 d, \end{cases} \tag{7.16}$$

where a_1, a_2, a_3 are some constants such that the matrix

$$A = \begin{bmatrix} -a_1 & 1 & 0 \\ -a_2 & 0 & 1 \\ -a_3 & 0 & 0 \end{bmatrix} \tag{7.17}$$

is Hurwitz and $\varphi(d)$ is defined by

$$\varphi(d) = \begin{cases} -\frac{1}{4}, & d \in (-\infty, -\frac{\pi}{2}), \\ \frac{1}{4}\sin(d), & d \in [-\frac{\pi}{2}, \frac{\pi}{2}], \\ \frac{1}{4}, & d \in (\frac{\pi}{2}, +\infty). \end{cases} \tag{7.18}$$

Define a function $P : \mathbb{R}^3 \to \mathbb{R}$ as

$$P(d) = \langle \tilde{P} d, d \rangle + \int_0^{d_1} \varphi(x) dx \tag{7.19}$$

where the matrix $\tilde{P} > 0$ satisfies

$$\tilde{P} A + A^T \tilde{P} = -I. \tag{7.20}$$

By choosing the functions $h_i(d)$ and $\varphi(d)$, there exists a nonnegative-definite function $S : \mathbb{R}^3 \to \mathbb{R}$ satisfying

$$\begin{cases} \mu_1 \|d\|^2 \le P(d) \le \mu_2 \|d\|^2, \\ \mu_3 \|d\|^2 \le S(d) \le \mu_4 \|d\|^2, \\ \sum_{i=1}^{2} \frac{\partial P}{\partial d_i}(d_{i+1} - h_i(d_1)) - \frac{\partial P}{\partial d_3} h_3(d_3) \le -S(d), \\ \left\| \frac{\partial P}{\partial d} \right\| \le \varpi_1 \|d_1\| \end{cases} \tag{7.21}$$

where μ_1, μ_2, μ_3, μ_4, and ϖ_1 are positive constants. Note that $S(d) = \mu_k \|d\|^2$, where μ_k is a positive constant, is an adoptable form and is used in our design.

The control signal u is designed as

$$u = -\frac{k_1}{\rho_5}\hat{x}_1 - \frac{k_2}{\rho_5}\hat{x}_2 - \frac{1}{\rho_5}\hat{x}_3 \tag{7.22}$$

where k_i are some constants such that the matrix H is Hurwitz:

$$H = \begin{bmatrix} 0 & 1 \\ -k_1 & -k_2 \end{bmatrix}. \tag{7.23}$$

Obviously, there is a matrix $\tilde{Q} > 0$ such that

$$\tilde{Q}H + H^T\tilde{Q} = -I \tag{7.24}$$

then we define $Q(d)$ as

$$Q(d) = d^T\tilde{Q}d \tag{7.25}$$

for which we have

$$\begin{cases} \iota_1\|d\|^2 \leq Q(d) \leq \iota_2\|d\|^2, \\ \left\|\dfrac{\partial Q}{\partial d}\right\| \leq \varpi_2\|d\| \end{cases} \tag{7.26}$$

for some positive constants ι_1, ι_2, and ϖ_2. Next, $\tilde{x}_i$ and e_j are defined to describe the performance of the ESO in (7.14)

$$\begin{cases} \tilde{x}_i = x_i - \hat{x}_i, \\ e_j = \dfrac{\tilde{x}_i}{\varepsilon^{3-i}}, \end{cases} \tag{7.27}$$

for $i = 1,2,3$, based on which we define the sampling error $\sigma(t)$ as

$$\sigma(t) = \frac{x_1(t_k) - x_1(t)}{\varepsilon^2}, \tag{7.28}$$

where $t \in [t_k, t_{k+1})$. Furthermore, define $\alpha_j(e_1, \sigma) = h_j(e_1 + \sigma) - h_j(e_1)$ for $j = 1, 2, 3$.

In this chapter, the electric cylinder system is modeled as (7.6), which is a second-order SISO system with the input u and the output θ_1. Since the LuGre friction model from (7.3) to (7.5) is adopted to represent the most friction behaviors (e.g., viscous friction, Coulomb friction, static friction, and Stribeck friction) of the electric cylinders, the system function $\eta(\theta)$ in (7.7) is nonlinear, which brings challenges to the stability proof for the ET-ESO and the convergence analysis for the closed-loop system. For the remainder of this chapter, the following problems are discussed:

- Considering $\sigma(t)$ and $\dot{\sigma}(t)$, can we propose a $\Gamma(t)$ with efficient performance in terms of communication reduction without causing Zeno phenomenon and asymptotic boundedness in observation error?
- Based on the improved event-triggered ESO, can we propose an ADRC algorithm such that $\limsup_{t\to\infty} \|x\| \leq B_e$ for the control system?

7.3 Event-Triggered ADRC of Electric Cylinders

In this section, an event-triggered ADRC scheme for electric cylinders is introduced in detail. The boundedness of the event-triggered ESO and the position tracking error is guaranteed theoretically.

The block diagram of the controller is as shown in Fig. 7.2. From (7.22), the control signal u can be rewritten as

$$u = -\sum_{i=1}^{2} \frac{k_i}{\rho_5} x_i + \sum_{i=1}^{2} \frac{k_i}{\rho_5} \varepsilon^{3-i} e_i - \frac{1}{\rho_5} \hat{x}_3. \tag{7.29}$$

For notational brevity, we define K as

$$K := \max\{k_i, k_i \varepsilon^{3-i}, 1\}, i \in \{1,2\}. \tag{7.30}$$

From (7.7), (7.9) and (7.29), we obtain that

$$|u| + |\eta| + |\dot{\lambda}| + \left|\frac{d\eta}{dt}\right| + \left|\frac{d\eta}{d\theta}\right| \leq c_0 + \sum_{i=1}^{2} c_i |\theta_i|^c, \tag{7.31}$$

where c and $c_i (i = 0,1,2)$ are positive constants. So, we have

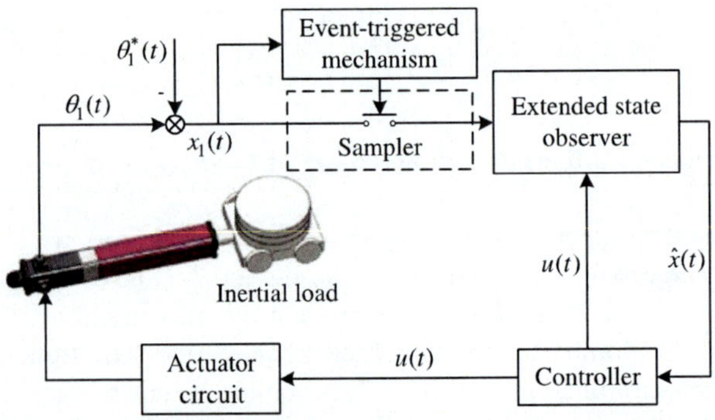

Fig. 7.2 Block diagram of an event-triggered ADRC for the electric cylinders

$$
\begin{aligned}
\left|\frac{d}{dt}(\eta+\lambda)\right| &\leq \left|\frac{d\eta}{dt}\right| + |\dot{\lambda}| \\
&\leq c_0 + \sum_{i=1}^{2} c_i|\theta_i|^c - |u| - |\eta| - \left|\frac{d\eta}{d\theta}\right|.
\end{aligned} \tag{7.32}
$$

There exist a positive constant b_2 and a domain Ξ such that the $u(t)$ and $\theta(t)$ are bounded by

$$
\sum_{i=1}^{2} c_i|\theta_i|^c - |u| \leq b_2, \tag{7.33}
$$

for the initial value $\theta(t_0) \in \Xi$. This implies that there exists an $M > 0$ such that

$$
\left|\frac{d}{dt}(\eta(\theta(t)) + \lambda(t))\right| < M. \tag{7.34}
$$

Then, an event-triggered mechanism is proposed as

$$
\Gamma(t) = \begin{cases} 0, \text{ if } \frac{\varpi_1\mu_2}{\mu_1\mu_3} \sum\limits_{i=1}^{3} L_i \dot{\sigma}(t)\sigma(t) \leq \left(E - \frac{\varpi_1\mu_2 M\varepsilon}{\mu_1\mu_3}\right)|\dot{\sigma}(t)|, \\ 1, \text{ otherwise,} \end{cases} \tag{7.35}
$$

where E is a positive constant and $\dot{\sigma}(t)$ represents the left derivative of $\sigma(t)$. In the next result, we show the non-existence of Zeno phenomenon for this event-triggering condition, which guarantees the existence of the solution of the control system when $t \to \infty$.

Theorem 7.1 *Consider the closed-loop system in (7.12), the ESO in (7.14), and the control signal in (7.22). There exist $\varepsilon^* > 0$ and $\tau > 0$ such that for $\varepsilon \in (0, \varepsilon^*)$ and for any $k > 0$*

$$
\min\{t_{k+1} - t_k\} \geq \tau. \tag{7.36}
$$

Proof Define $\varepsilon^* := \min\{\frac{E\mu_1\mu_3}{\varpi_1\mu_2 M}, 1\}$ such that for any $\varepsilon \in (0, \varepsilon^*)$ we have $E - \frac{\varpi_1\mu_2 M\varepsilon}{\mu_1\mu_3} > 0$. Then the event-triggering condition is equivalent to

$$
\Gamma(t) = \begin{cases} 0, \text{ if } \dot{\sigma}(t)\sigma(t) \leq \left(E - \frac{\varpi_1\mu_2 M\varepsilon}{\mu_1\mu_3}\right)|\dot{\sigma}(t)|/(\frac{\varpi_1\mu_2}{\mu_1\mu_3}\sum\limits_{i=1}^{3} L_i), \\ 1, \text{ otherwise.} \end{cases} \tag{7.37}
$$

According to (7.28), when $t \in [t_k, t_{k+1})$, we have

$$
\sigma(t) = \frac{1}{\varepsilon^2}\int_{t_k}^{t} \dot{x}_1(\tau)d\tau = \frac{1}{\varepsilon^2}\int_{t_k}^{t} x_2(\tau)d\tau. \tag{7.38}
$$

From the definition of the event-triggering condition, we obtain

$$\sigma(t)\dot{\sigma}(t) = \dot{\sigma}(t)\frac{1}{\varepsilon^2}\int_{t_k}^{t} x_2(\tau)d\tau \leq |\dot{\sigma}(t)|\frac{(t-t_k)B}{\varepsilon^2} \tag{7.39}$$

for some $B > 0$. Thus, there exists

$$\tau := \left(E - \frac{\varpi_1\mu_2 M\varepsilon}{\mu_1\mu_3}\right)\varepsilon^2 / \left(\frac{B\varpi_1\mu_2}{\mu_1\mu_3}\sum_{i=1}^{3} L_i\right) \tag{7.40}$$

such that $\sigma(t)\dot{\sigma}(t) \leq \left(E - \frac{\varpi_1\mu_2 M\varepsilon}{\mu_1\mu_3}\right)|\dot{\sigma}(t)| / \left(\frac{\varpi_1\mu_2}{\mu_1\mu_3}\sum_{i=1}^{3} L_i\right)$ holds for $t = t_k + \tau$. This completes the proof. □

Based on the above result, we further analyze the stability of the closed-loop system.

Theorem 7.2 *Consider the closed-loop system in (7.12), the ESO in (7.14), the control signal in (7.22) and the event-triggering condition $\Gamma(t)$ in (7.37). For $E > 0$ and any initial values of x and $\hat{x}$, there exist $\varepsilon^* > 0$, such that for any $\varepsilon \in (0, \varepsilon^*)$, and $i \in \{1,2,3\}$ it holds that*

$$\begin{cases} \limsup\limits_{t\to\infty} \|x\| \leq \sqrt{\frac{\varpi_1 \varepsilon^{\frac{1}{3}}(F_0+\Psi)}{2\iota_1\gamma}}, \\ \limsup\limits_{t\to\infty} |x_i - \hat{x}_i| \leq \varepsilon^{3-i}\sqrt{\frac{\varpi_1 \varepsilon^{\frac{1}{3}}(F_0+\Psi)}{2\mu_1\gamma}} \end{cases} \tag{7.41}$$

where $\Psi = \max\left\{\frac{E\mu_1\mu_3}{\varepsilon\varpi_1\mu_2} - M,\ 0\right\}$.

Proof For the closed-loop system proposed in (7.12),(7.14), and (7.22), considering the definition of the observation error in (7.27), we have

$$\begin{aligned} \dot{\tilde{x}}_j(t) &= \tilde{x}_{i+1}(t) - \varepsilon^{2-i}h_j\left(\frac{\xi(t)-\hat{x}_1(t)}{\varepsilon^2}\right), \quad i \in \{1,2\} \\ &= \tilde{x}_{i+1}(t) - \varepsilon^{2-i}h_j(e_1(t)) - \varepsilon^{2-i}\alpha_i(e_1(t),\sigma(t)), \\ \dot{\tilde{x}}_3(t) &= \frac{d}{dt}(\eta(\theta(t)) - \upsilon(t) + \lambda(t)) - \frac{1}{\varepsilon}h_3\left(\frac{\xi(t)-\hat{x}_1(t)}{\varepsilon^2}\right) \\ &= \frac{d}{dt}(\eta(\theta(t)) - \upsilon(t) + \lambda(t)) - \frac{h_3(e_1(t))}{\varepsilon} - \frac{\alpha_3(e_1(t),\sigma(t))}{\varepsilon}. \end{aligned} \tag{7.42}$$

From (7.27), the dynamic of e_i is obtained in the following form:

$$\begin{aligned}\dot{e}_i(t) &= \frac{1}{\varepsilon}(e_{i+1}(t) - h_i(e_1(t)) - \alpha_i(e_1(t), \sigma(t))), i \in \{1,2\}, \\ \dot{e}_3(t) &= \dot{x}_3(t) - \frac{1}{\varepsilon}h_3(e_1(t)) - \frac{1}{\varepsilon}\alpha_3(e_1(t), \sigma(t)).\end{aligned} \tag{7.43}$$

Consider a nonnegative-definite function $V(x, e)$:

$$V(x, e) := Q(x) + P(e) \tag{7.44}$$

where $Q(x)$ and $P(e)$ are provided in (7.19) and (7.25), respectively. For this function, we observe that

$$\begin{aligned}\frac{dV}{dt} &= \frac{dQ}{dt} + \frac{dP}{dt} \\ &= \frac{\partial Q}{\partial x_1}x_2 + \frac{\partial Q}{\partial x_2}(x_3 + \rho_5 u) + \sum_{i=1}^{3}\frac{\partial P}{\partial e_i}\dot{e}_i.\end{aligned} \tag{7.45}$$

From the dynamic of e_i in (7.43), we have

$$\begin{aligned}\frac{dV}{dt} &= \frac{\partial Q}{\partial x_1}x_2 + \frac{\partial Q}{\partial x_2}\left(x_3 - \sum_{i=1}^{2}k_i x_i + \sum_{i=1}^{2}k_i\varepsilon^{3-i}e_i - \hat{x}_3\right) \\ &\quad + \sum_{i=1}^{2}\frac{\partial P}{\partial e_i}\left[\frac{1}{\varepsilon}(e_{i+1} - h_i(e_1) - \alpha_i(e_1, \sigma))\right] \\ &\quad + \frac{\partial P}{\partial e_3}\left[\dot{x}_3 - \frac{1}{\varepsilon}(h_3(e_1) + \alpha_3(e_1, \sigma))\right] \\ &\le \frac{\partial Q}{\partial x_1}x_2 - \frac{\partial Q}{\partial x_2}\sum_{i=1}^{2}k_i x_i + \frac{\partial Q}{\partial x_2}\left(\sum_{i=1}^{2}k_i\varepsilon^{3-i}e_i + e_3\right) + \left|\frac{\partial P}{\partial e_3}\right| \cdot |\dot{x}_3| \\ &\quad + \frac{1}{\varepsilon}\left[\sum_{i=1}^{2}\frac{\partial P}{\partial e_i}(e_{i+1} - h_i(e_1)) - \frac{\partial P}{\partial e_3}h_3(e_1)\right] - \frac{1}{\varepsilon}\sum_{i=1}^{3}\frac{\partial P}{\partial e_i}\alpha_i(e_1, \sigma).\end{aligned} \tag{7.46}$$

According to the definition of $Q(b)$ in (7.25), we observe

$$\frac{\partial Q}{\partial x_1}x_2 - \frac{\partial Q}{\partial x_2}\sum_{i=1}^{2}k_i x_i = 2x^T\tilde{Q}Hx = x^T(\tilde{Q}H + H^T\tilde{Q})x = -\|x\|^2. \tag{7.47}$$

For the function $P(\cdot)$, we have

$$\frac{1}{\varepsilon}\left[\sum_{i=1}^{2}\frac{\partial P}{\partial e_i}(e_{i+1} - h_i(e_1)) - \frac{\partial P}{\partial e_3}h_3(e_1)\right] \le -\frac{1}{\varepsilon}S(e) \le -\frac{\mu_3}{\varepsilon}\|e\|^2. \tag{7.48}$$

According to inequality (7.25),

$$\begin{aligned}\frac{\partial Q}{\partial x_2}\left(\sum_{i=1}^{2}k_i\varepsilon^{3-i}e_i+e_3\right)&\le\varpi_2\|x\|\left|\sum_{i=1}^{2}k_i\varepsilon^{3-i}e_i+e_3\right|\\&=\frac{1}{\varepsilon}\varpi_2\left[\varepsilon^{\frac{2}{3}}\|x\|\varepsilon^{\frac{1}{3}}\left(\sum_{i=1}^{2}k_i\varepsilon^{3-i}|e_i|+|e_3|\right)\right]\\&\le\frac{3}{2}\varepsilon^{\frac{1}{3}}\varpi_2\|x\|^2+\frac{\varpi_2}{2\varepsilon}\left(\sum_{i=1}^{2}k_i^2\varepsilon^{\frac{20}{3}-2i}+\varepsilon^{\frac{2}{3}}\right)\|e\|^2,\\\left|\frac{\partial P}{\partial e_3}\right|\cdot|\dot{x}_3|&\le\varpi_1\|e\|\cdot|\dot{x}_3|\\&=\frac{1}{\varepsilon}\varpi_1\varepsilon^{\frac{1}{3}}\|e\|\varepsilon^{\frac{2}{3}}|\dot{x}_3|\\&\le\frac{1}{2\varepsilon}\varpi_1\left(\varepsilon^{\frac{2}{3}}\|e\|^2+\varepsilon^{\frac{4}{3}}|\dot{x}_3|^2\right).\end{aligned}\tag{7.49}$$

Based on the control signal provided in (7.29), we have

$$\begin{aligned}|\dot{x}_3|^2&=|\dot{\eta}(\theta)-\dot{\upsilon}+\dot{\lambda}|^2\\&=\left|\frac{\partial\eta}{\partial\theta_1}\theta_2+\frac{\partial\eta}{\partial\theta_2}\left(e_3-\sum_{i=1}^{2}k_ix_i+\sum_{i=1}^{2}k_i\varepsilon^{3-i}e_i+\upsilon\right)-\dot{\upsilon}+\dot{\lambda}\right|^2\\&\le(A_F\|x\|+B_F\|e\|+C_F)^2\end{aligned}\tag{7.50}$$

where A_F, B_F, and C_F are defined as $A_F:=2b_0(K+1)$, $B_F:=b_0(2K+1)$, $C_F:=b_0C_{\upsilon1}+C_{\upsilon2}+b_1$. Then, we choose $F_0:=C_F^2+A_FC_F+B_FC_F$, $F_1:=A_F^2+A_FB_F+A_FC_F$, $F_2:=B_F^2+A_FB_F+B_FC_F$, so that

$$|\dot{x}_3|^2\le F_0+F_1\|x\|^2+F_2\|e\|^2.\tag{7.51}$$

From (7.48)–(7.50), we have

$$\begin{aligned}\left|\frac{\partial P}{\partial e_3}\right||\dot{x}_3|&\le\frac{\varpi_1}{2\varepsilon}\left[\varepsilon^{\frac{2}{3}}\|e\|^2+\varepsilon^{\frac{4}{3}}(F_0+F_1\|x\|^2+F_2\|e\|^2)\right]\\&\le\frac{\varpi_1\varepsilon^{\frac{1}{3}}}{2}\left[(\varepsilon^{-\frac{2}{3}}+F_2)\|e\|^2+F_1\|x\|^2+F_0\right].\end{aligned}\tag{7.52}$$

According to the event-triggering condition in (7.35), if $\dot{\sigma}(t)\sigma(t)>0$, we have

$$\sum_{i=1}^{3}L_i|\sigma(t)|\le\frac{E\mu_1\mu_3}{\varpi_1\mu_2}-M\varepsilon.\tag{7.53}$$

If $\dot{\sigma}(t)\sigma(t) \le 0$, for every $t \in [t_k, t_{k+1})$, we have

$$|\sigma(t)| \le |\sigma(t_k)| = 0. \tag{7.54}$$

Considering the definition of Ψ, we obtain

$$\left| \frac{1}{\varepsilon} \sum_{i=1}^{3} L_i \sigma(t) \right| \le \Psi. \tag{7.55}$$

According to the definition in (7.16), we have

$$|\alpha_i(e_1, \sigma)| = |h_i(e_1 + \sigma) - h_i(e_1)| \le L_i |\sigma|. \tag{7.56}$$

Combining the fact that $\|\frac{\partial P}{\partial z}\| \le \varpi_1 \|z\|$, we obtain that

$$\begin{aligned} -\frac{1}{\varepsilon} \sum_{i=1}^{3} \frac{\partial P}{\partial e_i} \alpha_i(e_1, \sigma) &\le \frac{1}{\varepsilon} \sum_{i=1}^{3} \left| \frac{\partial P}{\partial e_i} \right| |\alpha_i(e_1, \sigma)| \\ &\le \frac{1}{\varepsilon} \varpi_1 \|e\| \Psi \varepsilon \\ &\le \left(\frac{\varepsilon^{-\frac{1}{3}} \|e\|^2}{2} + \frac{\varepsilon^{\frac{1}{3}}}{2} \right) \Psi \varpi_1. \end{aligned} \tag{7.57}$$

From (7.45)–(7.57), we have

$$\begin{aligned} \frac{dV}{dt} \le &- \|x\|^2 - \frac{\mu_3}{\varepsilon} \|e\|^2 + \frac{3}{2} \varepsilon^{\frac{1}{3}} \varpi_2 \|x\|^2 + \frac{\varpi_2}{2\varepsilon} \left(\sum_{i=1}^{2} k_i^2 \varepsilon^{\frac{20}{3} - 2i} + \varepsilon^{\frac{2}{3}} \right) \|e\|^2 \\ &+ \frac{\varpi_1 \varepsilon^{\frac{1}{3}}}{2} \left[(\varepsilon^{-\frac{2}{3}} + F_2) \|e\|^2 + F_1 \|x\|^2 + F_0 \right] + \left(\frac{\varepsilon^{-\frac{1}{3}} \|e\|^2}{2} + \frac{\varepsilon^{\frac{1}{3}}}{2} \right) \Psi \varpi_1. \end{aligned} \tag{7.58}$$

Write

$$\begin{aligned} r_1 &:= \mu_3 - \frac{\varpi_2}{2} \left(\sum_{i=1}^{2} k_i^2 \varepsilon^{\frac{20}{3} - 2i} + \varepsilon^{\frac{2}{3}} \right) - \frac{\varepsilon^{\frac{2}{3}}}{2} \left(\Psi \varpi_1 + \varpi_1 + \varpi_1 \varepsilon^{\frac{2}{3}} F_2 \right), \\ r_2 &:= 1 - \frac{3\varpi_2}{2} \varepsilon^{\frac{1}{3}} - \frac{\varpi_1}{2} \varepsilon^{\frac{1}{3}} F_1, \end{aligned} \tag{7.59}$$

and define ε^* as

$$\varepsilon^* := \max\{\varepsilon \in \mathbb{R} | r_1(\varepsilon) \ge 0, r_2(\varepsilon) \ge 0, \text{and } \varepsilon \ge 0\}. \tag{7.60}$$

For $\varepsilon \in (0, \varepsilon^*)$, from inequality (7.58), we have

$$\begin{aligned} \frac{dV}{dt} &\leq -r_2\|x\|^2 - \frac{r_1}{\varepsilon}\|e\|^2 + \frac{\varpi_1}{2}\varepsilon^{\frac{1}{3}}F_0 + \frac{\Psi\varpi_1}{2}\varepsilon^{\frac{1}{3}} \\ &\leq -\frac{r_2}{\iota_2}Q(x) - \frac{r_1}{\varepsilon\mu_2}P(e) + \frac{\varpi_1}{2}\varepsilon^{\frac{1}{3}}F_0 + \frac{\Psi\varpi_1}{2}\varepsilon^{\frac{1}{3}}. \end{aligned} \tag{7.61}$$

Writing $\gamma = \min\left\{\frac{r_2}{\iota_2}, \frac{r_1}{\varepsilon\mu_2}\right\}$ and $\delta(t, t_0) = \exp(-\gamma(t - t_0))$, we obtain that

$$V \leq V(x(t_0), e(t_0))\delta(t, t_0) + \frac{\varpi_1\varepsilon^{\frac{1}{3}}(F_0 + \Psi)}{2\gamma}(1 - \delta(t, t_0)). \tag{7.62}$$

When $t \to \infty$ and $\delta(t, t_0) \to 0$, we have

$$V(x, e) \leq \frac{\varpi_1\varepsilon^{\frac{1}{3}}(F_0 + \Psi)}{2\gamma}. \tag{7.63}$$

With the relationships in (7.21), (7.26), (7.27), and (7.44), we obtain that for $i \in \{1,2,3\}$

$$\begin{aligned} &\limsup_{t\to\infty} \|x\| \leq \sqrt{\frac{\varpi_1\varepsilon^{\frac{1}{3}}(F_0 + \Psi)}{2\iota_1\gamma}}, \\ &\limsup_{t\to\infty} |x_i - \hat{x}_i| \leq \varepsilon^{3-i}\sqrt{\frac{\varpi_1\varepsilon^{\frac{1}{3}}(F_0 + \Psi)}{2\mu_1\gamma}}. \end{aligned} \tag{7.64}$$

This completes the proof. □

The above results show the stability of the closed-loop control system under the proposed event-triggering condition, which depends on both the state and the trend of the tracking error and is called a "PD-type" of event-triggering condition. Specifically, for the PD-type event-triggering condition in (7.35) and $\sigma(t)$ in (7.28), we have $\dot{\sigma}(t) = -x_2(t)/\varepsilon^2$, which means the trend of position tracking error is under consideration. This development enables the predictive triggering ability of the event-triggering mechanism, but introduces additional challenges to ensure the stability of the closed-loop system. With the event-triggering mechanism, the ET-ESO obtains the value of the outputs only when an event occurs and a zero-order-hold module is applied at a no-event instant. As for the events, since $\Gamma(t) = 0$ for $\dot{\sigma}(t)\sigma(t) \leq 0$, the system states information is not available when $|\sigma(t)| \geq \left(E - \frac{\varpi_1\mu_2 M\varepsilon}{\mu_1\mu_3}\right)$ but $|\sigma(t)|$ is decreasing, which is the main difference of the ET-ESO adopted in this chapter from the ET-ESO considered in the previous chapter where the event-triggering condition depends on terms like $\sum_{j=1}^{3}|\alpha_j(e_1, \sigma)|$ determined by the tracking error. Therefore, we need to prove the convergence of the Lyapunov function considered here for every $\sigma(t)$ that satisfies $\dot{\sigma}(t)\sigma(t) \leq 0$.

On the other hand, recall that the event-triggering condition in Chap. 6 is provided as

$$\Gamma(t) = \begin{cases} 0, \text{ if } \sum_{j=1}^{3} \left|\alpha_j(e_1, \sigma)\right| \leq \Psi\varepsilon, \\ 1, \text{ otherwise.} \end{cases} \tag{7.65}$$

Considering the existence of e_1, the condition can be implemented by setting a local observer on the sensor side of their control system. Compared with (7.65), the event-triggering condition we introduced in (7.35) has the advantage of simplifying the controller structure, rendering the condition feasible for a robotic system in terms of energy consumption, engineering practice, and communication bandwidth.

From (7.31) and Theorem 7.2, we can obtain that the $x_i(t)$ of system (7.12) and the external disturbance $\lambda(t)$ are bounded. In particular, the value of B can be provided by

$$|x_i| + |\lambda(t)| \leq B, \ (i = 1, 2). \tag{7.66}$$

7.4 Experiment Results

To evaluate the proposed event-triggered control method, an electric cylinder control system shown in Fig. 7.3 is used in this section. This system mainly consists of a drive circuit, a control computer, two monitors, and an execution and loading mechanism. The control method can be programmed by the control computer. The drive circuit can convert digital control commands into analog control signals acceptable for electric cylinders. At the same time, the data of system operating status can be captured and displayed on two monitors. The execution and loading mechanism is presented in detail in Fig. 7.4. It consists of an electric cylinder (Exlar GSM20) which is the same as the electric cylinders used in a recently developed robotic system, a load device and additional fixing and connecting structural parts. With different numbers of inertia

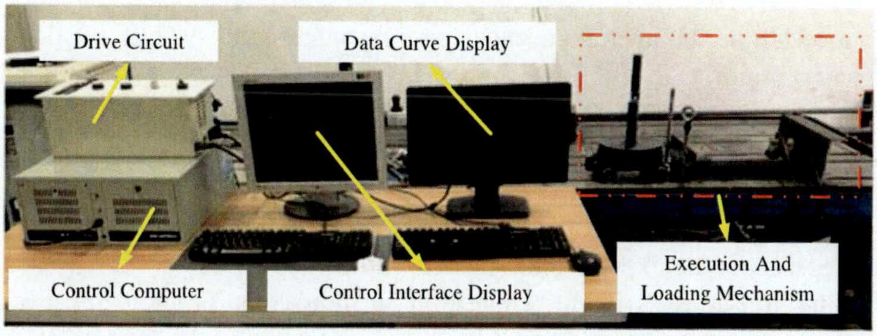

Fig. 7.3 An electric cylinder control system

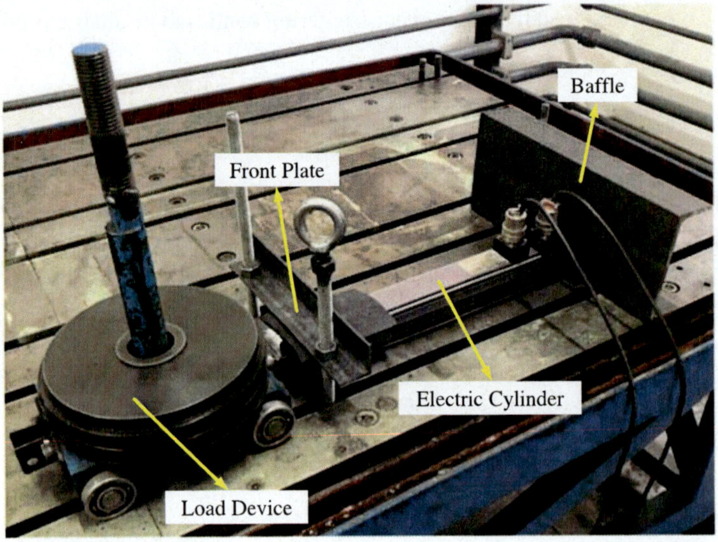

Fig. 7.4 Execution and loading mechanism

disks, the load device can simulate different inertia load conditions. Furthermore, the workspace of the electric cylinder is from -50.0 mm to 50.0 mm, the output torque is from -1.1 N·m to 1.1 N·m, and the inertia of the screw J is 1.3×10^{-4} kg·m^2. Since the control circuit outputs ± 10.0 V analog signal, the input-torque ratio K_p is $0.11\,\mathrm{N{\cdot}m{\cdot}V^{-1}}$, and it is obtained $\rho_5 = 846.2\,\mathrm{V^{-1}s^{-2}}$.

Two different kinds of reference trajectory inputs, namely, the squarewave input and the multitone sinusoid input, are chosen to carry out the experiments with the introduced ET-ADRC. A time-triggered continuous-time ADRC (CT-ADRC) is also implemented for comparison. The controllers are implemented by Euler approximation with sampling time $t_s = 2.2$ms. The differential term $\dot{\sigma}$ is approximated as

$$\dot{\sigma}(kt_s) = \frac{\sigma(kt_s) - \sigma((k-1)t_s)}{t_s}. \tag{7.67}$$

In order to evaluate the average event-triggering rate of the system, we define the average sampling time T_A as

$$T_A = \begin{cases} 2.2, & \text{for CT-ADRC,} \\ \frac{T}{N_s}, & \text{for ET-ADRC,} \end{cases} \tag{7.68}$$

where T represents the experimental time in milliseconds, and N_s is the total triggering counts in one experiment. Therefore, an increasing T_A indicates a reduced triggering rate.

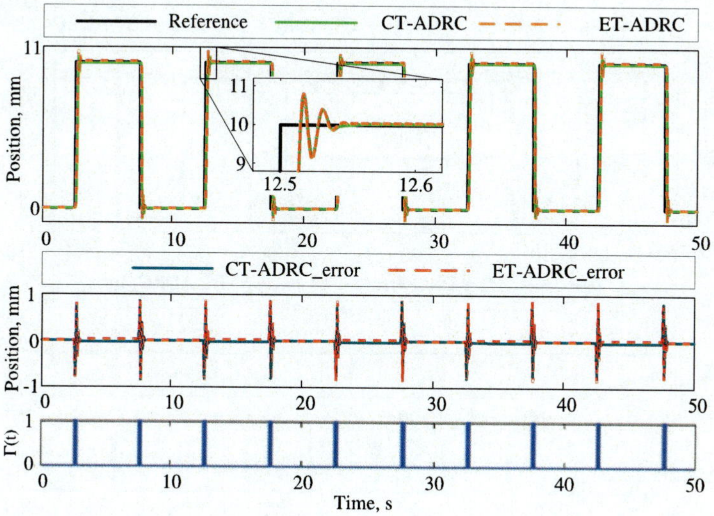

Fig. 7.5 Tracking performance without load

To quantitatively measure the efficiency of the proposed controller, the integral of absolute error (IAE) index is defined as

$$IAE = \frac{P}{2\pi} \int |\theta_1(t) - \theta_1^*(t)| dt. \tag{7.69}$$

7.4.1 Squarewave-Type Reference

The tracking performance of the system with a squarewave-type reference is first illustrated. For (7.16), the parameter $\varepsilon = 0.2$ and h_i is given as

$$h_1(d) = 2d + \varphi(d),\ h_2(d) = 3d,\ h_3(d) = d. \tag{7.70}$$

The control law and event-triggering condition are designed as

$$\rho_5 u_1 = -5\hat{x}_1 - \hat{x}_2 - \hat{x}_3, \tag{7.71}$$

and

$$\Gamma(kt_s) = \begin{cases} 0, & \text{if } \dot{\sigma}(kt_s)\sigma(kt_s) \leq 0.3|\dot{\sigma}(kt_s)|, \\ 1, & \text{otherwise}. \end{cases} \tag{7.72}$$

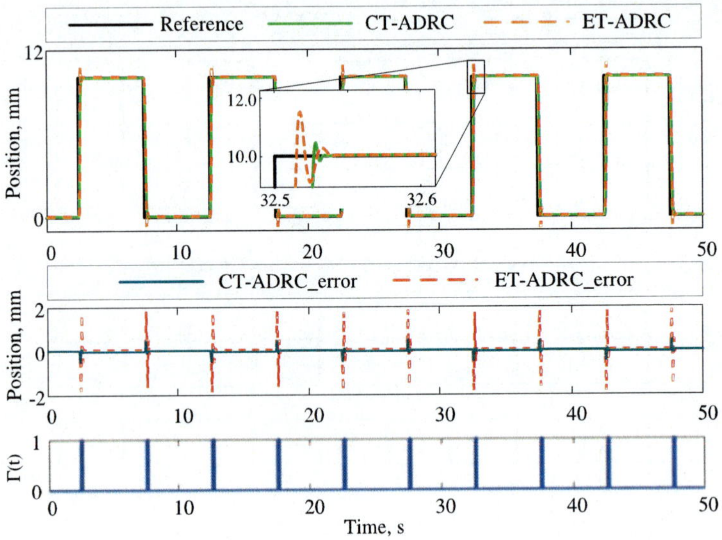

Fig. 7.6 Tracking performance with load of 25.6 kg

Table 7.2 Performance for Squarewave-Type Reference

Experimental schemes	Overshoot	T_A	IAE
CT-ADRC without load	0.77 mm	2.20 ms	26.9 m
ET-ADRC without load	0.77 mm	36.85 ms	31.2 m
CT-ADRC with load	0.55 mm	2.20 ms	15.6 m
ET-ADRC with load	1.84 mm	37.52 ms	41.8 m

The reference is a squarewave signal processed by a low-pass filter so that $|\upsilon(t)|$ and $|\dot{\upsilon}(t)|$ are both guaranteed to be bounded. Figures 7.5 and 7.6 show the tracking performance without load and with load of 25.6 kg. For these figures, the plot on the top shows the position tracking response, the plot in the middle shows the position tracking error curve, and the bottom plot shows the event-triggering instants. The overshoot, T_A, and IAE index are provided in Table 7.2.

The overshoots of the electric cylinder controlled by CT-ADRC and ET-ADRC without load are both less than 1 mm, but the IAE index for ET-ADRC (31.2 m) is higher than that for CT-ADRC (26.9 m). With a load of 25.6 kg, the overshoot of CT-ADRC is 0.55 mm with the IAE index equal to 15.6 m, while the overshoot of ET-ADRC is 1.84 mm with the IAE index being 41.8 m. These results indicate the ET-ADRC has decreased position tracking performance compared with the CT-ADRC. However, considering the electric cylinder is connected by joints for which the motion gap exceeds 2.0 mm in practical applications, the tracking performances are acceptable to ensure the stable operation of a robotic system. The average sampling time

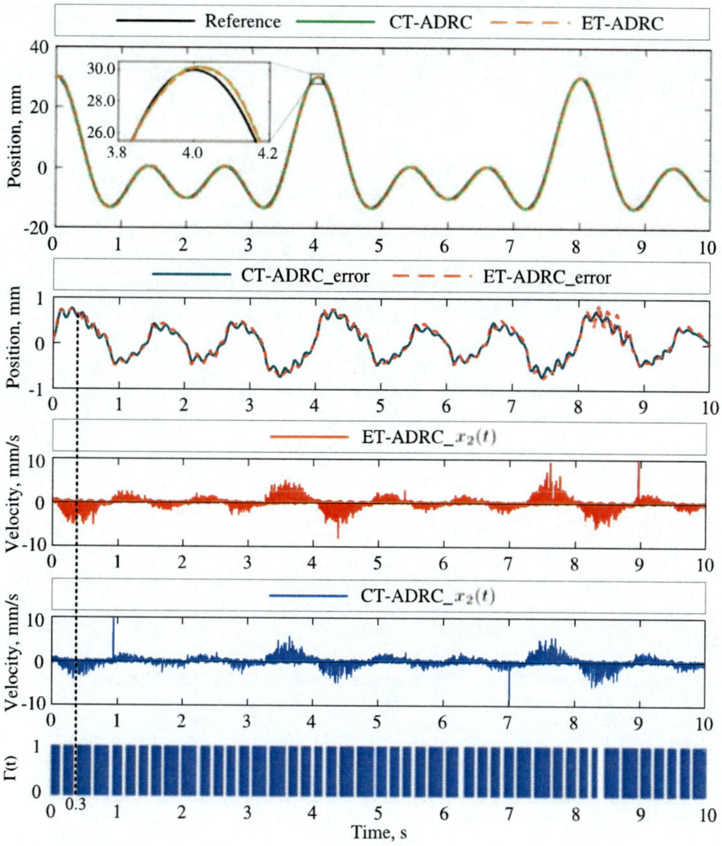

Fig. 7.7 Tracking performance without load

T_A of ET-ADRC increases up to 37.52 ms which is a sharp increase compared with $T_A = 2.20$ ms for CT-ADRC.

7.4.2 *Multitone Sinusoid Reference*

Next, the tracking performance of the system with multitone sinusoid reference is illustrated. In this case, for (7.16), h_i is given as

$$h_1(d) = 3d + \varphi(d),\ h_2(d) = 3d,\ h_3(d) = d \tag{7.73}$$

and the parameter $\varepsilon = 0.2$. The event-triggering condition is the same as (7.72). The control law is designed as

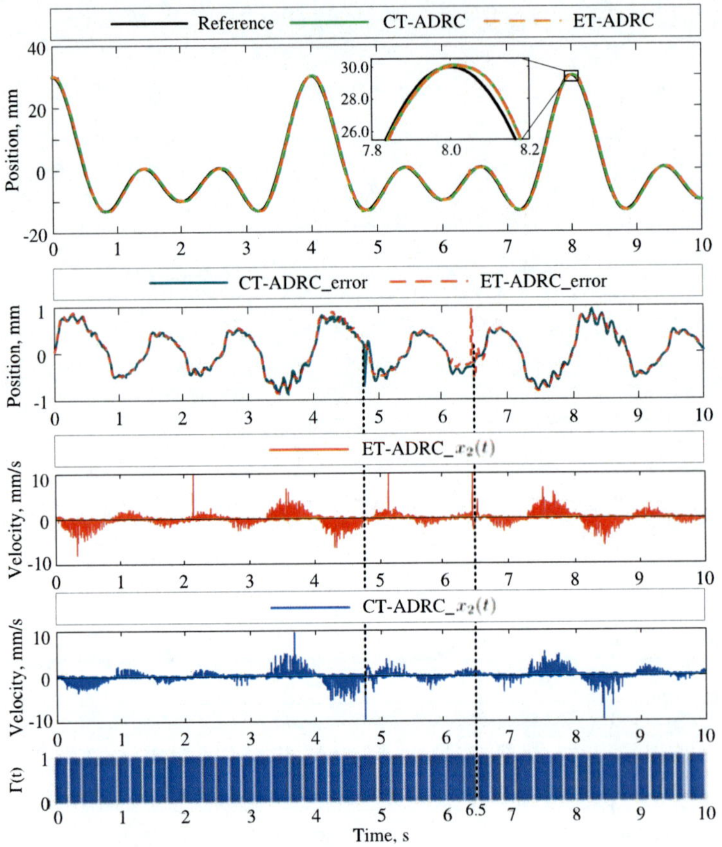

Fig. 7.8 Tracking performance with load of 25.6 kg

Table 7.3 Performance for Multitone Sinusoid Reference

Experimental schemes	Tracking error	T_A	IAE
CT-ADRC without load	0.78 mm	2.20 ms	34.1 m
ET-ADRC without load	0.83 mm	3.89 ms	36.1 m
CT-ADRC with load	0.94 mm	2.20 ms	39.5 m
ET-ADRC with load	0.94 mm	3.54 ms	40.5 m

$$\rho_5 u_1 = -4\hat{x}_1 - \hat{x}_2 - \hat{x}_3 \tag{7.74}$$

Since the reference trajectory is a multitone sinusoid signal, $|\upsilon(t)|$ and $|\dot{\upsilon}(t)|$ are bounded. Figures 7.8 and 7.7 show the results with and without a load of 25.6 kg.

The tracking error, T_A, and IAE index are shown in Table 7.3. From these results, we observe that the system can be controlled by both CT-ADRC and ET-ADRC with tracking error less than 1 mm for different loading conditions. Although the ET-ADRC leads to higher IAE indexes than CT-ADRC, the differences are less than 5.8%. As a compromise, ET-ADRC achieves a much larger T_A. To illustrate the performance of ESO, the trends of $x_2(t)$ are provided in Figs. 7.7 and 7.8. From Fig. 7.7, we can observe that the condition $\Gamma(t)$ is not triggered at 0.3 s when the tracking error exceeds the threshold but is decreasing. This characteristic highlights the benefit of having a PD-type of event-triggering condition.

7.5 Discussion

In this chapter, an event-triggered ADRC is adopted to control an electric cylinder, and a PD-type event-triggering condition considering both the size and trend of tracking error is proposed. With the introduced ET-ADRC, the observation error of the ESO and the tracking error of the system is shown to be asymptotically bounded. The performance of the controller is extensively evaluated through comparative experiments on an actual electric cylinder control system for different load conditions and reference signals.

References

1. Årzen, K.E.: A simple event-based pid controller. IFAC World Congress **32**, 8687–8692 (1999)
2. Åstrom, K.J., Bernhardsson, B.: Comparison of periodic and event based sampling for first order stochastic systems. IFAC World Congress **32**, 5006–5011 (1999)
3. Behzad, A., Amin, N.: Hardware implementation of an adrc controller on a gimbal mechanism. IEEE Trans. Control. Syst. Technol. **26**(6), 2268–2275 (2017)
4. Chaoui, H., Khayamy, M., Aljarboua, A.A.: Adaptive interval type-2 fuzzy logic control for pmsm drives with a modified reference frame. IEEE Trans. Ind. Electron. **64**(5), 3786–3797 (2017)
5. Chen, S., Xue, W., Huang, Y.: Analytical design of active disturbance rejection control for nonlinear uncertain systems with delay. Control Eng. Pract. **84**, 323–336 (2019)
6. Chu, Z., Sun, Y., Wu, C., Sepehri, N.: Active disturbance rejection control applied to automated steering for lane keeping in autonomous vehicles. Control Eng. Pract. **74**, 13–21 (2018). https://doi.org/10.1016/j.conengprac.2018.02.002
7. Ding, W., Deng, H., Xia, Y., Duan, X.: Tracking control of electro-hydraulic servo multi-closed-chain mechanisms with the use of an approximate nonlinear internal model. Control Eng. Pract. **58**, 225–241 (2017). https://doi.org/10.1016/j.conengprac.2016.11.003
8. Durand, S., Guerrero-Castellanos, J.F.: Event-based digital pid control. In: 2015 International Conference on Event-based Control, Communication, and Signal Processing (EBCCSP), pp. 1–7 (2015). https://doi.org/10.1109/EBCCSP.2015.7300654
9. Guo, J., Xue, W., Hu, T.: Active disturbance rejection control for pmlm servo system in cnc machining. J. Syst. Sci. Complex. **29**(1), 74–98 (2016). https://doi.org/10.1007/s11424-015-3258-2
10. Han, J.: From pid to active disturbance rejection control. IEEE Trans. Ind. Electron. **56**(3), 900–906 (2009). https://doi.org/10.1109/TIE.2008.2011621
11. Heemels, W.P.M.H., Donkers, M.C.F.: Model-based periodic event-triggered control for linear systems. Automatica **49**(3), 698–711 (2013)

12. Huang, Y., Wang, J., Shi, D., Shi, L.: Toward event-triggered extended state observer. IEEE Trans. Autom. Control **63**(6), 1842–1849 (2018)
13. Li, H., Shi, Y.: Event-triggered robust model predictive control of continuous-time nonlinear systems. Automatica **50**(5), 1507–1513 (2014)
14. Liu, X., Shan, L., Chen, H., Li, J.: The application of the improved udp based on the extrapolation algorithm in the network servo system. In: IEEE International Conference on Information and Automation (2017)
15. Matraji, I., Al-Durra, A., Haryono, A., Al-Wahedi, K., Abou-Khousa, M.: Trajectory tracking control of skid-steered mobile robot based on adaptive second order sliding mode control. Control Eng. Pract. **72**, 167–176 (2018). https://doi.org/10.1016/j.conengprac.2017.11.009
16. Mendoza-Mondragon, F., Hernandez-Guzman, V.M., Rodriguez-Resendiz, J.: Robust speed control of permanent magnet synchronous motors using two-degrees-of-freedom control. IEEE Trans. Ind. Electron. **65**(8), 6099–6108 (2018)
17. Plummer, A.: Model-based motion control for multi-axis servohydraulic shaking tables. Control Eng. Pract. **53**, 109–122 (2016). https://doi.org/10.1016/j.conengprac.2016.05.004
18. Puchta, E.D.P., Siqueira, H.V., dos Santos Kaster, M.: Optimization tools based on metaheuristics for performance enhancement in a gaussian adaptive pid controller. IEEE Trans. Cybern. **50**, 1–10 (2019). https://doi.org/10.1109/TCYB.2019.2895319
19. Rubaai, A., Castro-Sitiriche, J., Ofoli, A.R.: Design and implementation of parallel fuzzy pid controller for high-performance brushless motor drives: an integrated environment for rapid control prototyping. IEEE Trans. Ind. Appl. **44**(4), 1090–1098 (2008)
20. Ruiz-Carcel, C., Starr, A.: Data-based detection and diagnosis of faults in linear actuators. IEEE Trans. Instrum. Meas. **67**(9), 2035–2047 (2018). https://doi.org/10.1109/TIM.2018.2814067
21. Shi, D., Shi, L., Chen, T.: Event-Based State Estimation. Springer, Berlin (2016)
22. Shi, D., Xue, J., Zhao, L., Wang, J., Huang, Y.: Event-triggered active disturbance rejection control of dc torque motors. IEEE/ASME Trans. Mechatron. **22**(5), 2277–2287 (2017)
23. Sobczyk, M.R., Perondi, E.A., Cunha, M.A.B.: A continuous approximation of the lugre friction model. ABCM Symp. Ser. Mechatron. **4**, 218–228 (2010)
24. Sobczyk, M.R., Perondi, E.A., Cunha, M.A.B.: A continuous extension of the lugre friction model with application to the control of a pneumatic servo positioner. In: 2012 IEEE 51st IEEE Conference on Decision and Control (CDC), pp. 3544–3550 (2012). https://doi.org/10.1109/CDC.2012.6426406
25. Tabuada, P.: Event-triggered real-time scheduling of stabilizing control tasks. IEEE Trans. Autom. Control **52**(9), 1680–1685 (2007)
26. Wang, A., Mu, B., Shi, Y.: Consensus control for a multi-agent system with integral-type event-triggering condition and asynchronous periodic detection. IEEE Trans. Ind. Electron. **64**(7), 5629–5639 (2017). https://doi.org/10.1109/TIE.2017.2677312
27. Wang, Y., Wang, Y., Liu, X., Gu, Y.: Control research of network servo system based on smith predictor. In: Intelligent Control and Automation (2015)
28. Xue, W., Bai, W., Yang, S., Song, K., Huang, Y., Xie, H.: Adrc with adaptive extended state observer and its application to air-fuel ratio control in gasoline engines. IEEE Trans. Ind. Electron. **62**(9), 5847–5857 (2015). https://doi.org/10.1109/TIE.2015.2435004
29. Xue, W., Madonski, R., Lakomy, K., Gao, Z., Huang, Y.: Add-on module of active disturbance rejection for set-point tracking of motion control systems. IEEE Trans. Ind. Appl. **53**(4), 4028–4040 (2017). https://doi.org/10.1109/TIA.2017.2677360
30. Yang, J., Chen, W., Li, S., Guo, L., Yan, Y.: Disturbance/uncertainty estimation and attenuation techniques in pmsm drives-a survey. IEEE Trans. Ind. Electron. **64**(4), 3273–3285 (2017)
31. Yao, J., Deng, W., Jiao, Z.: Adaptive control of hydraulic actuators with lugre model-based friction compensation. IEEE Trans. Ind. Electron. **62**(10), 6469–6477 (2015)
32. Zheng, Q., Dong, L., Lee, D.H., Gao, Z.: Active disturbance rejection control for mems gyroscopes. IEEE Trans. Control Syst. Technol. **17**(6), 1432–1438 (2009). https://doi.org/10.1109/TCST.2008.2008638
33. Zhong, S., Huang, Y., Guo, L.: A parameter formula connecting PID and ADRC. In: Science China Information Sciences, pp. 1–13. (2020). https://doi.org/10.1007/s11432-019-2712-7

Chapter 8
Event-Triggered Attitude Tracking for Rigid Spacecraft

8.1 Background

In this chapter, we illustrate the applicability of ET-ADRC in attitude tracking for spacecraft systems. Since the kinematic and dynamic equations of rigid spacecraft possess highly coupled nonlinear characteristics and attitude dynamics may suffer from external disturbances, inertia-matrix uncertainties, and even actuator failures, control systems play an important role in achieving attitude maneuvering, attitude tracking, and high-precision pointing of spacecraft, and the attitude-control problem has attracted significant attention over the past few decades [1–5].

Different strategies have been investigated for attitude-control problems of rigid spacecraft under different restrictive conditions, including sliding mode control [6, 7], adaptive robust control [8, 9], and optimal control [10]. Considering the limitations of the onboard communication and computation resources for embedded computer systems on spacecraft, resource-aware control strategies need to be adopted, which is why the ET-ADRC approach is considered for attitude tracking in this chapter. In [11], an event-triggered control approach to reduce the control-updating frequency was proposed for spacecraft-attitude stabilization, but without considering the effect of disturbances. The same problem was addressed in [12], where two different types of triggering events, i.e., fixed- and relative-threshold strategies, were proposed. Attitude stabilization with an event-triggered mechanism and external disturbances was considered in [13]. In [14], an observer-based event-triggered control approach was proposed to cope with external disturbances and actuator faults and to reduce the control-updating frequency.

Unlike [11–14], which focused upon attitude stabilization, we consider an event-triggered attitude-tracking problem for a spacecraft with inertial uncertainties and external disturbances in this chapter. In particular, we consider the effect of inertial uncertainties that are coupled to the system states and make the attitude tracking more difficult to achieve with reduced sampling frequency. To attenuate these effects, we utilize an ET-ADRC approach to the control-system design. Specifically, we focus on how to utilize ESO to observe the coupled three-channel disturbances and how

D. Shi et al., *Event-Triggered Active Disturbance Rejection Control*,
Studies in Systems, Decision and Control 356,
https://doi.org/10.1007/978-981-16-0293-1_8

to design an event-triggered mechanism to guarantee the stability of the integrated tracking system without information about the disturbances. The main contents of this chapter are summarized as follows:

(1) An event-triggered mechanism is proposed for attitude tracking of a spacecraft under communication-resource restrictions. The event-triggered mechanism only needs measurements of spacecraft-attitude orientation and angular velocity, without information concerning parameter uncertainties and external disturbances.

(2) Using the designed event-triggered mechanism, an ET-ADRC scheme is proposed and the corresponding theoretical analysis is developed. The asymptotic upper bounds on the steady-state-observation errors with parameters in the event-triggering condition are provided. Moreover, the tracking system is proved to be asymptotically stable, indicating that the tracking problem can be solved by using ET-ADRC.

(3) The developed results are illustrated by numerical simulations, which show that compared with the time-triggered ADRC scheme, ET-ADRC achieves satisfactory tracking performance at a reduced average data transmission rate.

The remainder of this chapter is organized as follows. In Sect. 8.2, the attitude-tracking system for a rigid spacecraft is described and the problem is formulated. The theoretical analysis of the performance of the event-triggered ADRC scheme is discussed in Sect. 8.3. The numerical results are provided in Sects. 8.4, and 8.5 presents some concluding remarks.

Finally, to aid the analysis of attitude dynamics, we use the operator $a^\times$ to denote a skew-symmetric matrix of any vector $a = [a_1, a_2, a_3]^{\mathrm{T}}$ such that

$$a^\times = \begin{bmatrix} 0 & -a_3 & a_2 \\ a_3 & 0 & -a_1 \\ -a_2 & a_1 & 0 \end{bmatrix}.$$

This notation is used throughout this chapter.

8.2 Problem Formulation

Consider an event-triggered control scheme for attitude tracking in Fig. 8.1. Following [1], the attitude kinematics and dynamics of a rigid spacecraft is modeled in terms of the unit quaternion that is free of singularity:

$$\dot{q}_v = \frac{1}{2}(q_4 I_3 + q_v^\times)\omega,\ \dot{q}_4 = -\frac{1}{2}q_v^{\mathrm{T}}\omega,\ \text{and}\ J\dot{\omega} = -\omega^\times J\omega + u + d, \tag{8.1}$$

where $\omega \in \mathbb{R}^3$ is the angular velocity of the spacecraft, $u \in \mathbb{R}^3$ and $d \in \mathbb{R}^3$ are the control torques and external unknown disturbances, respectively, $J \in \mathbb{R}^{3\times 3}$ is the symmetric inertial matrix of the spacecraft, the unit quaternion $q = [q_v^{\mathrm{T}}, q_4]^{\mathrm{T}} \in \mathbb{R}^3 \times \mathbb{R}$ represents the attitude orientation of a body-fixed frame with respect to an inertial frame comprising a vector part $q_v = [q_1, q_2, q_3]^{\mathrm{T}} \in \mathbb{R}^3$ and scalar component $q_4 \in \mathbb{R}$, which satisfies the constraint $q_v^{\mathrm{T}} q_v + q_4^2 = 1$.

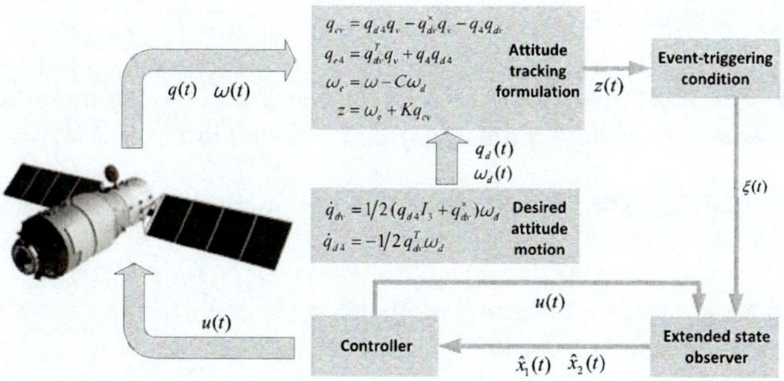

Fig. 8.1 Schematic of the event-triggered ADRC scheme

The attitude-tracking problem is formulated according to [7, 10]. We assume that the desired attitude motion is generated by

$$\dot{q}_{dv} = \frac{1}{2}(q_{d4}I_3 + q_{dv}^{\times})\omega_d,\ \dot{q}_{d4} = -\frac{1}{2}q_{dv}^{\mathrm{T}}\omega_d, \tag{8.2}$$

where $q_d = [q_{dv}^{\mathrm{T}}, q_{d4}]^{\mathrm{T}} \in \mathbb{R}^3 \times \mathbb{R}$ is the target-attitude unit quaternion satisfying $q_{dv}^{\mathrm{T}}q_{dv} + q_{d4}^2 = 1$, and ω_d is the target angular velocity. Following the discussions in [10], ω_d, $\dot{\omega}_d$, and $\ddot{\omega}_d$ are assumed to be bounded. Subsequently, the objective of attitude tracking can be expressed as

$$\limsup_{t\to\infty} \eta_1(t) = 0,\ \eta_1(t) = q(t) - q_d(t), \tag{8.3}$$

$$\limsup_{t\to\infty} \eta_2(t) = 0,\ \eta_2(t) = \omega(t) - \omega_d(t). \tag{8.4}$$

This problem can be converted into a stabilization problem by considering the attitude-tracking-error quaternion $q_e = [q_{ev}^{\mathrm{T}}, q_{e4}]^{\mathrm{T}} \in \mathbb{R}^3 \times \mathbb{R}$ introduced in [1]:

$$q_{ev} = q_{d4}q_v - q_{dv}^{\times}q_v - q_4q_{dv},\ q_{e4} = q_{dv}^{\mathrm{T}}q_v + q_4q_{d4},\ \text{and } \omega_e = \omega - C\omega_d, \tag{8.5}$$

where

$$C = (1 - 2q_{ev}^{\mathrm{T}}q_{ev})I_3 + 2q_{ev}q_{ev}^{\mathrm{T}} - 2q_{e4}q_{ev}^{\times}$$

with $\|C\| = 1$ and $\dot{C} = -\omega_e^{\times}C$. Note that according to [10], the error quaternion also satisfies the constraint $q_{ev}^{\mathrm{T}}q_{ev} + q_{e4}^2 = 1$. Then, the attitude-tracking system can be formulated as

$$\dot{q}_{ev} = \frac{1}{2}(q_{e4}I_3 + q_{ev}^{\times})\omega_e,\ \dot{q}_{e4} = -\frac{1}{2}q_{ev}^{\mathrm{T}}\omega_e, \tag{8.6}$$

$$J\dot{\omega}_e = -(\omega_e + C\omega_d)^{\times}J(\omega_e + C\omega_d) + J(\omega_e^{\times}C\omega_d - C\dot{\omega}_d) + u + d. \tag{8.7}$$

To design the control law, a few technical lemmas are first presented.

Lemma 8.1 *The objectives (8.4) can be achieved if there exists a control law for system (8.6) and (8.7) such that* $\lim_{t\to\infty} q_{ev}(t) = 0$ *and* $\lim_{t\to\infty} \omega_e(t) = 0$.

Proof This result can be proved based on the analysis of Sections II and III in [15]. □

Then, taking the coordinate transformation of $z = \omega_e + K q_{ev}$ introduced in [9], systems (8.6) and (8.7) can be rewritten as

$$\dot{q}_{ev} = \frac{1}{2}(q_{e4}I_3 + q_{ev}^{\times})\omega_e,\ \dot{q}_{e4} = -\frac{1}{2}q_{ev}^{\mathrm{T}}\omega_e, \tag{8.8}$$

$$\begin{aligned} J\dot{z} =& -(\omega_e + C\omega_d)^{\times} J(\omega_e + C\omega_d) + J(\omega_e^{\times} C\omega_d - C\dot{\omega}_d) \\ &+ \frac{1}{2} JK(q_{e4}I_3 + q_{ev}^{\times})\omega_e + u + d, \end{aligned} \tag{8.9}$$

where K is a positive definite matrix.

Lemma 8.2 ([7]) *Consider the rigid spacecraft governed by systems (8.8) and (8.9). Then, for any* $z(t)$ *satisfying* $\lim_{t\to\infty} z(t) = 0$, *it follows that* $\lim_{t\to\infty} q_{ev}(t) = 0$ *and* $\lim_{t\to\infty} \omega_e(t) = 0$, *respectively.*

Due to the fuel consumption and payload variations, parameter uncertainties of the inertial matrix invariably exist in the control system. Similar to [7], the inertial matrix containing parameter uncertainties has the form $J = J_0 + \Delta J$, where J_0 denotes the known nominal matrix that is selected to be nonsingular and ΔJ is the uncertainty associated with J. Thus, the dynamics given by (8.9) can be rewritten as

$$\begin{aligned} (J_0 + \Delta J)\dot{z} =& -(\omega_e + C\omega_d)^{\times}(J_0 + \Delta J)(\omega_e + C\omega_d) + (J_0 + \Delta J)(\omega_e^{\times} C\omega_d - C\dot{\omega}_d) \\ &+ \frac{1}{2}(J_0 + \Delta J)K(q_{e4}I_3 + q_{ev}^{\times})\omega_e + u + d. \end{aligned} \tag{8.10}$$

In fact, the disturbance forces caused by solar radiation, aerodynamics, and magnetic fields vary slowly in time, and the properties of the spacecraft (e.g., shape and mass) also vary slowly in time because of fuel consumption and payload variations. Based on these considerations, the following assumptions are adopted.

Assumption 8.1 The disturbance d in (8.1) is assumed to be bounded as $\|d\| \le D_1$ and to be differentiable with the bound, i.e., $\|\dot{d}\| \le D_2$, where D_1 and D_2 are unknown positive constants.

Assumption 8.2 The uncertainty part of the inertial matrix ΔJ is assumed to be bounded as $\|\Delta J\| \le J_c$, where J_c is an unknown positive constant. The inertial matrix is assumed to be slowly varying, that is, $\|\dot{J}\| \le J_d$ with J_d being an unknown positive constant.

According to [7, 16, 17], these assumptions are common requirements in robust attitude control of a rigid spacecraft.

On the other hand, utilizing the matrix inversion lemma, $(J_0 + \Delta J)^{-1}$ can be expressed as follows:

$$(J_0 + \Delta J)^{-1} = J_0^{-1} + \Delta \tilde{J}, \text{ with } \Delta \tilde{J} = -(I_3 + J_0^{-1}\Delta J)^{-1} J_0^{-1} \Delta J J_0^{-1},$$

where $\Delta \tilde{J}$ is an additive uncertainty. Then, following some simple algebraic transformations to (8.10), we obtain

$$\dot{z} = F + G + J_0^{-1} u + \bar{d}, \tag{8.11}$$

where

$$F = J_0^{-1}[-\omega^\times J_0 \omega + J_0(\omega_e^\times C\omega_d - C\dot{\omega}_d) + \frac{1}{2} J_0 K(q_{e4} I_3 + q_{ev}^\times)\omega_e], \tag{8.12}$$

$$\begin{aligned} G = {} & J_0^{-1}[-\omega^\times \Delta J \omega + \Delta J(\omega_e^\times C\omega_d - C\dot{\omega}_d) + \frac{1}{2} \Delta J K(q_{e4} I_3 + q_{ev}^\times)\omega_e] \\ & + \Delta \tilde{J}[-\omega^\times J\omega + J(\omega_e^\times C\omega_d - C\dot{\omega}_d) + \frac{1}{2} J K(q_{e4} I_3 + q_{ev}^\times)] + \Delta \tilde{J} u, \end{aligned} \tag{8.13}$$

$$\bar{d} = (J_0^{-1} + \Delta \tilde{J})d. \tag{8.14}$$

Equation (8.11) models the attitude-tracking system in the presence of parameter uncertainties and unknown bounded disturbances. Finally, according to Lemma 8.2, if there exists a dynamic state-feedback-control law for system (8.11) such that $\lim_{t\to\infty} z(t) = 0$, it follows that the tracking objectives in (8.4) are achieved. In this chapter, we introduce ET-ADRC to design one such control law. To do this, we define the extended state x_2 as

$$x_2 = F + G + \bar{d}. \tag{8.15}$$

Furthermore, we write $x_1 := z$; then the system (8.11) can be written as

$$\dot{x}_1 = x_2 + J_0^{-1} u, \; \dot{x}_2 = h(t), \tag{8.16}$$

where the function $h(t)$ is the derivative of the extended state, which is assumed to satisfy the following assumption.

Assumption 8.3 The derivative of the extended state is assumed to be bounded, namely, $|h_i(t)| \le M_i$, where $i \in \{1, 2, 3\}$, and M_i are some nonnegative constants.

The extended state x_2 in (8.15) represents the total disturbances, which integrates the information of external disturbances, actual and desired angular velocities and attitude unit quaternions. Suppose Assumptions 8.1–8.2 hold; then, from (8.14), it is clear that $\dot{\bar{d}}$, the derivative of $\bar{d}$, is bounded. Note that the spacecraft is always under smooth operation, which implies that angular velocity ω and its derivative are bounded. Moreover, ω_d, $\dot{\omega}_d$ and $\ddot{\omega}_d$ are the desired system states that are designed to be bounded, and q_{ev}, q_{e4} and C and their derivatives are bounded due to $\|q_e\| = \|C\| = 1$. Then, according to (8.12), we can conclude that $\dot{F}$ is bounded. The control law u and its derivative can be enforced to be bounded, and with the bounded conditions on $\dot{F}$ and $\ddot{d}$, we conclude that $\dot{G}$ is bounded based on (8.13). By combining the discussions above, the derivative of x_2 is bounded. The discussions here are consistent with those in [18, 19]. Note that F includes additional uncertainties due to the asynchronous information caused by the event-triggered sampling scheme adopted. Consequently, we treat F to be a part of the extended state in our design.

Now we introduce the ET-ESO:

$$\begin{aligned} \dot{\hat{x}}_1(t) &= \hat{x}_2(t) + g_1\left(\frac{1}{\varepsilon}[\xi(t) - \hat{x}_1(t)]\right) + J_0^{-1}u,\ \hat{x}_1(t_0) = \hat{x}_{10}, \\ \dot{\hat{x}}_2(t) &= \frac{1}{\varepsilon}g_2\left(\frac{1}{\varepsilon}[\xi(t) - \hat{x}_1(t)]\right),\ \hat{x}_2(t_0) = \hat{x}_{20}, \end{aligned} \tag{8.17}$$

where the above observer contains three sub-observers. Let the suffix i denote the ith element of the vector state x. With this notation, $[\hat{x}_{1i}, \hat{x}_{2i}]^{\mathrm{T}} \in \mathbb{R}^2, i \in \{1, 2, 3\}$ denotes the state of the ith observer, $[\hat{x}_{1i0}, \hat{x}_{2i0}]^{\mathrm{T}} \in \mathbb{R}^2$ denotes the initial value of $[\hat{x}_{1i}, \hat{x}_{2i}]$, ε_i is the high-gain parameter of the ith observer and $\frac{1}{\varepsilon} = \mathrm{diag}\{\frac{1}{\varepsilon_1}, \frac{1}{\varepsilon_2}, \frac{1}{\varepsilon_3}\}$. An operator $g_j(a) \in \mathbb{R}^3$, $j \in \{1, 2\}$ on a vector $a = [a_1, a_2, a_3]^{\mathrm{T}} \in \mathbb{R}^3$ is defined as $g_j(a) = [g_{j1}(a_1), g_{j2}(a_2), g_{j3}(a_3)]^{\mathrm{T}}$, where the detailed expressions of $g_{ji}(\cdot)$ are provided later in our design. Let $\xi(t)$ denote the previously received output x_1, which is given by

$$\xi(t) = \begin{cases} x_1(t_k), & \text{if } \Gamma(t) = 0, \\ x_1(t), & \text{otherwise}, \end{cases} \tag{8.18}$$

where $\Gamma(t)$ is the event-triggering condition to be designed and t_k is the transmission instant determined by the event-triggering condition. The value of $\xi(t)$ will be updated only when $\Gamma(t) = 1$.

For the ith observer, $g_{1i}(\cdot)$ and $g_{2i}(\cdot)$ are selected as

$$g_{1i}(y_1) = b_i y_1 + \varphi(y_1), \tag{8.19}$$

$$g_{2i}(y_1) = c_i y_1 \tag{8.20}$$

with $\varphi(\theta)$ defined as

$$\varphi(\theta) = \begin{cases} -\frac{1}{4}, & \theta < -\frac{\Pi}{2}, \\ \frac{1}{4}\sin\theta, & -\frac{\Pi}{2} \le \theta \le \frac{\Pi}{2}, \\ \frac{1}{4}, & \theta > \frac{\Pi}{2}. \end{cases} \tag{8.21}$$

Moreover, b_i and c_i are some constants such that the matrix $A := \begin{bmatrix} -b_i & 1 \\ -c_i & 0 \end{bmatrix}$ is Hurwitz; then there exists a matrix $\tilde{P} > 0$ satisfying

$$\tilde{P}A + A^{\mathrm{T}}\tilde{P} = -I. \tag{8.22}$$

We define functions $P_i : \mathbb{R}^2 \to \mathbb{R}$ as

$$P_i(y) = \langle \tilde{P}y, y \rangle + \int_0^{y_1} \varphi(s)ds.$$

By choosing the functions g_{1i}, g_{2i}, and φ in this form with b_i and c_i satisfying the inequality of [20]

$$(1 + c_i)/b_i + b_i < 4,$$

there exist nonnegative-definite functions $\Theta_i : \mathbb{R}^2 \to \mathbb{R}$ such that

$$\lambda_{1i}\|y\|^2 \le P_i(y) \le \lambda_{2i}\|y\|^2, \tag{8.23}$$

$$\lambda_{3i}\|y\|^2 \le \Theta_i(y) \le \lambda_{4i}\|y\|^2, \tag{8.24}$$

$$\frac{\partial P_i}{\partial y_1}(y_2 - g_{1i}(y_1)) - \frac{\partial P_i}{\partial y_2} g_{2i}(y_1) \le -\Theta_i(y), \tag{8.25}$$

$$\|\frac{\partial P_i}{\partial y}\| \le \tau_i \|y\|, \tag{8.26}$$

where λ_{1i}, λ_{2i}, λ_{3i}, λ_{4i}, and τ_i are some positive constants. The minimum and maximum values of the corresponding parameters of the three sub-observers are denoted with the suffixes min and max, respectively, e.g., $\{\lambda_{1i}\}$ with $\lambda_{1\min}$ and $\lambda_{1\max}$. Note that the choice of $\varphi(\theta)$ is not unique as long as it satisfies the above requirements.

With the disturbances estimated by the ET-ESO, a state-feedback control law is designed as

$$u = -J_0\tilde{K}\hat{x}_1 - J_0\hat{x}_2, \tag{8.27}$$

where $\tilde{K} = \mathrm{diag}\{\tilde{K}_1, \tilde{K}_2, \tilde{K}_3\}$, and $\tilde{K}_i$ are some positive constants.

For convenience of analysis, we define $\tilde{x}_j$, e_j and $\tilde{e}_i$ as

$$\tilde{x}_j := x_j - \hat{x}_j,\ e_j := [e_{j1}, e_{j2}, e_{j3}]^{\mathrm{T}} = \left[\frac{\tilde{x}_{j1}}{\varepsilon_1^{2-j}}, \frac{\tilde{x}_{j2}}{\varepsilon_2^{2-j}}, \frac{\tilde{x}_{j3}}{\varepsilon_3^{2-j}}\right]^{\mathrm{T}},\ j \in \{1, 2\},$$

$$\tilde{e}_i := [e_{1i}, e_{2i}]^{\mathrm{T}},\ i \in \{1, 2, 3\}, \tag{8.28}$$

where the $\tilde{e}_i$ is related to the observation error of the ith observer consisting of the ith elements of e_1 and e_2. To aid our discussion, define the sampling error $\sigma(t)$ as

$\sigma(t) := \frac{1}{\varepsilon}[x_1(t_k) - x_1(t)], t \in [t_k, t_{k+1})$ and $\alpha_j(e_1, \sigma)$ as $\alpha_j(e_1, \sigma) := g_j(e_1 + \sigma) - g_j(e_1),\ j \in \{1, 2\}$.

In the remainder of this chapter, we mainly discuss the following problems:

- Is there an event-triggering condition such that the observation errors of three sub-observers can be guaranteed to be bounded?
- Based on the proposed event-triggering condition and the designed control law, is it possible to stabilize the system (8.11) in the presence of inertial uncertainties and unknown disturbances?

8.3 ET-ADRC Design

In this section, the problems outlined in the previous section are investigated. By definition, the control law u in (8.27) can also be expressed as

$$u = -J_0\tilde{K}x_1 + J_0\tilde{K}(x_1 - \hat{x}_1) - J_0\hat{x}_2 = -J_0\tilde{K}x_1 + J_0\tilde{K}\varepsilon e_1 - J_0\hat{x}_2, \tag{8.29}$$

where $\varepsilon = \text{diag}\{\varepsilon_1, \varepsilon_2, \varepsilon_3\}$. The minimum and maximum values of $\{\varepsilon_i\}$ are denoted as $\varepsilon_{\min}$ and $\varepsilon_{\max}$, respectively. The event-triggering condition of the observer is proposed to have the following form:

$$\Gamma(t) = \begin{cases} 0, \text{ if} \sum_{i=1}^{3}\sum_{j=1}^{2} |\alpha_{ji}(e_{1i}, \sigma_i)| \le \Psi\varepsilon_{\min}, \\ 1, \text{ otherwise}, \end{cases} \tag{8.30}$$

where $\alpha_{ji}(\cdot)$ is the ith component of vector $\alpha_j(\cdot)$ and Ψ is a given constant. In the following theorem, we analyze the performance of the proposed event-triggered control scheme.

Theorem 8.1 *Consider the closed-loop system in (8.8) and (8.9), the ESO in (8.17), and the control law in (8.27). Suppose that Assumptions 8.1–8.3 hold. For any initial values of x_1, x_2 and $\hat{x}_1, \hat{x}_2$, and a given $\Psi > 0$, there exist $\varepsilon^* > 0$ and an event-triggering condition in (8.30) of the observer such that for any $\varepsilon_i \in (0, \varepsilon^*)$, $i \in \{1, 2, 3\}$, it holds that, for $j \in \{1, 2\}$,*

$$\limsup_{t\to\infty} \|x_j - \hat{x}_j\| \le \varepsilon_{\max}^{2-j}\sqrt{\frac{3\varepsilon_{\max}^{\frac{1}{2}}}{2\gamma\lambda_{1\min}}[(\tau M)_{\max} + (\tau\Psi)_{\max}]}, \tag{8.31}$$

$$\limsup_{t\to\infty} \|z\| \le \sqrt{\frac{3\varepsilon_{\max}^{\frac{1}{2}}}{\gamma}[(\tau M)_{\max} + (\tau\Psi)_{\max}]} \tag{8.32}$$

where $(\tau M)_{\max}$ and $(\tau\Psi)_{\max}$ are the maximum values of $\{\tau_i M_i\}$ and $\{\tau_i\Psi\}$, $i \in \{1, 2, 3\}$, respectively.

Proof From (8.16), (8.17) and the definitions of $\tilde{x}_j$ and e_j in (8.28), we observe that

$$\begin{aligned}\dot{\tilde{x}}_1 &= x_2 - \hat{x}_2 - g_1\left(\frac{1}{\varepsilon}[\xi(t) - \hat{x}_1(t)]\right)\\ &= \tilde{x}_2 - g_1\left(\frac{1}{\varepsilon}[x_1(t) - \hat{x}_1(t)] + \frac{1}{\varepsilon}[\xi(t) - x_1(t)]\right)\\ &= \tilde{x}_2 - g_1(e_1(t)) - \alpha_1(e_1(t), \sigma(t)),\\ \dot{\tilde{x}}_2 &= h(t) - g_2\left(\frac{1}{\varepsilon}[\xi(t) - \hat{x}_1(t)]\right)\\ &= h(t) - \frac{1}{\varepsilon}g_2(e_1(t)) - \frac{1}{\varepsilon}\alpha_2(e_1(t), \sigma(t)).\end{aligned}$$

Then, through a simple computation, we obtain the dynamics of e_j as

$$\begin{aligned}\dot{e}_1(t) &= \frac{1}{\varepsilon}e_2(t) - \frac{1}{\varepsilon}g_1(e_1(t)) - \frac{1}{\varepsilon}\alpha_1(e_1(t), \sigma(t)),\\ \dot{e}_2(t) &= h(t) - \frac{1}{\varepsilon}g_2(e_1(t)) - \frac{1}{\varepsilon}\alpha_2(e_1(t), \sigma(t)).\end{aligned} \tag{8.33}$$

Consider a nonnegative-definite function $V(x_1, e_1, e_2)$ defined as

$$V(x_1, e_1, e_2) = \frac{1}{2}x_1^{\mathrm{T}}x_1 + \sum_{i=1}^{3} P_i(\tilde{e}_i) \tag{8.34}$$

where $P_i(\tilde{e}_i)$ and $\tilde{e}_i$ are provided in (8.2) and (8.28), respectively. From the dynamics of e_j in (8.33), we observe that

$$\begin{aligned}\frac{dV}{dt} =& x_1^{\mathrm{T}}\dot{x}_1 + \sum_{i=1}^{3}\frac{\partial P_i}{\partial e_{1i}}\dot{e}_{1i} + \sum_{i=1}^{3}\frac{\partial P_i}{\partial e_{2i}}\dot{e}_{2i}\\ =& x_1^{\mathrm{T}}\dot{x}_1 + \sum_{i=1}^{3}\frac{\partial P_i}{\partial e_{1i}}\left[\frac{1}{\varepsilon_i}e_{2i}(t) - \frac{1}{\varepsilon_i}g_{1i}(e_{1i}(t)) - \frac{1}{\varepsilon_i}\alpha_{1i}(e_{1i}(t), \sigma_i(t))\right]\\ &+ \sum_{i=1}^{3}\frac{\partial P_i}{\partial e_{2i}}\left[h_i(t) - \frac{1}{\varepsilon_i}g_{2i}(e_{1i}(t)) - \frac{1}{\varepsilon_i}\alpha_{2i}(e_{1i}(t), \sigma_i(t))\right]\\ \le& x_1^{\mathrm{T}}\dot{x}_1 + \sum_{i=1}^{3}|\frac{\partial P_i}{\partial e_{2i}}||h_i(t)| + \sum_{i=1}^{3}\frac{1}{\varepsilon_i}\left[\frac{\partial P_i}{\partial e_{1i}}(e_{2i} - g_{1i}(e_{1i})) - \frac{\partial P_i}{\partial e_{2i}}g_{2i}(e_{1i})\right]\\ &- \sum_{i=1}^{3}\frac{1}{\varepsilon_i}\left[\frac{\partial P_i}{\partial e_{1i}}\alpha_{1i}(e_{1i}, \sigma_i) + \frac{\partial P_i}{\partial e_{2i}}\alpha_{2i}(e_{1i}, \sigma_i)\right].\end{aligned} \tag{8.35}$$

According to (8.16) and control law u in (8.29), we have

$$
\begin{aligned}
x_1^{\mathrm{T}}\dot{x}_1 =& x_1^{\mathrm{T}}[x_2 - \tilde{K}x_1 + \tilde{K}(x_1 - \hat{x}_1) - \hat{x}_2] = x_1^{\mathrm{T}}(-\tilde{K}x_1 + \tilde{K}\varepsilon e_1 + e_2) \\
\le & -\tilde{K}_{\min}\|x_1\|^2 + \tilde{K}_{\max}\varepsilon_{\max}\|x_1\|\|e_1\| + \|x_1\|\|e_2\| \\
\le & -\tilde{K}_{\min}\|x_1\|^2 + \frac{1}{2}\left(\tilde{K}_{\max}^2\varepsilon_{\max}^{\frac{5}{2}}\|x_1\|^2 + \varepsilon_{\max}^{-\frac{1}{2}}\|e_1\|^2\right) + \frac{1}{2}\left(\varepsilon_{\max}^{\frac{1}{2}}\|x_1\|^2 + \varepsilon_{\max}^{-\frac{1}{2}}\|e_2\|^2\right).
\end{aligned}
\tag{8.36}
$$

Considering the properties of functions $P_i(\cdot)$ and $\Theta_i(\cdot)$ in (8.24) and (8.25), we find that

$$
\begin{aligned}
\sum_{i=1}^{3}\frac{1}{\varepsilon_i}\left[\frac{\partial P_i}{\partial e_{1i}}(e_{2i} - g_{1i}(e_{1i})) - \frac{\partial P_i}{\partial e_{2i}}g_{2i}(e_{1i})\right] \le & -\sum_{i=1}^{3}\frac{1}{\varepsilon_i}\Theta(\tilde{e}_i) \le -\sum_{i=1}^{3}\frac{\lambda_{3i}}{\varepsilon_i}\|\tilde{e}_i\|^2 \\
\le & -\frac{\lambda_{3\min}}{\varepsilon_{\max}}(\|e_1\|^2 + \|e_2\|^2).
\end{aligned}
\tag{8.37}
$$

According to (8.26) and the event-triggering condition in (8.30), we observe that

$$
-\sum_{i=1}^{3}\frac{1}{\varepsilon_i}\left[\frac{\partial P_i}{\partial e_{1i}}\alpha_{1i}(e_{1i},\sigma_i) + \frac{\partial P_i}{\partial e_{2i}}\alpha_{2i}(e_{1i},\sigma_i)\right] \tag{8.38}
$$

$$
\le \sum_{i=1}^{3}\frac{1}{\varepsilon_i}\left[|\frac{\partial P_i}{\partial e_{1i}}\alpha_{1i}(e_{1i},\sigma_i)| + |\frac{\partial P_i}{\partial e_{2i}}\alpha_{2i}(e_{1i},\sigma_i)|\right]
$$

$$
\le \sum_{i=1}^{3}\frac{1}{\varepsilon_i}\tau_i\|\tilde{e}_i\|\Psi\varepsilon_i \le \sum_{i=1}^{3}\frac{\tau_i\Psi}{2}(\varepsilon_{\max}^{-\frac{1}{2}}\|\tilde{e}_i\|^2 + \varepsilon_{\max}^{\frac{1}{2}}) \tag{8.39}
$$

$$
\le \frac{(\tau\Psi)_{\max}}{2}(\varepsilon_{\max}^{-\frac{1}{2}}\|e_1\|^2 + \varepsilon_{\max}^{-\frac{1}{2}}\|e_2\|^2 + 3\varepsilon_{\max}^{\frac{1}{2}}). \tag{8.40}
$$

Similarly, according to Assumption 8.3, we obtain that

$$
\begin{aligned}
\sum_{i=1}^{3}|\frac{\partial P_i}{\partial e_{2i}}||h_i(t)| \le \sum_{i=1}^{3}\tau_i\|\tilde{e}_i\|M_i \le & \sum_{i=1}^{3}\frac{\tau_i M_i}{2}(\varepsilon_{\max}^{-\frac{1}{2}}\|\tilde{e}_i\|^2 + \varepsilon_{\max}^{\frac{1}{2}}) \\
\le & \frac{(\tau M)_{\max}}{2}(\varepsilon_{\max}^{-\frac{1}{2}}\|e_1\|^2 + \varepsilon_{\max}^{-\frac{1}{2}}\|e_2\|^2 + 3\varepsilon_{\max}^{\frac{1}{2}}).
\end{aligned}
\tag{8.41}
$$

Combining the results in (8.33)–(8.41), we finally obtain

$$
\begin{aligned}
\frac{dV}{dt} \le & -\frac{1}{\varepsilon_{\max}}\left[\lambda_{3\min} - \frac{(\tau M)_{\max} + (\tau\Psi)_{\max}}{2}\varepsilon_{\max}^{\frac{1}{2}} - \frac{1}{2}\varepsilon_{\max}^{\frac{1}{2}}\right](\|e_1\|^2 + \|e_2\|^2) \\
& -\left(\tilde{K}_{\min} - \frac{1}{2}\varepsilon_{\max}^{\frac{1}{2}} - \frac{\tilde{K}_{\max}^2}{2}\varepsilon_{\max}^{\frac{5}{2}}\right)\|x_1\|^2 + \frac{3}{2}\varepsilon_{\max}^{\frac{1}{2}}[(\tau M)_{\max} + (\tau\Psi)_{\max}].
\end{aligned}
$$

Write

$$\beta_1 := \tilde{K}_{\min} - \frac{1}{2}\varepsilon_{\max}^{\frac{1}{2}} - \frac{\tilde{K}_{\max}^2}{2}\varepsilon_{\max}^{\frac{5}{2}},$$
$$\beta_2 := \lambda_{3\min} - \frac{(\tau M)_{\max} + (\tau\Psi)_{\max}}{2}\varepsilon_{\max}^{\frac{1}{2}} - \frac{1}{2}\varepsilon_{\max}^{\frac{1}{2}}, \tag{8.42}$$

and define ε^* as

$$\varepsilon^* = \max\{\varepsilon_{\max} \in \mathbb{R} | \beta_1 \geq 0,\ \beta_2 \geq 0,\ \text{and } \varepsilon_{\max} > 0\}. \tag{8.43}$$

It is obvious that β_1 and β_2 are monotonically decreasing functions of $\varepsilon_{\max}$, that $\lim_{\varepsilon_{\max}\to 0}\beta_1 > 0$, $\lim_{\varepsilon_{\max}\to 0}\beta_2 > 0$, and that $\beta_1 < 0$, $\beta_2 < 0$ when $\varepsilon_{\max}$ becomes sufficiently large. Thus, ε^* is well defined. For $\varepsilon_{\max} \in (0, \varepsilon^*)$, we have

$$\begin{aligned}\frac{dV}{dt} &\leq -\beta_1\|x_1\|^2 - \frac{\beta_2}{\varepsilon_{\max}}(\|\tilde{e}_1\|^2 + \|\tilde{e}_2\|^2 + \|\tilde{e}_3\|^2) + \frac{3}{2}\varepsilon_{\max}^{\frac{1}{2}}[(\tau M)_{\max} + (\tau\Psi)_{\max}] \\ &\leq -2\beta_1\frac{1}{2}x_1^{\mathrm{T}}x_1 - \frac{\beta_2}{\varepsilon_{\max}\lambda_{2\max}}\sum_{i=1}^{3}P_i(\tilde{e}_i) + \frac{3}{2}\varepsilon_{\max}^{\frac{1}{2}}[(\tau M)_{\max} + (\tau\Psi)_{\max}].\end{aligned}$$

Furthermore, we define

$$\gamma := \min\left\{2\beta_1, \frac{\beta_2}{\varepsilon_{\max}\lambda_{2\max}}\right\},\ \phi(t, t_0) := \exp(-\gamma(t - t_0)); \tag{8.44}$$

then $\frac{dV}{dt}$ can be rewritten as

$$\frac{dV}{dt} \leq -\gamma V + \frac{3}{2}\varepsilon_{\max}^{\frac{1}{2}}[(\tau M)_{\max} + (\tau\Psi)_{\max}]. \tag{8.45}$$

From (8.34), it is clear that $V \geq 0$, and then according to the comparison lemma in [21], we obtain that

$$V(x_1(t), e_1(t), e_2(t)) \leq W(t), \tag{8.46}$$

where $W(t)$ is the solution of the differential equation

$$\dot{W} = -\gamma W + \frac{3}{2}\varepsilon_{\max}^{\frac{1}{2}}[(\tau M)_{\max} + (\tau\Psi)_{\max}],\ W(t_0) = V(x_1(t_0), e_1(t_0), e_2(t_0)). \tag{8.47}$$

Solving the differential equation in (8.47), we have

$$W(t) = V(x_1(t_0), e_1(t_0), e_2(t_0))\phi(t, t_0) + \frac{3}{2}\varepsilon_{\max}^{\frac{1}{2}}[(\tau M)_{\max} + (\tau\Psi)_{\max}] \int_{t_0}^{t} \phi(t, v)dv. \tag{8.48}$$

Then, we obtain

$$\begin{aligned} V(x_1(t), e_1(t), e_2(t)) \leq W(t) \leq V(x_1(t_0), e_1(t_0), e_2(t_0))\phi(t, t_0) \\ + \frac{3}{2\gamma}\varepsilon_{\max}^{\frac{1}{2}}[(\tau M)_{\max} + (\tau\Psi)_{\max}](1 - \phi(t, t_0)). \end{aligned} \tag{8.49}$$

Due to the fact that $\phi(t, t_0) \to 0$ when $t \to \infty$, we have that when $t \to \infty$

$$V(x_1, e_1, e_2) \leq \frac{3}{2\gamma}\varepsilon_{\max}^{\frac{1}{2}}[(\tau M)_{\max} + (\tau\Psi)_{\max}]. \tag{8.50}$$

Using the fact in (8.23), when $t \to 0$ we have

$$\frac{1}{2}\|x_1\| + \lambda_{1\min}(\|e_1\|^2 + \|e_2\|^2) \leq V(x_1, e_1, e_2). \tag{8.51}$$

From (8.50) and (8.51), we observe that

$$\limsup_{t\to\infty} \|z\| = \limsup_{t\to\infty} \|x_1\| \leq \sqrt{\frac{3\varepsilon_{\max}^{\frac{1}{2}}}{\gamma}[(\tau M)_{\max} + (\tau\Psi)_{\max}]}. \tag{8.52}$$

Similarly, we obtain that, for $j \in \{1, 2\}$

$$\limsup_{t\to\infty} \|e_j\| \leq \sqrt{\frac{3\varepsilon_{\max}^{\frac{1}{2}}}{2\gamma\lambda_{1\min}}[(\tau M)_{\max} + (\tau\Psi)_{\max}]}. \tag{8.53}$$

From the definitions of e_j, we conclude that

$$\limsup_{t\to\infty} \|x_j - \hat{x}_j\| \leq \varepsilon_{\max}^{2-j}\sqrt{\frac{3\varepsilon_{\max}^{\frac{1}{2}}}{2\gamma\lambda_{1\min}}[(\tau M)_{\max} + (\tau\Psi)_{\max}]}. \tag{8.54}$$

This completes the proof. □

The above results indicate that the ESO in (8.17), the control law in (8.27), and the event-triggering mechanism proposed in (8.30) can guarantee the asymptotic boundedness of the observation error and stabilize the system (8.11). According to Lemmas 8.1–8.2, the stability of system (8.11) implies that the states of the rigid spacecraft system governed by (8.1) can track the given desired attitudinal motion (8.2) in the presence of parameter uncertainties and unknown disturbances. Moreover, from (8.52) and (8.54), the system state z and observation errors can converge to an arbitrarily small bounded range around 0 for properly chosen values of $\{\varepsilon_i\}$ and Ψ, even though the bound of the derivative of the extended state is unknown. The input-to-state stability of the ith ESO is ensured by the parameterization of the functions g_{ji}, $j \in \{1, 2\}$, and $\varphi(\theta)$, as chosen in Sect. 8.2, which satisfy the requirements in (8.22) and (8.23)–(8.26). The function $\varphi(\theta)$ is the nonlinear part of the ESO. The proposed event-triggering condition in (8.30) is related to the value of the sampling errors $\sigma(t)$ and Ψ, and only when $\sigma(t)$ becomes sufficiently large will the value of $\xi(t)$ be updated; In this way, the larger the Ψ is, the larger the tolerance range, and the lower the transmission cost will be. Furthermore, note that the tracking performance can be adjusted by changing the value of parameter Ψ according to the quantitative relations in (8.52) and (8.54).

8.4 Numerical Evaluation

In this section, the effectiveness of the introduced event-triggered control scheme is evaluated by numerical simulations. Consider the following nominal inertia matrix of the spacecraft model [7]:

$$J_0 = \begin{bmatrix} 20 & 1.2 & 0.9 \\ 1.2 & 17 & 1.4 \\ 0.9 & 1.4 & 15 \end{bmatrix} \text{kg} \cdot \text{m}^2. \tag{8.55}$$

The parameter uncertainties of the inertial matrix are described as

$$\Delta J = \text{diag}[\sin(0.1t), 2\sin(0.2t), 3\sin(0.3t)] \ \text{kg} \cdot \text{m}^2.$$

The external disturbances are described as

$$d(t) = [0.1\sin(0.1t), 0.2\sin(0.2t), 0.3\sin(0.2t)]^{\text{T}} \ \text{N} \cdot \text{m}.$$

All the initial values of the parameters in the two groups stay the same. The initial attitude orientation of the unit quaternion is

$$q(0) = [0.3, -0.2, -0.3, 0.8832]^{\text{T}},$$

and the initial target unit quaternion is $q_d(0) = [0, 0, 0, 1]^{\text{T}}$. The initial value of the angular velocity is $\omega(0) = [0, 0, 0]^{\text{T}}$ rad/s. To illustrate the tracking performance, we perform two groups of simulations with different desired angular velocities. The corresponding desired unit quaternion to be tracked is generated by (8.2).

For comparison purposes, a time-triggered continuous-time ADRC (CT-ADRC) is implemented for each group, in addition to the ET-ADRC controller. In these simulations, the ADRC controllers are implemented by Euler approximation with a sampling time $t_s = 1$ ms. To evaluate the tracking performance of the system, we define the tracking errors of the angular velocity (E_v), unit quaternion (E_q), and average sampling time (T_A) as follows:

$$E_v = \frac{1}{3T}\int_0^T \sum_{i=1}^{3} |\omega_i - \omega_{di}|dt, \; E_q = \frac{1}{4T}\int_0^T \sum_{i=1}^{4} |q_i - q_{di}|dt, \tag{8.56}$$

$$T_A = \begin{cases} 10, \text{ for CT-ADRC scheme,} \\ \frac{T}{N_s}, \text{ for ET-ADRC scheme,} \end{cases} \tag{8.57}$$

where we recall that the suffix d represents the desired target state, the suffix i denotes the ith element of the vector state, T is the total simulation time in milliseconds, and N_s is the total triggering counts in one simulation. Moreover, since energy consumption in attitude control is an extremely important fact to consider, comparison of energy consumption between the time- and event-triggered approaches is performed based on the energy-consumption model in [22] as follows:

$$E_c = \sum_{k=1}^{N}\sum_{i=1}^{3} |\frac{1}{2}\omega_i^2(k) - \frac{1}{2}\omega_i^2(k-1)|, \tag{8.58}$$

where $\omega_i(k)$ is the ith angular velocity at step k, and N denotes the total number of time steps in the whole simulation. Note that the difference here is that the mass of the spacecraft is not considered as it does not affect the comparison.

8.4.1 Desired Angular Velocity in Sinusoidal Form

In this section, the attitude-tracking problem for a spacecraft with the desired angular velocity in a sinusoidal form is simulated. The desired angular velocity is supposed to have the form

$$\omega_d(t) = 0.05\left[\sin\left(\frac{\pi t}{20}\right), \sin\left(\frac{2\pi t}{20}\right), \sin\left(\frac{3\pi t}{20}\right)\right]^{\mathrm{T}} \text{ rad/s}. \tag{8.59}$$

Without loss of generality, the parameters of three sub-ESOs in (8.17) are selected to be the same, such that for any $i \in \{1, 2, 3\}$, $g_{1i}(y_1) = 2y_1 + \varphi(y_1)$ and $g_{2i}(y_1) = y_1$. The high-gain parameters are chosen as $\varepsilon_1 = \varepsilon_2 = \varepsilon_3 = 0.4$. The parameters K in (8.9) and $\tilde{K}$ in (8.27) are chosen as $K = 2I_3$ and $\tilde{K} = 4I_3$, respectively.

For comparison purpose, we perform the simulations considering the utilization of sampling periods equal to 1 ms and 400 ms for the time-triggered cases, and the utilization of event-triggering parameters Ψ that lead to average sampling periods of 60.9 ms and 402 ms for the event-triggered case. The tracking results of the attitude quaternion and the angular velocity component for the CT-ADRC scheme with 1 ms

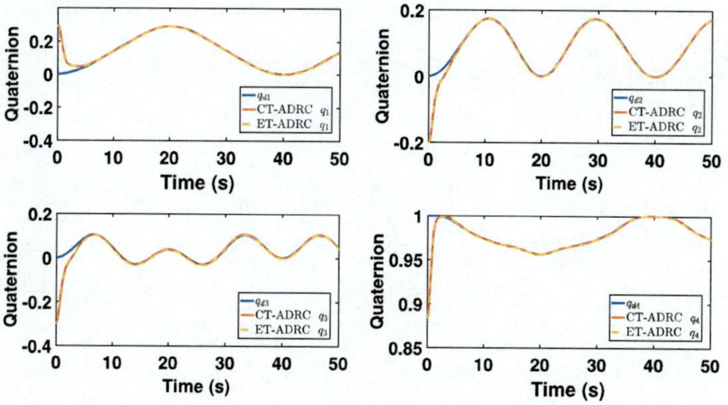

Fig. 8.2 Attitude quaternion tracking performance of sinusoidal form

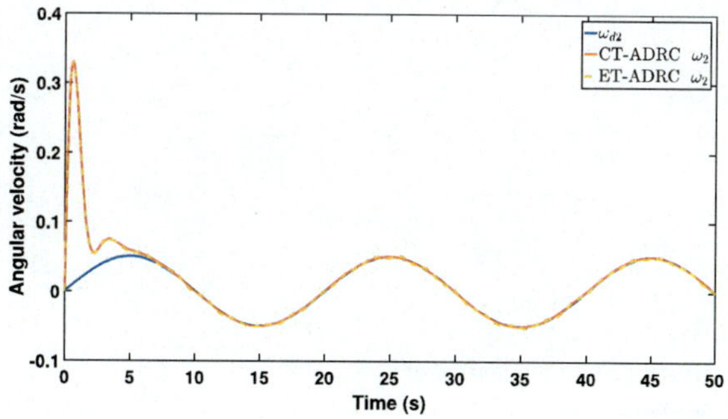

Fig. 8.3 Angular velocity tracking performance of sinusoidal form

sampling periods and the ET-ADRC scheme with $\Psi = 0.1$ are shown in Figs. 8.2 and 8.3, respectively. Moreover, the corresponding performance of the ESO for these schemes is shown in Figs. 8.4 and 8.5, respectively. The tracking errors E_v, E_q, average sampling time T_A, and qualitative energy consumption (QEC) E_c are summarized in Table 8.1. From Table 8.1, we observe that compared with CT-ADRC with a small sampling period (1 ms), the proposed scheme can maintain a comparable control performance and almost identical energy consumption at a much smaller average sampling rate. Moreover, we also observe that when the (average) sampling periods are both increased to larger values (approximately 400 ms), the proposed scheme leads to a much smaller energy consumption index. The implication is that the ET-ADRC scheme is more energy efficient in certain extreme cases (e.g., sensor faults or communication bandwidth outrage) and bears improved fault-tolerant robustness in terms of performance and energy efficiency. Additionally, the results of sampling intervals between two consecutive event-triggering instants are provided in Fig. 8.6, where we can observe that nonzero sampling intervals can be guaranteed to avoid Zeno behavior. With longer inter-triggering intervals, the cost of data transmission and computation can be reduced.

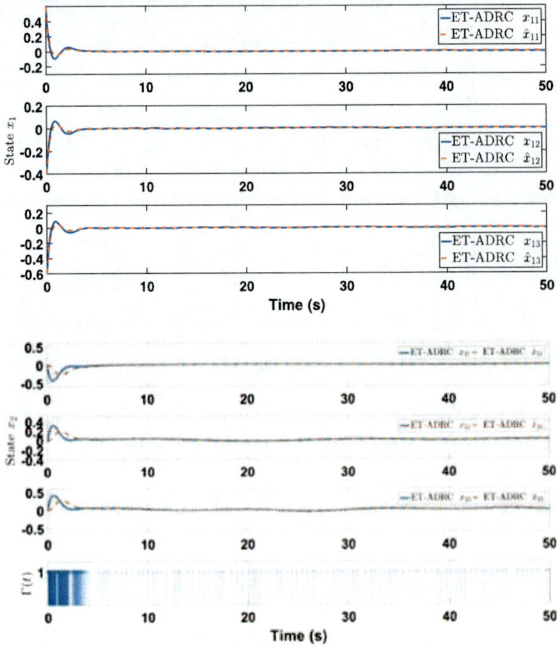

Fig. 8.4 Observation performance of ESO with event-triggered mechanism of sinusoidal form

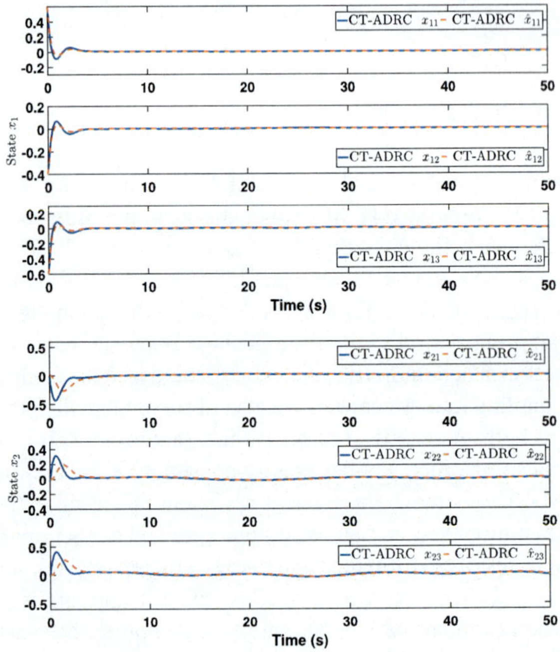

Fig. 8.5 Observation performance of ESO with time-triggered mechanism of sinusoidal form

Table 8.1 Performances of two control schemes of sinusoidal form

Simulation schemes	Average sampling time (T_A)	Tracking errors (E_v/E_q)	QEC (E_c)
Time-triggered	1 ms	0.0115/0.0059	0.58
	400 ms	0.0407/0.0073	3.41
Event-triggered	101.21 ms ($\Psi = 0.1$)	0.0122/0.0059	0.59
	402.13 ms ($\Psi = 1.5$)	0.0268/0.0075	0.84

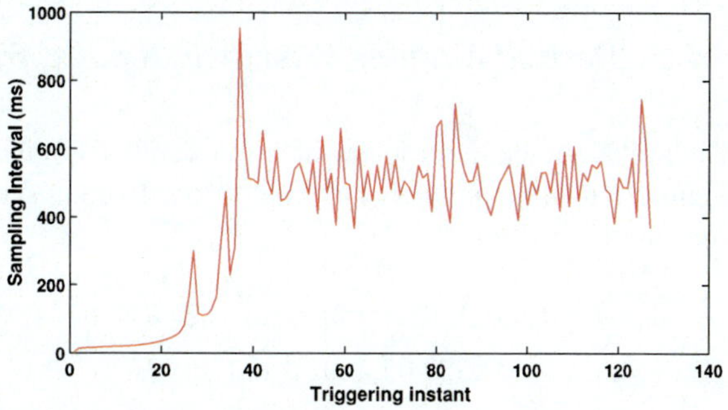

Fig. 8.6 Sampling intervals of ET-ADRC scheme with $\Psi = 1.5$ for sinusoidal form

Table 8.2 The effect of uncertainties upon the performance of the ET-ADRC scheme with $\Psi = 0.1$

References	Simulation scenarios	Average inputs signal (N*m)	Tracking errors E_v/E_q
Sinusoid form	With uncertainties	1.49	0.0122/0.0059
	Without uncertainties	1.35	0.0123/0.0059
Squarewave form	With uncertainties	1.90	0.0158/0.0063
	Without uncertainties	1.67	0.0157/0.0063

For the ET-ADRC scheme with $\Psi = 0.1$, we perform comparative analysis concerning the effects of uncertainties in terms of tracking errors and average input signal. In particular, the uncertainties considered in the simulations include the external disturbances and parameter uncertainties of the inertial matrix. The results are presented in Table 8.2; we observe that the ET-ADRC scheme with uncertainties has similar tracking performance in terms of tracking errors but requires larger inputs in comparison with the cases without uncertainties. This is because of the disturbance-rejection capability of ADRC. Thus, the ET-ADRC scheme is required to produce larger input signals to eliminate the effects of uncertainties.

Table 8.3 Performances of two control schemes of square wave form

Simulation schemes	Average sampling time (T_A)	Tracking errors (E_v/E_q)	QEC (E_c)
Time-triggered	1 ms	0.0152/0.0063	0.60
	400 ms	0.0575/0.0094	3.77
Event-triggered	60.90 ms ($\Psi = 0.1$)	0.0158/0.0063	0.61
	401.52 ms ($\Psi = 2.5$)	0.0402/0.0092	1.16

8.4.2 *Desired Angular Velocity in Square Wave Form*

The attitude-tracking problem of a spacecraft with abrupt changes in the desired angular velocity is simulated in this section. To do this, consider the desired angular velocity in the form

$$\omega_d(t) = [0.05, 0.03, 0.02]^{\mathrm{T}} \text{ rad/s}, \ 2kT < t < (2k+1)T, \tag{8.60}$$

$$\omega_d(t) = -[0.05, 0.03, 0.02]^{\mathrm{T}} \text{ rad/s}, \ (2k+1)T < t < (2k+2)T, \tag{8.61}$$

where $T = 10$ s is the switching period.

The parameters needed for CT-ADRC and ET-ADRC are all the same as described in the previous subsection. Comparative analysis results concerning the effects of uncertainties on the ET-ADRC scheme are summarized in Table 8.2. The simulation results of the ET-ADRC and CT-ADRC schemes with different average sampling periods (T_A) in terms of tracking errors (E_v) and (E_q) and QEC (E_c) are summarized in Table 8.3. The results of the two control schemes are shown in Figs. 8.7, 8.8, 8.9, 8.10, and 8.11. Consistent with the results observed in the previous subsection, the proposed ET-ADRC scheme achieves similar performance in terms of attitude quaternion and angular velocity tracking except for the switching points (see Figs. 8.8 and 8.9) and similar energy consumptions but with a lower sampling rate, compared with the CT-ADRC scheme. Moreover, with the increase of average sam-

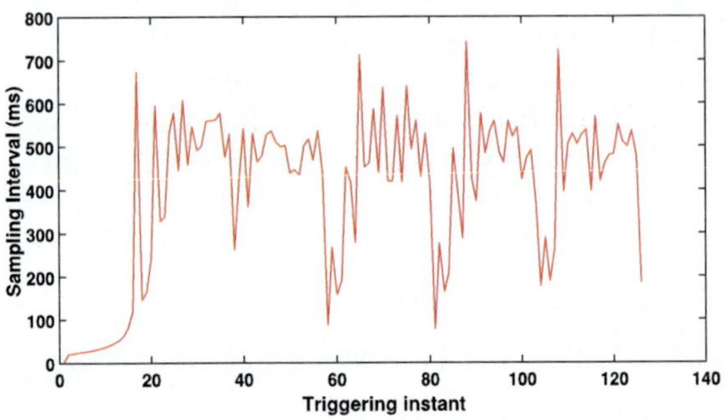

Fig. 8.7 Sampling intervals of ET-ADRC scheme with $\Psi = 2.5$ for square wave form

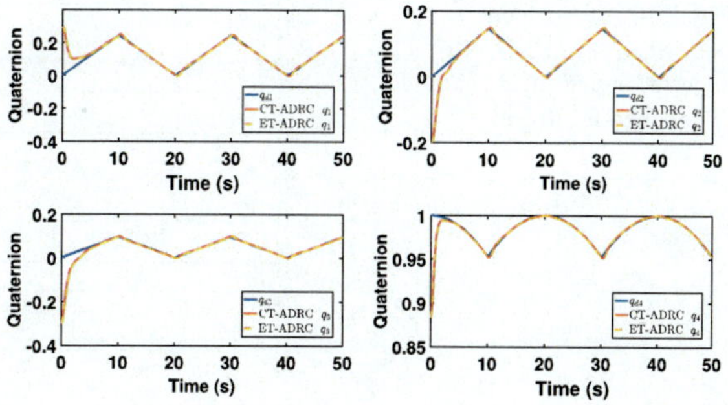

Fig. 8.8 Attitude quaternion tracking performance of square wave form

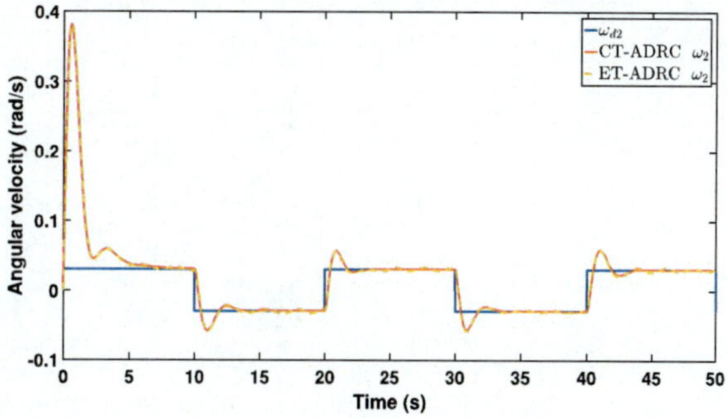

Fig. 8.9 Angular velocity tracking performance of square wave form

pling times, the performance of the CT-ADRC scheme becomes worse than that of the ET-ADRC scheme (see Table 8.3); in particular, the ET-ADRC scheme achieves improved angular velocity tracking performance but with lower sampling rates and energy consumption.

To summarize, compared with the CT-ADRC scheme, the ET-ADRC scheme with two kinds of desired angular velocity generally can guarantee similar performance in terms of tracking errors with reduced average data transmission rate.

8.5 Summary

In this chapter, a problem of attitude tracking for spacecraft models with inertial uncertainties and external disturbances is considered using the ET-ADRC approach. An event-triggered mechanism is designed to achieve the asymptotic boundedness of observation errors and the asymptotic stabilization of the attitude-tracking system. Simulation results with different types of desired angular velocity are employed to illustrate the effectiveness of the control approach. Note that the performance of the system largely relies on the selection of parameters, especially the parameters $\tilde{K}$ of the control law, ε_i and Ψ of the sub-ESOs, and event-triggered mechanism. Larger Ψ can reduce resource consumption with lower average sampling rates but

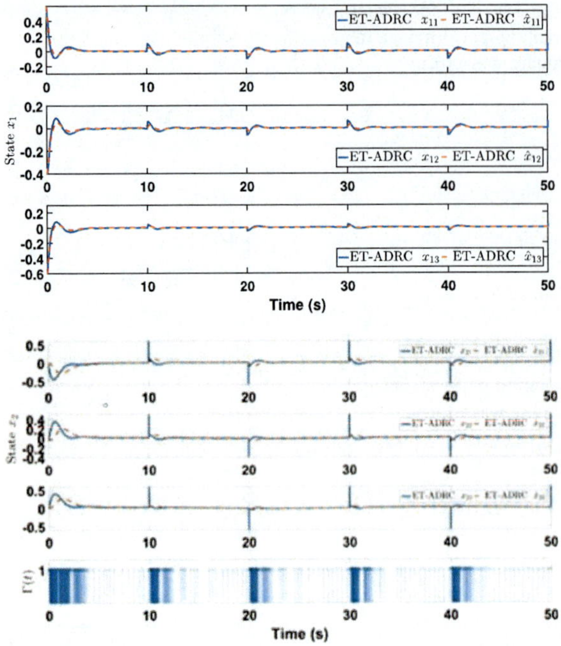

Fig. 8.10 Observation performance of ESO with event-triggered mechanism of square wave form

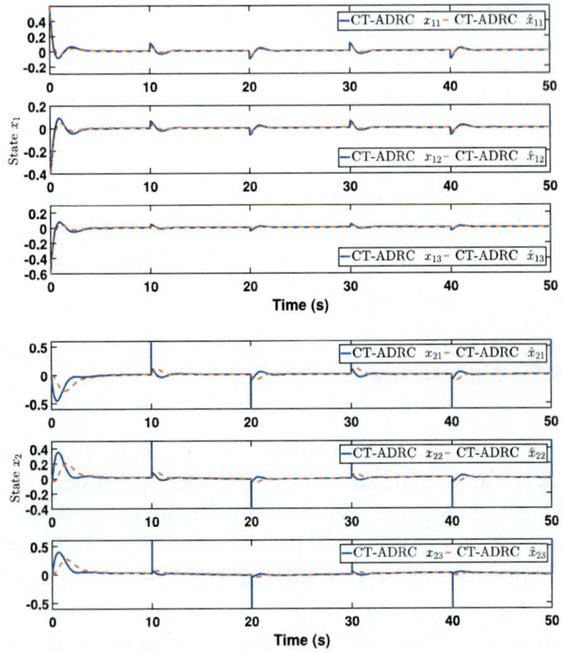

Fig. 8.11 Observation performance of ESO with time-triggered mechanism of square wave form

the observation errors are more likely to be bigger, causing larger tracking errors. A smaller ε_i can speed up the convergence rate of sub-ESOs and reduce the undesired chattering, but requires larger control inputs and energy consumption. Larger $\tilde{K}$ can improve the performance of attitude tracking and reduce undesired chattering, but will also require larger control inputs. Interestingly, the numerical results indicate through introducing event-triggered control, the energy efficiency of the controller can also be improved, which deserves further investigation as energy/fuel efficiency is crucially important in aerospace applications.

References

1. Sidi, M.J.: Spacecraft Dynamics and Control. Cambridge University Press, Cambridge (2000)
2. Crouch, P.: Spacecraft attitude control and stabilization: applications of geometric control theory to rigid body models. IEEE Trans. Autom. Control **29**, 321–331 (2003)
3. Wang, N., Zhang, T.W., Xu, J.Q.: Formation control for networked spacecraft in deep space: with or without communication delays and with switching topology. Sci. China Inf. Sci. **54**, 469–481 (2011)
4. Jin, Y.Q., Liu, X.D., Qiu, W., et al.: Time-varying sliding mode control for a class of uncertain MIMO nonlinear system subject to control input constraint. Sci. China Inf. Sci. **53**, 89–100 (2010)
5. Pukdeboon, C.: Extended state observer-based third-order sliding mode attitude tracking controller for rigid spacecraft. Sci. China Inf. Sci. (2017)
6. Pukdeboon, C., Zinober, A.S.I., Thein, M.W.L.: Quasi-continuous higher order sliding-mode controllers for spacecraft attitude tracking maneuvers. IEEE Trans. Ind. Electron. **57**, 1436–1444 (2010)
7. Xia, Y.Q., Zhu, Z., Fu, M.Y., et al.: Attitude tracking of rigid spacecraft with bounded disturbances. IEEE Trans. Ind. Electron. **58**, 647–659 (2011)
8. Li, Z.K., Duan, Z.S.: Distributed adaptive attitude synchronization of multiple spacecraft. Sci. China Technol. Sci. **54**, 1992–1998 (2011)
9. Chen, Z.Y., Huang, J.: Attitude tracking and disturbance rejection of rigid spacecraft by adaptive control. IEEE Trans. Autom. Control **54**, 600–605 (2009)
10. Luo, W.C., Chu, Y.C., Ling, K.V.: Inverse optimal adaptive control for attitude tracking of spacecraft. IEEE Trans. Autom. Control **50**, 1639–1654 (2005)
11. Sun, S., Yang, M.F., Wang, L.: Event-triggered nonlinear attitude control for a rigid spacecraft. In: Chinese Control Conference, pp. 7582–7586 (2017)
12. Xing, L.T., Wen, C.Y., Liu, Z.T., et al.: An event-triggered design scheme for spacecraft attitude control. In: IEEE Conference on Industrial Electronics and Applications, pp. 1552–1557 (2017)
13. Wu, B.L., Shen, Q., Cao, X.B.: Event-triggered attitude control of spacecraft. Adv. Space Res. **61**, 927–934 (2018)
14. Zhang, C.X., Wang, J.H., Zhang, D.X., et al.: Learning observer based and event-triggered control to spacecraft against actuator faults. Aerosp. Sci. Technol. **78**, 522–530 (2018)
15. Yuan, J.S.: Closed-loop manipulator control using quaternion feedback. IEEE Trans. Autom. Control **4**, 434–440 (1988)
16. Lu, K.F., Xia, Y.Q., Fu, M.Y.: Controller design for rigid spacecraft attitude tracking with actuator saturation. Inf. Sci. **220**, 343–366 (2013)
17. Yang, H.J., You, X., Xia, Y.Q., et al.: Adaptive control for attitude synchronisation of spacecraft formation via extended state observer. IET Control Theory Appl. **8**, 2171–2185 (2014)
18. Li, B., Hu, Q.L., Ma, G.F.: Extended state observer based robust attitude control of spacecraft with input saturation. Aerosp. Sci. Technol. **50**, 173–182 (2016)

19. Li, B., Hu, Q.L., Yu, Y.B., et al.: Observer-based fault-tolerant attitude control for rigid spacecraft. IEEE Trans. Aerosp. Electron. Syst. **53**, 2572–2582 (2017)
20. Cai, D.H., Zou, H.G., Wang, J.Z., et al.: Supplementary discussions on "Event-triggered attitude tracking for rigid spacecraft" (2019). www.escience.cn/people/dshi/
21. Khalil, H.K.: Nonlinear Systems. Prentice Hall, Upper Saddle River (2002)
22. Li, P., Yue, X.K., Chi, X.B., et al.: Optimal relative attitude tracking control for spacecraft proximity operation. In: 2013 25th Chinese Control and Decision Conference (2013)

Chapter 9
Event-Triggered Adaptive Disturbance Rejection for Artificial Pancreas

9.1 Background

In this chapter, we show how the ideas of ET-ADRC can be generalized and employed to achieve closed-loop glucose regulation for patients with type 1 diabetes mellitus (T1DM). In particular, we diverge a little bit from our previous ET-ADRC design approaches and show how a different controller with scenario-dependent adaptability can be designed through event-based parameter adaptation. This is very important and interesting since the controller needs to behave drastically differently to achieve safe and satisfactory glucose regulation performance and more importantly, this approach appears to be mimicking how the human brain reacts against different scenarios.

T1DM is caused by autoimmune destruction of the beta cells in the pancreas and can result in chronic health problems, e.g., retinopathy and nephropathy [9]. The commercialization of continuous glucose monitoring (CGM) sensors and continuous subcutaneous insulin infusion (CSII) pumps enable the development of artificial pancreas (AP) systems for patients with T1DM, which determines real-time insulin infusion rates based on CGM-driven feedback control algorithms.

The move to develop home-use-safe wearable AP systems calls for low-complexity, reliable, and efficient control algorithms. In this chapter, we introduce a design that simultaneously exhibits the performance and flexibility of Model Predictive Control (MPC) for AP and the simplicity of a PID controller. Our controller is designed on the basis of an ADRC structure, which can be understood as a generalization of the classic PID control. The main contribution of our approach is the development of glucose- and velocity-dependent event-triggered control parameter adaptation laws that determine the response of the proposed controller, which is inspired by the adaptive zone MPC design in [33] and the ET-ADRC approach introduced in this book. To ensure the safety of the proposed controller, insulin on board (IOB) constraints and limitation on the maximum allowable insulin infusion rates are incorporated. Note that glucose- and velocity-based controller design was also adopted in fuzzy logic controllers (FLC) for AP [1, 27, 29]. The main difference

D. Shi et al., *Event-Triggered Active Disturbance Rejection Control*,
Studies in Systems, Decision and Control 356,
https://doi.org/10.1007/978-981-16-0293-1_9

of our approach from FLC is that the glucose- and velocity information is mainly used to characterize different controller operation modes and build explicit ADRC parameter adaptation laws for each of these modes, whereas FLC encodes clinician expertise to develop a pre-determined dosing matrix based on all possible combinations of glucose, velocity, and even acceleration [29]. This approach is also related with the switched linear parameter varying (LPV) control approach to AP design [5, 6], but the approach in this chapter focuses on designing glucose- and velocity-triggered parameter adaptation mechanisms, instead of designing sub-controllers and switching logics in the LPV approach.

The performance of the proposed controller is evaluated through the 10-adult cohort of the US Food and Drug Administration (FDA) accepted Universities of Virginia (UVA)/Padova T1DM simulator [7] for different *in silico* protocols. The proposed controller achieves similar and satisfactory performance compared with the adaptive zone MPC introduced in [33], in terms of mean glucose level and percentage time in the safe range for both announced and unannounced meals, without increasing the risk of hypoglycemia. We also show that the proposed controller can behave properly in the scenarios of moderate meal-bolus over/underestimates and basal-rate mismatches.

9.2 Related Literature

Multiple efforts have been devoted to designing AP control algorithms in the literature [10, 18, 38]. The main approaches include proportional-integral-differentiation (PID) control and model predictive control (MPC) [2, 34]. PID control was argued to have the ability of emulating the pancreatic β-cell insulin response [37], and different PID-based AP algorithms have been evaluated in clinical studies [11, 36, 42]. Gopakumaran et al. proposed a fading memory PD control (FMPD) method [14] by exploring the plateaus phenomenon of pancreatic beta-cell response after sustained hyperglycemia. Based on the FMPD and an adaptive system for estimating changing sensitivity to insulin, Jacobs et al. [25] developed an adaptive PD control algorithm for a bihormonal (glucagon and insulin) AP. To improve the safety of the algorithms, Steil et al. [35] proposed a PID control algorithm for AP systems, in which an insulin feedback term was adopted to avoid controller-induced hypoglycemia. In Huyett et al. [24], a robust PID controller for a fully implantable AP with insulin feedback and anti-reset windup strategy was developed. A PID control algorithm with IOB constraints was proposed in Rossetti et al. [30] to prevent hypoglycemia caused by insulin over-delivery.

Since MPC can easily handle practical constraints in engineering systems by formulating the controller design problem into a constrained optimization problem, MPC-based AP algorithms became popular in recent years and have also been evaluated extensively in clinical trials [21, 31, 39]. Utilizing a compartment model determined by a Bayesian parameter estimation approach, Hovorka et al. [22] proposed an MPC with safety constraints on insulin injection doses. Cameron et al.

[4] studied a stochastic predictive control algorithm through adopting a multiple model approach. Toffanin et al. [40] introduced an unconstrained MPC algorithm with multiple safety constraints designed based on clinical experience. With the glucose metabolism modeled as an autoregressive moving average model, Turksoy et al. [41] proposed a generalized predictive control method for AP. Since euglycemia is defined as a range of blood glucose (BG) values (e.g. 70–180 mg/dL), Doyle, Dassau and co-authors [8, 12, 13, 15] developed and clinically evaluated a zone MPC approach featuring a diurnal zone cost function, IOB constraints, velocity-weighting and velocity penalty; in a recent update of zone MPC, Shi et al. [32, 33] introduced control penalty adaptation based on glucose and velocity information. Moreover, Boiroux et al. [3] investigated an adaptive MPC with an asymmetric cost function and time-varying reference signals; the benefit of adopting asymmetric cost functions in AP MPC design was also explored by Lee et al. [26]. In Paolo Incremona et al. [28], an integral error term was introduced into the cost function to eliminate the static error of the unconstrained MPC algorithm.

9.3 Controller Structure Design

In this section, the structure of the proposed ADRC is introduced. The controller builds on an ADRC scheme (see Fig. 9.1), which inherits the simple error-driven model-free structure from PID control, but takes the benefit of the state observer from modern control theory [19, 43]. In particular, ADRC is robust to the model-plant mismatch by treating the internal uncertainties and external disturbances, namely total disturbances, as an additional state of the system, which is the so-called extended state.

The proposed controller operates around the subject-specific time-dependent-basal insulin infusion rate $u_{\rm b}$ that can virtually achieve a steady-state glucose level $G_{\rm b}$, and assumes a *virtual* nonlinear insulin-glucose model of the form:

$$\ddot{G} = f(G, \dot{G}, d(t), t) + bu, \tag{9.1}$$

$$y = G, \tag{9.2}$$

where the input and output of the model are $u = u_{\rm I} - u_{\rm b}$ and $G = G_{\rm BG} - G_{\rm b}$, respectively, $u_{\rm I}$ is the absolute insulin infusion rate, $G_{\rm BG}$ is the subject's BG value, $d(t)$ represents the disturbance introduced by the meals, exercise, or other unmeasured effects, b denotes the input gain in the form of $b = \Delta b + b_0$, with b_0 representing the nominal gain selected as $b_0 = -k/u_{\rm TDI}$ and Δb being the uncertain part, and $u_{\rm TDI}$ denotes the subject specific total daily insulin amount. Note that the exact knowledge of $f(\cdot)$ is not needed in controller design and that the model in (9.1) is consistent with the structure of the control relevant model introduced in [20].

The proposed controller includes three basic components of ADRC (see green blocks in Fig. 9.1): TD, ESO, and nonlinear feedback. To ensure the safety of the control algorithm, insulin on board (IOB) constraints and constraints on minimum and maximum insulin infusion rates (see blue blocks in Fig. 9.1) are incorporated.

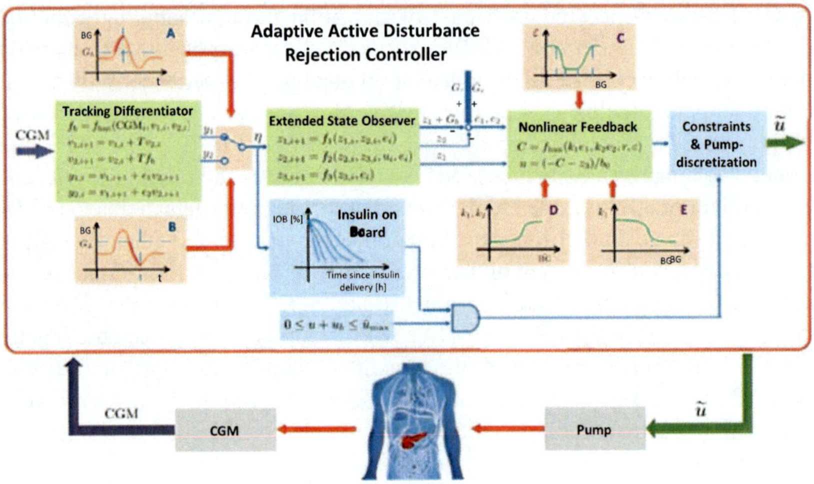

Fig. 9.1 Schematic of the adaptive active disturbance rejection controller. The basic ADRC modules (green blocks) include a tracking differentiator, an extended state observer, and a nonlinear feedback module. To ensure the safety of the controller, IOB constraints and upper/lower limits for insulin infusion are incorporated (blue blocks). Adaptation mechanisms are designed for ESO, IOB, and nonlinear feedback (orange blocks), to enable different controller behavior in ascending and descending BG situations (see Sect. 9.4 for details)

9.3.1 Time-Optimal Control Synthesis Function

Before introducing the controller, the function $f_{\text{han}}(n_1, n_2, r, \varepsilon)$ introduced earlier in this book is firstly reviewed, which is mathematically defined as

$$\begin{cases} d = r\varepsilon^2,\ a_0 = \varepsilon n_2,\ y = n_1 + a_0, \\ a_1 = \sqrt{d(d + 8|y|)}, \\ a_2 = a_0 + \text{sign}(y)(a_1 - d)/2, \\ s_y = (\text{sign}(y + d) - \text{sign}(y - d))/2, \\ a = (a_0 + y - a_2)s_y + a_2, \\ s_a = (\text{sign}(a + d) - \text{sign}(a - d))/2, \\ f_{\text{han}} = -r(\frac{a}{d} - \text{sign}(a))s_a - r\text{sign}(a). \end{cases} \tag{9.3}$$

Note that this function is a time-optimal control synthesis function for the discrete-time stabilization of two integrators in cascade with bounded input [19]:

$$n_1(i + 1) = n_1(i) + Tn_2(i), \tag{9.4}$$

$$n_2(i + 1) = n_2(i) + Tu,\ |u| \leq r, \tag{9.5}$$

where T is sampling period, and u is control law. Importantly, letting v be a desired tracking signal, we can obtain the time-optimal discrete-time TD used in our design (see Sect. 9.3.2) using $f_{\text{han}}(n_1 - v, n_2, r, \varepsilon)$. A rigorous proof on the convergence of the TD is provided in [17].

Besides a linear approximate expression $f_{\text{han}}^{\text{a}}(\cdot)$ of $f_{\text{han}}(\cdot)$ in (9.3) was explored in the literature, which is in the form of

$$\begin{cases} q = n_1 + \varepsilon n_2, \\ p = n_2 + \frac{q}{\varepsilon}, \\ m = n_1 + \frac{\varepsilon}{2} n_2 + \frac{|n_2| n_2}{2r}, \\ f_{\text{han}}^{\text{a}} = -r \text{sat}(p, \varepsilon r), \quad |p| \leq \varepsilon r \text{ and } |q| \leq \varepsilon^2 r, \\ f_{\text{han}}^{\text{a}} = -r \text{sign}(m), \quad \text{otherwise.} \end{cases} \tag{9.6}$$

where $\text{sat}(p, \varepsilon r)$ is a linear saturation function with the form of

$$\text{sat}(p, \varepsilon r) = \begin{cases} \text{sign}(p), & |p| > \varepsilon r, \\ \frac{p}{\varepsilon r}, & |p| \leq \varepsilon r. \end{cases} \tag{9.7}$$

This approximate function is used and analyzed in parameter design of our controller.

9.3.2 Tracking Differentiator

A TD attempts to estimate the differential signal of a given signal [17, 19], and has been adopted in a wide range of engineering applications [16, 23]. Here, driven by the arrival of a CGM measurement CGM(i), TD calculates the glucose estimate $y_1(i)$, the glucose prediction $y_2(i)$, and the approximate differential signal $v_2(i)$ (namely, glucose velocity). Specifically, the TD module utilized in the controller has the form

$$f_{\text{h}}(i) = f_{\text{han}}(v_1(i) - \text{CGM}(i), v_2(i), r_1, h_0), \tag{9.8}$$

$$v_1(i+1) = v_1(i) + T v_2(i), \tag{9.9}$$

$$v_2(i+1) = v_2(i) + T f_{\text{h}}(i), \tag{9.10}$$

$$y_1(i) = v_1(i+1) + \epsilon_1 v_2(i+1), \tag{9.11}$$

$$y_2(i) = v_1(i+1) + \epsilon_2 v_2(i+1), \tag{9.12}$$

where $T := 5$ [min] is sampling period, r_1 determines the tracking speed, h_0 is a filter coefficient, ϵ_1 and ϵ_2 are compensation coefficients satisfying $\epsilon_2 > \epsilon_1 > 0$, and $f_{\text{han}}(\cdot)$ denotes a time-optimal control synthesis function that enables fast tracking of the targeted CGM signal without overshoot (see Sect. 9.3.1 for the definition and [19, Sect. III.A] for the detailed explanations). In principle, $v_1(i)$ is a signal that tracks CGM(i), and the glucose estimate $y_1(i)$ of CGM(i) and glucose prediction $y_2(i)$ are calculated using the differential signal $v_2(i)$, which helps to compensate the phase delay and amplitude decay. The detailed design of ϵ_1, ϵ_2, r_1, and h_0 are discussed in Sect. 9.4.3.

9.3.3 Extended State Observer

For the model in (9.1) and (9.2), let $x_1 := G$ and $x_2 := \dot{G}$, and we define an extended state according to the standard ESO formulation [19, 43] as

$$x_3 := f(G, \dot{G}, d(t), t) + \Delta bu. \tag{9.13}$$

The ESO is implemented in discrete form according to [19] to estimate the total disturbances:

$$e(i) = z_1(i) - (\eta(i) - G_b), \tag{9.14}$$
$$f_2(i) = f_e(e(i), \sigma_1, \mu), \;\; f_3(i) = f_e(e(i), \sigma_2, \mu), \tag{9.15}$$
$$z_1(i+1) = z_1(i) + T(z_2(i) - \lambda_1 e(i)), \tag{9.16}$$
$$z_2(i+1) = z_2(i) + T(z_3(i) + b_0 u(i) - \lambda_2 f_2(i)), \tag{9.17}$$
$$z_3(i+1) = z_3(i) + T(-\lambda_3 f_3(i)), \tag{9.18}$$

where z_1 and z_2 correspond to the estimates of x_1 and x_2, respectively, z_3 is the estimation of extended state x_3, and η is either the glucose estimate y_1 or the glucose prediction y_2 determined by a condition to be discussed in Sect. 9.4.1. Here λ_1, λ_2 and λ_3 are the observer gains, σ_1 and σ_2 are design parameters, and the nonlinear function $f_e(\cdot)$ has the form

$$f_e(e, \sigma, \mu) = \begin{cases} \frac{e}{\mu^{1-\sigma}}, & |e| \le \mu, \\ |e|^\sigma \operatorname{sign}(e), & |e| > \mu. \end{cases} \tag{9.19}$$

Parameter design will be discussed in Sect. 9.4.2.

9.3.4 Nonlinear Feedback

ADRC usually employs a nonlinear feedback control law to improve the closed-loop control performance [43]. With the state estimates provided by ESO, we consider the following feedback control law:

$$u(i) = (-u_0(i) - z_3(i))/b_0, \tag{9.20}$$
$$u_0(i) = f_{\text{han}}(k_1 e_1(i), k_2 e_2(i), r, \varepsilon), \tag{9.21}$$
$$e_1 = G_r - (z_1 + G_b), \;\; e_2 = \dot{G}_r - z_2, \tag{9.22}$$

where G_r and $\dot{G}_r$ are the reference value of blood glucose and its rate of change, respectively, and k_1, k_2, r, and ε are positive controller adaptation parameters. These parameters will be designed in Sect. 9.4.2. Note that the term $-z_3(i)$ in (9.20) is used to compensate the total disturbances at step i, so that the transfer function from $u_0(i)$ to $G(i)$ (c.f. Eq. (9.1)) will be approximately two cascaded integrators in discrete time [43]. In this regard, u_0 is designed in the form of $f_{\text{han}}(\cdot)$, which is the time-optimal control synthesis function for two cascaded discrete-time integrators.

9.3.5 *Safety Constraints*

For safety concern, the controller enforces several constraints at each step i. The first type of constraint provides upper and lower limits for the insulin infusion rate:

$$0 \leq u(i)T + \frac{u_b(i)T}{60 \text{ min/h}} \leq u_{\max}, \tag{9.23}$$

where $u_{\max} := 0.5$ U denotes an upper bound on the dose of a micro-bolus. From (9.23), the minimum value of $u(i)$ is set to $-\frac{u_b(i)}{60 \text{ min/h}}$, since the insulin-glucose system is a positive system and it is impossible to remove the insulin from the subject.

The controller also enforces an IOB constraint based on insulin delivery history to prevent over-bolus when much insulin has been recently delivered. Let

$$\tau(\eta) := \max\{\min\{-\eta/30 + 12, 8\}, 2\}, \tag{9.24}$$

where the design of η is introduced in Sect. 9.4.1. A revised IOB constraint based on the version used in [13] is adopted in the controller. In particular, it employs IOB decay curves of length $\tau(\eta)$ based on either the estimated CGM y_1 or CGM prediction y_2, instead of the current CGM level in [13]. The reason for this is discussed in Sect. 9.4.1. At each step i, the IOB constraint is given as

$$u(i)T \leq \overline{u}_{\text{IOB}}(i), \tag{9.25}$$

$$\overline{u}_{\text{IOB}}(i) := \max\{\Gamma(i) - \Phi(i), 0\}, \tag{9.26}$$

where $\Phi(i) \in \mathbb{R}$ is the estimated IOB, and $\Gamma(i) \in \mathbb{R}$ is the required IOB at time step i depending on the current $\eta(i)$. The detailed calculation of $\Phi(i)$ and $\Gamma(i)$ can be found in [13]. Note that $\overline{u}_{\text{IOB}}(i)$ is nonnegative and $\overline{u}_{\text{IOB}}(i) = 0$ implies the controller is allowed to deliver no more than the basal rate $u_b(i)T$; therefore the IOB constraint does not affect insulin delivery to below the basal rate.

9.3.6 *Pump-Discretization*

Similar to [12], as a CSII pump has a delivery resolution of δ [U], the final absolute insulin infusion $\widetilde{u}(i)$ commanded to the pump is calculated according to a so-called carry-over module:

$$\widetilde{u}(i) := \delta\left\lfloor\frac{\hat{u}(i)}{\delta}\right\rfloor, \quad \hat{u}(i) := u(i)T + u_{\text{carray}}(i) + \frac{u_b(i)T}{60\ \text{min/h}},$$
$$u_{\text{carray}}(i) := \hat{u}(i-1) - \widetilde{u}(i-1) \geq 0, \quad u_{\text{carray}}(0) := 0,$$

where $\lfloor\cdot\rfloor$ denotes the floor operator, and $u(i)T$ is the control input at time step i after enforcing insulin delivery constraints and IOB constraints.

9.4 Event-Triggered Parameter Adaptation for ADRC

Due to the asymmetric risks of hyper- and hypoglycemia, the AP controller is required to react differently in ascending and descending BG situations. To ensure dynamic glucose management performance, event-triggered parameter adaptation is incorporated for ESO, IOB, and nonlinear feedback modules of the proposed controller.

9.4.1 Adaptation of ESO and IOB

To enable a safe insulin infusion strategy when the glucose is decreasing toward hypoglycemia and a responsive strategy when the glucose level is increasing toward hyperglycemia, an event-triggered signal switching policy (see orange blocks A and B in Fig. 9.1) is proposed for ESO and IOB, which has the form:

$$\eta(i) := \begin{cases} y_2(i), & y_1(i) > G_h, \ \text{sign}(v_2(i)) \geq 0, \\ y_2(i), & y_1(i) < G_L, \ \text{sign}(v_2(i)) < 0, \\ y_1(i), & \text{otherwise}. \end{cases} \tag{9.27}$$

Specifically, when the glucose is ascending and its measurement value is higher than a threshold G_h, or when the glucose is descending and its measurement value is lower than G_L, ESO and IOB will use the glucose prediction $y_2(i)$ to calculate their output signals, otherwise the glucose estimate $y_1(i)$ is used. Note that $y_1(i)$, $y_2(i)$ and $v_2(i)$ are provided by TD at each time step i.

From (9.11) and (9.12), we observe that when the glucose level is steadily increasing, the predicted values are higher than the filtered values, which would introduce larger estimates z_1 and z_2 in (9.16) and (9.17), and lead to more responsive controller behavior. Meanwhile, the predicted values for ascending glucose traces would lead to the shorter length of IOB decay curves than the filtered values (c.f., Eq. (9.24)), which loosens the insulin infusion constraints so that a more responsive controller behavior is encouraged. Following this idea, a relatively more conservative insulin infusion policy will be observed for decreasing glucose traces.

9.4.2 Parameter Adaptation for Nonlinear Feedback

Parameter adaptation is designed for k_1, k_2, r, and ε in the nonlinear feedback module, which determines the aggressiveness of the control response. In particular, according to (9.6) in Sect. 9.3.1, if the following inequalities hold:

$$|n_1 + \varepsilon n_2| \leq \varepsilon^2 r, \ |n_2 + \frac{n_1 + \varepsilon n_2}{\varepsilon}| \leq \varepsilon r, \tag{9.28}$$

then a linear approximate expression of $f_{\text{han}}(n_1, n_2, r, \varepsilon)$

$$f_{\text{han}}^{\text{approx}}(n_1, n_2, r, \varepsilon) = -\frac{2n_2 + n_1/\varepsilon}{\varepsilon} \tag{9.29}$$

can be obtained. Based on this relationship, we fix r and design ε to make (9.28) always hold, so that the parameter design problem can be simplified. In this way, the control law in (9.20) can be approximated as

$$u(i) \approx \left((2k_2 e_2 + \frac{k_1 e_1}{\varepsilon})/\varepsilon - z_3(i) \right) / b_0. \tag{9.30}$$

From (9.30), we observe that ε is a coefficient that determines the overall aggressiveness of controller (in terms of increasing insulin infusion above basal rate); a small value for ε will increase the aggressiveness of controller and vice versa. On the other hand, k_1 and k_2 adjust the weight of tracking error and its rate of change in control law, and thus determine the aggressiveness of controller with respect to glucose excursion and its rate of change, respectively.

To assist our design, we first introduce a basic function $f_a(x, \Upsilon)$ that is utilized to build the adaptation laws for ε, k_1 and k_2. Specifically, $f_a(x, \Upsilon)$ is designed as an exponential function with saturation:

$$f_a(x, \Upsilon) := \min\{a_1, \exp\{\alpha x^{\beta}\} + a_2 - 1\}, \ \ x \geq 0, \tag{9.31}$$

where $\Upsilon := [a_1, a_2, \alpha, \beta]$ is a quadruple that determines the shape of this function (see Fig. 9.2). In particular, the maximum (saturation) and minimum values of the curve is determined by a_1 and a_2, respectively, and the steepness of the curve is controlled by α and β.

To achieve parameter adaptation, we define different scenarios based on the value and trend of glucose prediction. Specifically, when the glucose is ascending, we divide the glucose range into three regions based on the glucose estimate y_1 provided by TD: $y_1 \leq G_h$, $G_h < y_1 < G_H$ and $y_1 \geq G_H$. Similarly, when glucose is descending, we divide the glucose range into two regions: $y_1 \geq G_L$ and $y_1 < G_L$, respectively. Glucose- and velocity-triggered parameter adaptation laws are designed for each region as follows (see orange blocks C-E in Fig. 9.1):

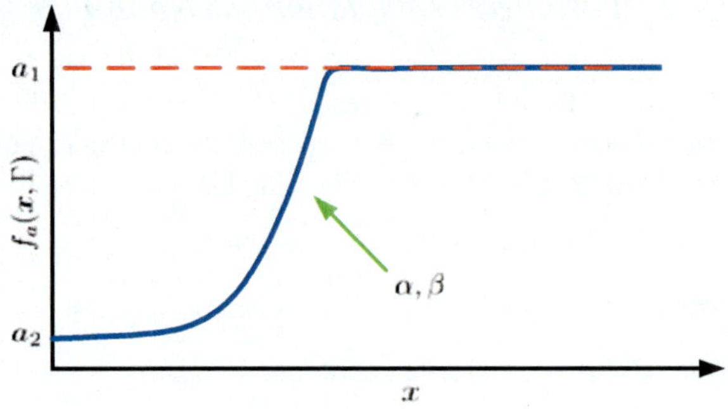

Fig. 9.2 An illustration of the basic function defined in (9.31)

$$
k_1 := \begin{cases} f_a(v_2, \Upsilon_1^{k_1}), & y_1 \le G_h, \ \mathrm{sign}(v_2) \ge 0, \\ b_1, & G_h < y_1 < G_H, \ \mathrm{sign}(v_2) \ge 0, \\ b_1, & y_1 \ge G_H, \ \mathrm{sign}(v_2) \ge 0, \\ b_2, & y_1 \ge G_L, \ \mathrm{sign}(v_2) < 0, \\ b_3, & y_1 < G_L, \ \mathrm{sign}(v_2) < 0, \end{cases}
$$

$$
k_2 := \begin{cases} f_a(v_2, \Upsilon_1^{k_2}), & y_1 \le G_h, \ \mathrm{sign}(v_2) \ge 0, \\ c_1, & G_h < y_1 < G_H, \ \mathrm{sign}(v_2) \ge 0, \\ c_1, & y_1 \ge G_H, \ \mathrm{sign}(v_2) \ge 0, \\ c_2, & y_1 \ge G_L, \ \mathrm{sign}(v_2) < 0, \\ f_a(|v_2|, \Upsilon_2^{k_2}), & y_1 < G_L, \ \mathrm{sign}(v_2) < 0, \end{cases}
$$

$$
\varepsilon := \begin{cases} f_a(G_h - y_1, \Upsilon_1^{\varepsilon}), & y_1 \le G_h, \ \mathrm{sign}(v_2) \ge 0, \\ A - f_a(G_H - y_1, \Upsilon_2^{\varepsilon}), & G_h < y_1 < G_H, \ \mathrm{sign}(v_2) \ge 0, \\ d_1, & y_1 \ge G_H, \ \mathrm{sign}(v_2) \ge 0, \\ d_2, & y_1 \ge G_L, \ \mathrm{sign}(v_2) < 0, \\ d_2, & y_1 < G_L, \ \mathrm{sign}(v_2) < 0, \end{cases} \tag{9.32}
$$

where v_2 is the approximate differential signal of glucose measurements provided by TD in (9.10) to determine whether the glucose is ascending or descending; b_1 is the maximum (saturation) value of $f_a(v_2, \Upsilon_1^{k_1})$, c_1 is the maximum (saturation) value of $f_a(v_2, \Upsilon_1^{k_2})$, c_2 is the minimum value of $f_a(|v_2|, \Upsilon_2^{k_2})$, and d_1 is the maximum (saturation) value of $f_a(G_H - y_1, \Upsilon_2^{\varepsilon})$; b_2, b_3, and d_2 are positive constants; A is the sum of maximum (saturation) and minimal values of $f_a(G_H - y_1, \Upsilon_2^{\varepsilon})$. The design principles for the adaptation laws in each region are presented in the following, and the design of the parameters is presented in Sect. 9.4.3.

Region (1) $y_1 \le G_h, \ \mathrm{sign}(v_2) \ge 0$: In this case, when the glucose value is at a low level or rising slowly, the controller should keep a conservative insulin delivery policy for robustness to the CGM noise, and become gradually responsive when the glucose is rising. Consequently, according to (9.30), k_1 and k_2 are designed to gradually increase with the increase of change rate of glucose value, and ε is designed to gradually decrease with the increase of glucose value to determine the overall aggressiveness of controller. In this way, the controller can remain conservative when the glucose value is at a low level to ensure safety.

Region (2) $G_h < y_1 < G_H$, $\text{sign}(v_2) \geq 0$: This is a key region where the pump should deliver enough insulin to prevent early postprandial hyperglycemia so as to regulate the glucose level back into normal region as soon as possible. The controller is required to stay responsive until y_1 is approaching an upper threshold G_H to prevent postprandial hypoglycemia. To implement this principle, k_1 and k_2 keep the maximum (saturation) value in $f_a(v_2, \Upsilon_1^{k_1})$ and $f_a(v_2, \Upsilon_1^{k_2})$ unchanged, respectively, and ε is designed to increase when y_1 is approaching upper the threshold G_H.

Region (3) $y_1 \geq G_H$, $\text{sign}(v_2) \geq 0$: As discussed above, a suitably designed controller would have delivered enough insulin in Region 2 so that only moderate insulin need to be delivered in this region. To do this, k_1 and k_2 still take the values in Region 2, and ε is set as the maximum (saturation) value in $f_a(G_H - y_1, \Upsilon_2^{\varepsilon})$. In this way, according to (9.30), the controller is designed to be insensitive to extremely high glucose value but is sensitive to its rate of change, which allows the controller to deliver a moderate amount of insulin when glucose is high but still rising rapidly.

Region (4) $y_1 \geq G_L$, $\text{sign}(v_2) < 0$: In this region, when the glucose is at a high level but is falling slowly, a moderate amount of insulin (above basal rate) could be delivered to alleviate hyperglycemia, but pump suspension should be adopted when glucose is falling rapidly. From (9.30), insulin delivery will increase with the increase of glucose value but decrease with the increase of its absolute rate of change. Based on this observation and the design principle, we assign constant values to ε, k_1 and k_2 so that the weight of the tracking error is lower than that of the glucose rate of change in the control law.

Region (5) $y_1 < G_L$, $\text{sign}(v_2) < 0$: When glucose is descending into this region, the controller should be extremely conservative and sensitive to the change rate of glucose value to prevent over-delivery. To do this, ε still takes the value in Region 4, k_1 is fixed to a designed constant, and k_2 is designed to gradually increase with the increase of the absolute glucose velocity. Moreover, when glucose value y_1 is lower than threshold G_L, TD provides glucose prediction values to both IOB and ESO, which will also make the controller act conservatively.

Table 9.1 Parameters for basic ADRC

Parameters for nominal input gain and TDIFF	$b_0 = -6/u_{\text{TDI}}$	$K_1 = 3$ (Filter signal)
	$K_2 = 7$(Glucose ascending ($y_1 > G_h$))	
	$r_1 = 20$	$h_0 = 2T$
	$K_2 = 5$ (Glucose descending ($y_1 < G_L$))	
Parameters for ESO	$\lambda_1 = 1/T$	$\lambda_2 = 1/(2T^2)$
	$\lambda_3 = 1/(160T^3)$	
	$\sigma_1 = 0.5$	$\sigma_2 = 0.25$
	$\mu = T$	

In this table, T denotes sampling period, $G_h = 250$ mg/dL and $G_L = 135$ mg/dL

Table 9.2 Parameters for nonlinear feedback

Glucose ascending ($G_h = 135$, $G_H = 250$)		
$y_1 \leq G_h$	$G_h < y_1 < G_H$	$y_1 \geq G_H$
$\Upsilon_1^{k_1} = \{0.01, 0.005, 0.01, 5\}$	$k_1 = 0.01$	$k_1 = 0.01$
$\Upsilon_1^{k_2} = \{0.05, 0.01, 0.01, 5\}$	$k_2 = 0.05$	$k_2 = 0.05$
$\Upsilon_1^{\varepsilon} = \{25, 12, 0.35, 0.8\}$	$\Upsilon_2^{\varepsilon} = \{40, 12, 2, 0.5\}$	$\varepsilon = 40$
Glucose descending ($G_L = 180$)		Other parameters
$y_1 \geq G_L$	$y_1 < G_L$	
$k_1 = 0.002$	$k_1 = 0.005$	$G_r = 110$
$k_2 = 0.02$	$\Upsilon_2^{k_2} = \{0.1, 0.02, 0.002, 2\}$	$\dot{G}_r = 0$
$\varepsilon = 8$	$\varepsilon = 8$	$r = 10$

9.4.3 Parameter Design

The parameters of the proposed controller are designed and evaluated using the UVA/Padova T1DM metabolic simulator[7]. The "average patient" is selected as the "test set", and the 10-patient cohort is selected as the "verification set". In the parameter design phase, a 12-hour *in silico* protocol starting from 7:00 with one meal of 60 g carbohydrate (CHO) at 9:00 is employed. The parameters are designed using a trial-and-error approach based on the glucose data obtained from the "average patient" for both announced and unannounced meals, so that satisfactory glucose regulation performance in terms of average glucose, percent time in [70, 180] mg/dL, and percent time below 70 mg/dL can be achieved. The obtained parameters are summarized in Tables 9.1 and 9.2. In the verification phase, the proposed control algorithm with the obtained parameters is evaluated on the 10-patient cohort of the simulator for different scenarios. The results are reported in Sect. 9.5.

Note that the 12-hour protocol is used as a "test protocol" for controller design, and a different protocol with multiple meals of different amounts of CHOs is used as the "verification protocol" to evaluate the performance of the designed controller (see Sect. 9.5).

9.5 In Silico Performance Analysis

As is mentioned above, the proposed controller is evaluated on the 10-adult cohort of the UVA/Padova simulator. For comparison purpose, the protocol designed in [32] is introduced to evaluate the performance of the controller. The protocol starts from 7:00 on Day one and lasts for two days (48 h) where on each day, breakfast (50 g CHO), lunch (75 g CHO), and dinner (75 g CHO) are consumed at 8:00, 12:00, and 19:00, respectively.

The proposed controller is compared with the adaptive zone MPC controller introduced in [33] (denoted as "Control"). For both controllers, the pump resolution is selected as 0.05 [U]. Based on the introduced protocol, *in silico* experiments are performed on the whole 10-adult cohort using additive CGM noises (with random seeds 1–10) for scenarios considering different conditions on meal boluses and basal rates, and thus a total of 100 simulations are performed for each scenario and each controller. Note that the results of the adaptive zone MPC reported in [33] were obtained on the 100-adult cohort, which is why the results reported here are slightly different. The results for the scenario of nominal basal rate with announced meals and unannounced meals are presented in Sects. 9.5.1 and 9.5.2, respectively. The discussion for the scenarios of basal-rate mismatches (130 and 70% of the nominal basal rate) and over/underestimated meal boluses (130 and 70% of the nominal meal boluses) are provided in Sect. 9.5.3. The statistical results are summarized in Tables 9.3, 9.4, 9.5, and 9.6, and the performance of the proposed controller in terms of glucose regulation is illustrated in Figs. 9.3, 9.4, 9.5, 9.6, and 9.7, where the 5, 25, 75, and 95% quartile curves together with the median curves are presented.

9.5.1 *Announced Meals with Nominal Basal Rate*

From Table 9.3, compared with the adaptive zone MPC, the proposed non-optimization-based controller achieves similar (and slightly improved) performance for glucose regulation with announced meals without increasing the risk of hyperglycemia (percent time < 70 mg/dL, 0.0% versus 0.0%, $p = 0.441$). Satisfactory performance for hyperglycemia control is achieved, which is reflected in percent time in the euglycemic range of 70–180 mg/dL (93.8% versus 92.4%, $p < 0.001$), mean glucose (134.9 mg/dl versus 135.4 mg/dL, $p < 0.001$), hyperglycemia

Table 9.3 Comparison results with nominal basal rate

Scenario (#Simu = 100)	Announced meals			Unannounced meals		
Metric	Control	Proposed	p value	Control	Proposed	p value
% time						
<54 mg/dL	0.0 (0.0)	0.0 (0.0)	0.500	0.0 (0.0)	0.0 (0.0)	1.000
<70 mg/dL	0.0 (0.0)	0.0 (0.0)	0.441	0.0 (0.0)	0.0 (0.0)	0.211
70–180 mg/dL	92.4 (10.2)	**93.8 (9.2)**	**<0.001**	72.4 (13.4)	**76.0 (11.3)**	**<0.001**
>180 mg/dL	7.6 (10.0)	**6.2 (9.2)**	**<0.001**	27.6 (13.4)	**24.0 (11.0)**	**<0.001**
>250 mg/dL	0.0 (0.0)	0.0 (0.0)	0.500	1.2 (9.5)	**1.0 (7.9)**	**<0.001**
Mean glucose (mg/dL)	135.4 (6.1)	**134.9 (6.4)**	**<0.001**	151.7 (19.9)	**149.7 (16.8)**	**<0.001**
SD glucose (mg/dL)	28.2 (8.2)	**25.8 (4.8)**	**<0.001**	44.9 (14.1)	**43.0 (14.4)**	**<0.001**
Mean glucose at 07:00	121.0 (15.2)	**118.5 (17.0)**	**<0.001**	121 (15.5)	**113.0 (15.2)**	**<0.001**

Data in this table are shown as median (inter quartile range). Statistically significant ($p < 0.05$) changes are highlighted in bold

Table 9.4 Comparison results with 130% of nominal basal rate

Scenario (#Simu = 100)	Announced meals			Unannounced meals		
Metric	Control	Proposed	*p* value	Control	Proposed	*p* value
% time						
<54 mg/dL	0.0 (0.0)	0.0 (0.0)	0.201	0.0 (0.0)	0.0 (0.0)	0.537
<70 mg/dL	0.5 (1.8)	**0.0 (1.5)**	**0.002**	0.5 (3.0)	**0.0 (0.7)**	**<0.001**
70–180 mg/dL	94.7 (7.8)	**95.6 (6.8)**	**<0.001**	77.6 (14.4)	**81.4 (13.4)**	**<0.001**
>180 mg/dL	4.0 (7.4)	**3.3 (6.6)**	**<0.001**	19.0 (13.4)	**18.2 (11.3)**	**<0.001**
>250 mg/dL	0.0 (0.0)	0.0 (0.0)	1.000	0.4 (3.0)	0.0 (3.3)	0.141
Mean glucose (mg/dL)	122.7 (4.8)	**125.1 (6.6)**	**<0.001**	135.7 (15.8)	135.5 (15.8)	0.640
SD glucose (mg/dL)	28.2 (8.2)	**25.8 (4.8)**	**<0.001**	45.1 (13.2)	**41.6 (13.2)**	**<0.001**
Mean glucose at 07:00	107.5 (16.2)	**111.2 (18.5)**	**0.001**	107.2 (15.8)	**103.0 (19.8)**	**<0.001**

Data in this table are shown as median (interquartile range). Statistically significant ($p < 0.05$) changes are highlighted in bold

Table 9.5 Comparison results with 70% of nominal basal rate

Scenario (#Simu = 100)	Announced meals			Unannounced meals		
Metric	Control	Proposed	*p* value	Control	Proposed	*p* value
% time						
<54 mg/dL	0.0 (0.0)	0.0 (0.0)	1.000	0.0 (0.0)	0.0 (0.0)	1.000
<70 mg/dL	0.0 (0.0)	0.0 (0.0)	1.000	0.0 (0.0)	0.0 (0.0)	1.000
70–180 mg/dL	82.8 (9.3)	**85.2 (9.5)**	**<0.001**	56.5 (24.3)	**65.6 (15.8)**	**<0.001**
>180 mg/dL	17.2 (9.3)	**14.8 (9.5)**	**<0.001**	43.5 (24.3)	**34.4 (15.8)**	**<0.001**
>250 mg/dL	0.0 (0.0)	0.0 (0.0)	0.633	7.4 (19.3)	**7.5 (14.0)**	**<0.001**
Mean glucose (mg/dL)	153.2 (6.6)	**150.9 (5.4)**	**<0.001**	178.0 (36.6)	**170.6 (22.2)**	**<0.001**
SD glucose (mg/dL)	26.2 (7.2)	25.5 (5.0)	0.065	45.1 (15.4)	45.3 (16.1)	0.620
Mean glucose at 07:00	142.5 (14.0)	**137.8 (17.0)**	**<0.001**	141.5 (15.0)	**135.5 (16.0)**	**<0.001**

Data in this table are shown as median (inter quartile range). Statistically significant ($p < 0.05$) changes are highlighted in bold

Table 9.6 Comparison results with nominal basal rate for over/underestimated meal boluses

Scenario (#Simu = 100)	130% of nominal meal boluses			70% of nominal meal boluses		
Metric	Control	Proposed	*p* value	Control	Proposed	*p* value
% time						
<54 mg/dL	0.0 (0.0)	0.0 (0.0)	0.500	0.0 (0.0)	0.0 (0.0)	0.250
<70 mg/dL	0.0 (0.0)	0.0 (0.0)	0.985	0.0 (0.0)	0.0 (0.0)	0.484
70–180 mg/dL	96.4 (6.8)	**97.4 (4.9)**	**<0.001**	83.9 (15.9)	**86.2 (12.4)**	**<0.001**
>180 mg/dL	3.5 (6.4)	**2.6 (4.5)**	**<0.001**	16.1 (15.5)	**13.8 (12.2)**	**<0.001**
>250 mg/dL	0.0 (0.0)	0.0 (0.0)	1.000	0.0 (0.0)	**0.0 (0.0)**	**0.022**
Mean glucose (mg/dL)	128.9 (4.3)	**128.3 (4.8)**	**<0.001**	141.2 (8.8)	**140.2 (9.0)**	**<0.001**
SD glucose (mg/dL)	25.5 (7.4)	**22.7 (6.2)**	**<0.001**	32.7 (9.6)	**31.0 (8.0)**	**<0.001**
Mean glucose at 07:00	121.0 (14.0)	120.5 (18.2)	0.118	120.8 (14.8)	**116.5 (17.8)**	**<0.001**

Data in this table are shown as median (inter quartile range). Statistically significant ($p < 0.05$) changes are highlighted in bold

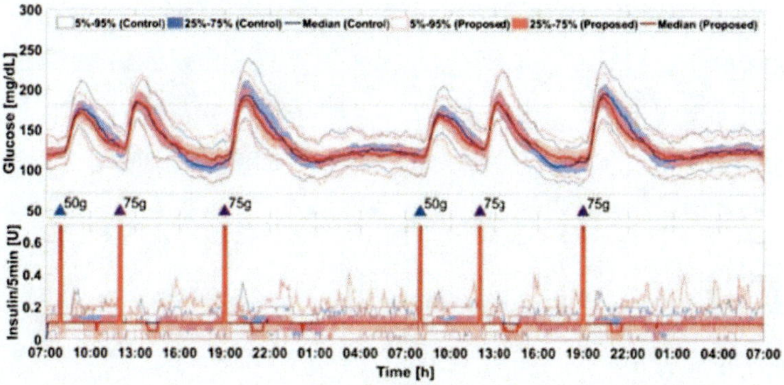

Fig. 9.3 Performance comparison with a nominal basal rate for unannounced meals in terms of glucose regulation and insulin deliver action. Blue and purple triangles denote meals of 50 g and 75 g CHO, respectively

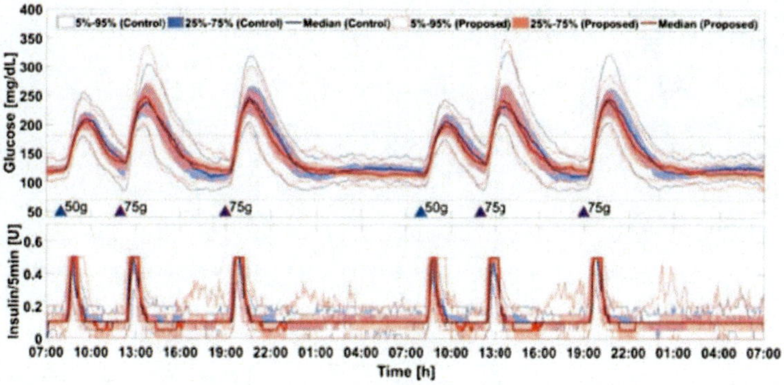

Fig. 9.4 Performance comparison with nominal basal rate for unannounced meals in terms of glucose regulation and insulin deliver action. Blue and purple triangles denote meals of 50 g and 75 g CHO, respectively

(percent time > 250 mg/dL, 0.0% versus 0.0%, $p = 0.500$), and glucose variability (SD glucose, 25.8 versus 28.2, $p < 0.001$). The discussions of controller performance are consistent with the quartile curves in Fig. 9.5.

9.5.2 Unannounced Meals with Nominal Basal Rate

Similar to the case of announced meals, the proposed controller achieves comparable performance to the adaptive zone MPC in terms of percent time in the euglycemic range of 70–180 mg/dL (76.0% versus 72.4%, $p < 0.001$), mean glucose (149.7 mg/dl versus 151.7 mg/dL, $p < 0.001$), hyperglycemia (percent time >

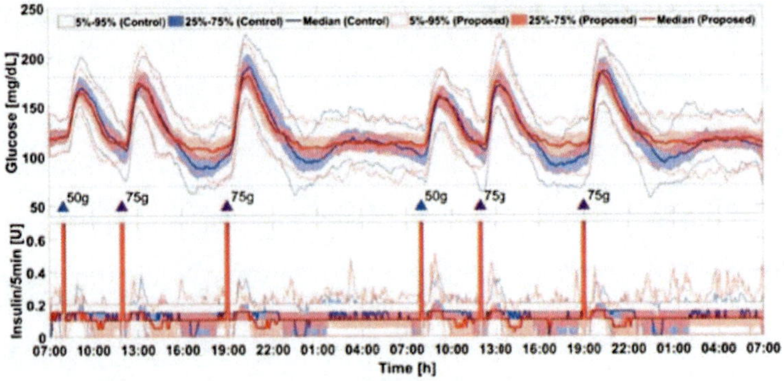

(a) Announced meals

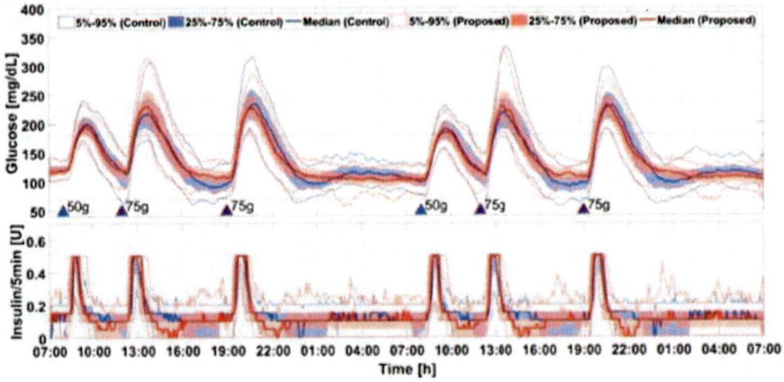

(b) Unannounced meals

Fig. 9.5 Performance comparison for the scenario of over-estimated basal rates

250 mg/dL, 1.0% versus 1.2%, $p < 0.001$), glucose variability (SD glucose, 43.0 versus 44.9, $p < 0.001$). From Fig. 9.3, we can observe that the controller is able to reduce insulin infusion when the glucose concentration is rapidly decreasing to ensure safety. Again, no increased risk of hypoglycemia is observed (percent time < 70 mg/dL, 0.0% versus 0.0%, $p = 0.211$).

9.5.3 *Robust Performance Evaluation*

To evaluate the robustness of the designed controller, extensive stress tests are performed considering realistic scenarios of moderate meal-bolus over/underestimates and basal-rate mismatches. The results are also compared with those obtained for the adaptive zone MPC. From Tables 9.4, 9.5, and 9.6 and Figs. 9.4, 9.6, and 9.7, we observe that the proposed controller behaves properly for these scenarios and

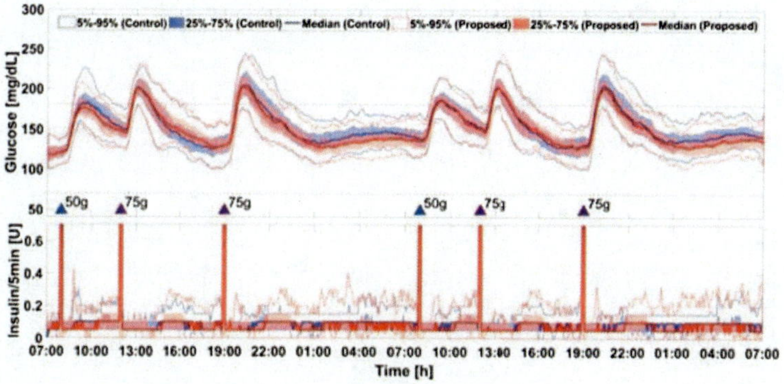

(a) Announced meals

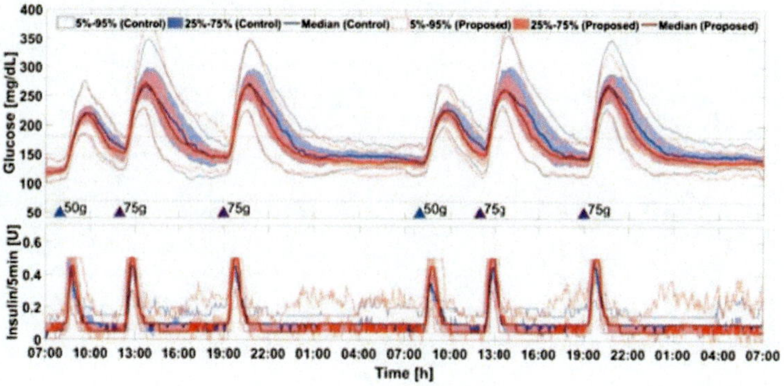

(b) Unannounced meals

Fig. 9.6 Performance comparison for the scenario of underestimated basal rates

again achieves satisfactory and comparable performance in terms of mean glucose and percentage time in [70, 180] mg/dL. Statistically significant improvements are observed for the scenario of underestimated basal rates (see Table 9.5), which is partly attributed to the offset-free tracking property inherited from ADRC [19].

9.6 Summary

In this chapter, we generalize the idea of ET-ADRC to design an adaptive ADRC controller for closed-loop glucose regulation. The core method is to enable different event-triggered controller behavior for different glucose situations, which is important to ensure safe and efficient glucose regulation. The proposed controller is able to responsively command insulin when blood glucose is rapidly increasing above the euglycemic range, and safely turn off or turn down insulin infusion when the glucose

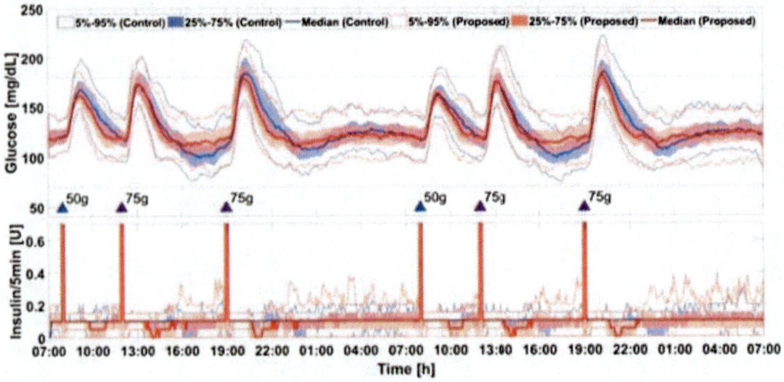

(a) Over-estimated meal boluses

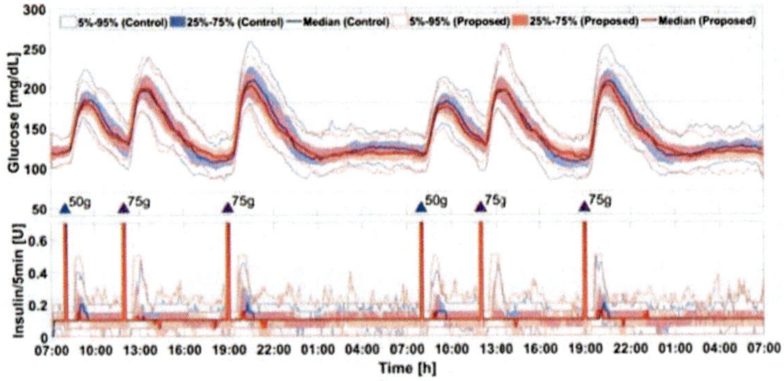

(b) Under-estimated meal boluses

Fig. 9.7 Performance comparison for the scenarios of over/underestimated meal boluses

is low or rapidly decreasing. The effectiveness and robustness of the controller is evaluated on the 10-adult cohort of the UVA/Padova simulator through comparisons with the adaptive zone MPC [33]. Although the usage of event-triggers to adjust controller behavior is implemented heuristically in this chapter, the idea behind opens the door of event-based parameter adaptation and learning for sampled-data control systems, which motivates and deserves further investigations. Finally, access to the distributed version of the University of Virginia (UVA)/Padova metabolic simulator for research purposes under a license agreement between the Epsilon Group Analytics and the Beijing Institute of Technology is acknowledged.

References

1. Atlas, E., Nimri, R., Miller, S., Grunberg, E.A., Phillip, M.: Md-logic artificial pancreas system: a pilot study in adults with type 1 diabetes. Diabetes Care **33**(5), 1072–1076 (2010)
2. Bequette, B.W.: Algorithms for a closed-loop artificial pancreas: the case for model predictive control. J. Diabetes Sci. Technol. **7**(6), 1632–1643 (2013)
3. Boiroux, D., Duun-Henriksen, A.K., Schmidt, S., Norgaard, K., Poulsen, N.K., Madsen, H., Jorgensen, J.B.: Adaptive control in an artificial pancreas for people with type 1 diabetes. Control Eng. Pract. **58**, 332–342 (2017)
4. Cameron, F., Niemeyer, G., Bequette, B.W.: Extended multiple model prediction with application to blood glucose regulation. J. Process Control **22**(8), 1422–1432 (2012)
5. Colmegna, P.H., Sanchez-Pena, R.S., Gondhalekar, R., Dassau, E., Doyle III, F.J.: Switched LPV glucose control in type 1 diabetes. IEEE Trans. Biomed. Eng. **63**(6), 1192–1200 (2016)
6. Colmegna, P.H., Sanchez-Pena, R.S., Gondhalekar, R., Dassau, E., Doyle III, F.J.: Reducing glucose variability due to meals and postprandial exercise in T1DM using switched LPV control: in silico studies. J. Diabetes Sci. Technol. **10**(3), 744–753 (2016)
7. Dalla Man, C., Micheletto, F., Lv, D., Breton, M., Kovatchev, B., Cobelli, C.: The UVA/PADOVA type 1 diabetes simulator: new features. J. Diabetes Sci. Technol. **8**(1), 26–34 (2014)
8. Dassau, E., Pinsker, J., Kudva, Y., Brown, S., R., G., C., D., Patek, S., Schiavon, M., Dadlani, V., Dasanayake, I., Church, M.M., Carter, R.E., Bevier, W.C., Huyett, L.M., Hughes, J., Anderson, S., Lv, D., Schertz, E., Emory, E., McCrady-Spitzer, S.K., Jean, T., Bradley, P.K., Hinshaw, L., Sanz, A.J.L., Basu, A., Kovatchev, B., Cobelli, C., Doyle III, F.J.: Twelve week 24/7 ambulatory artificial pancreas with weekly adaptation of insulin delivery settings: effect on hemoglobin A1c and hypoglycemia. Diabetes Care **40**(12), 1719–1726 (2017)
9. Diabetes Control and Complications Trial Research Group: The effect of intensive treatment of diabetes on the development and progression of long-term complications in insulin-dependent diabetes mellitus. N. Engl. J. Med. **1993**(329), 977–986 (1993)
10. Doyle III, F.J., Huyett, L.M., Lee, J.B., Zisser, H.C., Dassau, E.: Closed-loop artificial pancreas systems: engineering the algorithms. Diabetes Care **37**(5), 1191–1197 (2014)
11. Garg, S.K., Weinzimer, S.A., Tamborlane, W.V., Buckingham, B.A., Bode, B.W., Bailey, T.S., Brazg, R.L., Ilany, J., Slover, R.H., Anderson, S.M., Bergenstal, R.M., Grosman, B., Roy, A., Cordero, T.L., Shin, J., Lee, S.W., Kaufman, F.R.: Glucose outcomes with the in-home use of a hybrid closed-loop insulin delivery system in adolescents and adults with type 1 diabetes. Diabetes Technol. Ther. **19**(3), 155–163 (2017)
12. Gondhalekar, R., Dassau, E., Doyle III, F.J.: Periodic zone-mpc with asymmetric costs for outpatient-ready safety of an artificial pancreas to treat type 1 diabetes. Automatica **71**, 237–246 (2016)
13. Gondhalekar, R., Dassau, E., Doyle III, F.J.: Velocity-weighting & velocity-penalty MPC of an artificial pancreas: improved safety & performance. Automatica **91**, 105–117 (2018)
14. Gopakumaran, B., Duman, H.M., Overholser, D.P., Federiuk, I.F., Quinn, M.J., Wood, M.D., Ward, W.K.: A novel insulin delivery algorithm in rats with type 1 diabetes: the fading memory proportional-derivative method. Artif. Organs **29**(8), 599–607 (2005)
15. Grosman, B., Dassau, E., Zisser, H., Jovanovic, L., Doyle III, F.J.: Multi-zone-MPC: clinical inspired control algorithm for the artificial pancreas. In: 18th IFAC World Congress, pp. 7120 – 7125 (2011)
16. Guo, B., Han, J., Xi, F.: Linear tracking-differentiator and application to online estimation of the frequency of a sinusoidal signal with random noise perturbation. Int. J. Syst. Sci. **33**(5), 351–358 (2002)
17. Guo, B., Zhao, Z.: On convergence of tracking differentiator. Int. J. Control **84**(4), 693–701 (2011)
18. Haidar, A.: The artificial pancreas: how closed-loop control is revolutionizing diabetes. IEEE Control Syst. **36**(5), 28–47 (2016)

19. Han, J.: From PID to active disturbance rejection control. IEEE Trans. Ind. Electron. **56**(3), 900–906 (2009)
20. van Heusden, K., Dassau, E., Zisser, H.C., Seborg, D.E., Doyle III, F.J.: Control-relevant models for glucose control using a priori patient characteristics. IEEE Trans. Biomed. Eng. **59**(7), 1839–1849 (2012)
21. Hovorka, R., Allen, J.M., Elleri, D., Chassin, L.J., Harris, J., Xing, D., Kollman, C., Hovorka, T., Larsen, A.M.F., Nodale, M., Palma, A.D., Wilinska, M.E., Acerini, C.L., Dunger, D.B.: Manual closed-loop insulin delivery in children and adolescents with type 1 diabetes: a phase 2 randomised crossover trial. The Lancet **375**(9716), 743–751 (2010)
22. Hovorka, R., Canonico, V., Chassin, L.J., Haueter, U., Massi-Benedetti, M., Federici, M.O., Pieber, T.R., Schaller, H.C., Schaupp, L., Vering, T., Wilinska, M.E.: Nonlinear model predictive control of glucose concentration in subjects with type 1 diabetes. Physiol. Meas. **25**(4), 905–920 (2004)
23. Hua, C.C., Wang, K., Chen, J.N., You, X.: Tracking differentiator and extended state observer-based nonsingular fast terminal sliding mode attitude control for a quadrotor. Nonlinear Dyn. **94**(1), 1–12 (2018)
24. Huyett, L.M., Dassau, E., Zisser, H.C., Doyle III, F.J.: Design and evaluation of a robust PID controller for a fully implantable artificial pancreas. Ind. Eng. Chem. Res. **54**(42), 10311–10321 (2015)
25. Jacobs, P.G., Youssef, J.E., Castle, J., Bakhtiani, P., Branigan, D., Breen, M., Bauer, D., Preiser, N., Leonard, G., Stonex, T., Ward, W.K.: Automated control of an adaptive bihormonal, dual-sensor artificial pancreas and evaluation during inpatient studies. IEEE Trans. Biomed. Eng. **61**(10), 2569–2581 (2014)
26. Lee, J.B., Dassau, E., Gondhalekar, R., Seborg, D.E., Pinsker, J.E., Doyle III, F.J.: Enhanced model predictive control (eMPC) strategy for automated glucose control. Ind. Eng. Chem. Res. **55**(46), 11857–11868 (2016)
27. Mauseth, R., Wang, Y., Dassau Jr., E., R.K., Matheson, D., Zisser, H.: Proposed clinical application for tuning fuzzy logic controller of artificial pancreas utilizing a personalization factor. J. Diabetes Sci. Technol. **4**(4), 913–922 (2010)
28. Paolo Incremona, G., Messori, M., Toffanin, C., Cobelli, C., Magni, L.: Artificial pancreas: from control-to-range to control-to-target. In: Proceedings 20th IFAC World Congress, vol. 50, pp. 7737 – 7742 (2017)
29. Richard, M., Hirsch, I.B., Jennifer, B., Robert, K., Don, M., Srinath, S., Carla, G.: Use of a "fuzzy logi" controller in a closed-loop artificial pancreas. Diabetes Technol. Ther. **15**(8), 628–633 (2013)
30. Rossetti, P., Quiros, C., Moscardo, V., Comas, A., Gimenez, M., Ampudia-Blasco, F.J., Leon, F., Montaser, E., Conget, I., Bondia, J., Vehi, J.: Closed-loop control of postprandial glycemia using an insulin-on-board limitation through continuous action on glucose target. Diabetes Technol. Ther. **19**(6), 355–362 (2017)
31. Russell, S.J., El-Khatib, F.H., Sinha, M., Magyar, K.L., McKeon, K., Goergen, L.G., Balliro, C., Hillard, M.A., Nathan, D.M., Damiano, E.R.: Outpatient glycemic control with a bionic pancreas in type 1 diabetes. N. Engl. J. Med. **371**(4), 313–325 (2014)
32. Shi, D., Dassau, E., Doyle III, F.J.: Zone model predictive control with glucose- and velocity-dependent control penalty adaptation for an artificial pancreas. In: Proceedings of the American Control Conference (2018)
33. Shi, D., Dassau, E., Doyle III, F.J.: Adaptive zone model predictive control of artificial pancreas based on glucose and velocity-dependent control penalties. IEEE Trans. Biomed. Eng. **66**(4), 1045–1054 (2019)
34. Steil, G.M.: Algorithms for a closed-loop artificial pancreas: the case for proportional-integral-derivative control. J. Diabetes Sci. Technol. **7**(6), 1621–1631 (2013)
35. Steil, G.M., Palerm, C.C., Kurtz, N., Voskanyan, G., Roy, A., Paz, S., Kandeel, F.R.: The effect of insulin feedback on closed loop glucose control. J. Clin. Endocrinol. Metab. **96**(5), 1402–1408 (2011)

36. Steil, G.M., Rebrin, K., Darwin, C., Hariri, F., Saad, M.F.: Feasibility of automating insulin delivery for the treatment of type 1 diabetes. Diabetes **55**(12), 3344–3350 (2006)
37. Steil, G.M., Rebrin, K., Janowski, R., Darwin, C., Saad, M.F.: Modeling β-cell insulin secretion - implications for closed-loop glucose homeostasis. Diabetes Technol. Ther. **5**(6), 953–964 (2003)
38. Thabit, H., Hovorka, R.: Coming of age: the artificial pancreas for type 1 diabetes. Diabetologia **59**(9), 1795–1805 (2016)
39. Thabit, H., Tauschmann, M., Allen, J.M., Leelarathna, L., Hartnell, S., Wilinska, M.E., Acerini, C.L., Dellweg, S., Benesch, C., Heinemann, L., Mader, J.K., Holzer, M., Kojzar, H., Exall, J., Yong, J., Pichierri, J., Barnard, K.D., Kollman, C., Cheng, P., Hindmarsh, P.C., Campbell, F.M., Arnolds, S., Pieber, T.R., Evans, M.L., Dunger, D.B., Hovorka, R.: Home use of an artificial beta cell in type 1 diabetes. N. Engl. J. Med. **373**(22), 2129–2140 (2015)
40. Toffanin, C., Messori, M., Palma, F.D., Nicolao, G.D., Cobelli, C., Magni, L.: Artificial pancreas: model predictive control design from clinical experience. J. Diabetes Sci. Technol. **7**(6), 1470–1483 (2013)
41. Turksoy, K., Quinn, L., Littlejohn, E., Cinar, A.: Multivariable adaptive identification and control for artificial pancreas systems. IEEE Trans. Biomed. Eng. **61**(3), 883–891 (2014)
42. Weinzimer, S.A., Steil, G.M., Swan, K.L., Dziura, J., Kurtz, N., Tamborlane, W.V.: Fully automated closed-loop insulin delivery versus semiautomated hybrid control in pediatric patients with type 1 diabetes using an artificial pancreas. Diabetes Care **5**(5), 934 (2008)
43. Yi, H., Zhao, C.: Active disturbance rejection control: methodology and theoretical analysis. ISA Trans. **53**(4), 963–976 (2014)

Chapter 10
Summary and Future Work

In this monograph, we present our recent developments on event-triggered ADRC design, starting from event-based ESO design and then to event-based closed-loop control. We also show that the developed results are highly practical using application examples with experimental results, due to the model-free nature of the ADRC approach. Also, we show that by carefully designing the event-triggering conditions, it would be possible to achieve improved performance with a restricted number of events. In addition, it seems that improved system performance can be obtained by an event-based ADRC through introducing event-triggered parameter adaptation rules, which is a different way of incorporating event-triggering mechanisms in a control system, particularly when the controller is required to behave differently under different scenarios.

Like we mentioned earlier, the approach we take here is mostly the first approach in sampled-data control system design, namely, the emulation approach. Although the discretization approach (which corresponds to the second approach in sampled-data control) was also explored in the literature, many problems remain to be addressed. Moreover, for both the particular event-triggered ADRC formulation and general event-based control frameworks, it is not yet clear how to systematically perform a direct sampled-data design. These point out the first class of problems for future research.

The majority of the content focused on the sampler case of systems with matched uncertainties. The more complicated but challenging case of systems with unmatched uncertainties, however, definitely deserves future attention. Fortunately, the attempts we made in dealing with matched uncertainties paved the general way for/revealed the power of ADRC in handling the additional uncertainty caused by event-triggering mechanisms, which would definitely aid the further exploration of enhanced ET-ADRC design to handle unmatched uncertainties.

D. Shi et al., *Event-Triggered Active Disturbance Rejection Control*, Studies in Systems, Decision and Control 356,
https://doi.org/10.1007/978-981-16-0293-1_10

Finally, ADRC attempts to deal with control tasks through simplifying the knowledge requirements on model information, which is good and advantageous when the model information is very difficult to obtain. However, in many application scenarios, prior knowledge on the model is also available and accumulates with the abundance of measurement data. In this regard, an important research question is how to design an event-triggered ADRC that is able to efficiently exploit this knowledge and adapt itself for improved performance with real-time data. This question seems to go beyond the topic of event-triggered ADRC but the experience we learnt in Chap. 9 appears to be motivating and encouraging.